DESIGNING WITH CREO® PARAMETRIC 6.0

Dr. Michael J. Rider Ph.D.
Ohio Northern University

SDC Publications
P.O. Box 1334
Mission, KS 66222
913-262-2664
www.SDCpublications.com
Publisher: Stephen Schroff

ISBN-13: 978-1-63057-300-3
ISBN-10: 1-63057-300-0

Printed and bound in the United States of America.

Many thanks to my loving wife, Debbie, for her patience and understanding while I was writing this book.

Textbook Conventions for Creo® Parametric 6.0

LMB	Use Left mouse button pick/select
RMB	Depress Right mouse button and hold so pop-up menu appears
MMB	Use Middle mouse button to locate (same as **Accept, OK,** or **Close**)
File>Open	Bold type with a > between the words indicates a pull down menu
OK	Single word OK means LMB selects OK box
Close	Single word close means LMB selects Close box
Any_Word	Single word to be selected with left mouse button such as **OK, Edit...**
"text"	Characters to be entered into a dialog box and followed by the enter key
<Enter>	Press the Enter key
<Delete>	Press the Delete key
<Ctrl>	Press and hold the Control key
<Shift>	Press and hold the Shift key
<Alt>	Press and hold the Alt key
Select	Use Left mouse button pick/select
Pick	Use Left mouse button pick/select
Double-click	Press the left mouse button twice in consecution

BRIEF CONTENTS

v

CONTENTS

PREFACE

Today's world is troubled by the fact that various people do not understand each other's language. However, the graphic language has been around since the dawn of man and, in general, is widely understood by all. A drawing is a graphical representation of a real object, idea, or proposed design for construction at a later time. Engineering or technical drawings are one such form of graphical communication. Artistic drawings are the other form. This textbook will focus on the prior.

This textbook is intended as a classroom text or a self-paced tutorial for the person who wants to learn Creo® Parametric 5.0. Chapters 1 and 2 describe the design process and may be skipped by the reader if desired. However, recovering from a program crash in Chapter 1 should not be skipped. The meat of this text, learning the basic Creo Parametric software, is found in Chapters 3 through 6. Chapters 7, 8, and 12 deal with dimensioning and tolerancing an engineering part. Chapters 9 and 10 deal with assemblies and assembly drawings. Chapter 11 deals with family tables used when similar parts are to be designed or used. Chapter 13 is an introduction to Creo® Simulate and finite element analysis and may be skipped without loss of course content. A series of Appendices are provided to save time for the reader when he/she needs some basic technical information.

This textbook provides an introduction to Creo Parametric 5.0, formerly known as Pro/ENGINEER® Wildfire 5.0, and Creo Simulate, formerly known as the MECHANICA® package that accompanied it. This 3D modeling and analysis software is provided by Parametric Technologies Corporation®. All assignments are meant to be printed on 8.5"×11"paper. Because it is easier to learn new information if you have a reason for learning it, this textbook discusses design intent while you are learning Creo Parametric. At the same time, it shows how knowledge covered in basic engineering courses such as statics, dynamics, strength of materials, and design of mechanical components can be applied to design. You do not need an engineering degree nor be working toward a degree in engineering to use this textbook. Although FEA (Finite Element Analysis) is used in this textbook, its theory is not covered here.

Since we are using Creo Parametric as a design tool, there are many ways to accomplish the same task. If you are new to Creo Parametric, follow the tutorials closely. If you are an experienced Creo Parametric user, feel free to deviate from the tutorials occasionally.

I would like to extend a special note of appreciation to the following reviewers who provided valuable suggestions for improving the manuscript.

Holly K. Ault—Worcester Polytechnic Institute
Brian Brady—Ferris State University
Marca Lam—Rochester Institute of Technology
Kate Leipold—Rochester Institute of Technology
William A. Ross—Professor Emeritus, Purdue University

The author invites all readers of this text to send their comments or suggestions to the author at: Michael J. Rider, Ohio Northern University, Ada, OH 45810 or m-rider@onu.edu.

CHAPTER

1

COMPUTER-AIDED DESIGN

Objectives

▶ History of Computer-aided design

▶ Benefits of using Creo Parametric

▶ Introduction to Creo Parametric

▶ Understand filename extensions

▶ Getting started

▶ Understand filename conventions

▶ How to recover from a program

 History of Computer-Aided Design

The first computer with a cathode-ray tube used as an output device was the Whirlwind I, which became operational in 1950. The beginnings of Computer-Aided Design (CAD) can be traced to 1957 when Dr. Patrick Hanratty developed the first numerically controlled programming system. In 1962, Ivan Sutherland created SKETCHPAD, a system that demonstrated the basic principles of computer created technical drawing.

The first CAD systems simply served as electronic drawing boards. The author of this textbook created a 2D drafting program for Rostone Corporation of Indiana in 1977. The design engineer worked in 2D to create technical drawings. The productivity of design increased only slightly since design engineers had to learn how to use computers and CAD software. With CAD the modifications and revisions were easier; however, the orthographic projections still had to be in the minds of the engineers before they were created on-screen. With cheaper computers, more industries got involved and CAD programs became more functional and user-friendly.

Syntha Vision, the first commercially available solid modeler program, was released by Mathematical Applications Group, Inc. in 1967. Solid modeling enhanced the 3D capabilities of some CAD systems. The author of this textbook created a 3D sculpturing program called CASASP-3D (Computer-Aided Simulation and Analysis of Sculptured Parts in 3D) for Union Special Corporation of Chicago, Illinois, in 1979 with the development continuing through 1985 when the company was bought by the Germans. In 1989 the NURBS software came to market on Silicon Graphics workstations. NURBS surfaces were functions of two parameters mapped to a surface in three-dimensional space. The shape of the surface was determined by the control points. This representation allowed designers to create smooth-flowing shapes.

In 1988 Parametric Technology Corporation (PTC) of the United States and in 1989 Top Systems Ltd. of Russia introduced CAD systems based on parametric parameters. Parametric modeling gets its name from the fact that the model is defined by parameters. A change in a dimension at any point in the design process changes the geometry of the model. Parametric models also use geometrical constraints such as horizontal, vertical, perpendicular, parallel, tangent, etc.

MCAD systems introduced the concept of constraints that enabled the designer to define relationships between parts in an assembly. Designers used a bottom-up approach where parts were created and then assembled. Modeling became more intuitive since the designer created parts and assemblies in the computer the same way he created them in the physical world.

Computer-aided design systems are now widely accepted and used throughout the industry. 3D modeling has become the norm in engineering design today. Creo Parametric 6.0 leads the pack as the best all-around feature-based, parametric modeling and design software for mechanical engineering.

Creo from PTC is a scalable suite of design applications built on a common platform. Its applications are the correct size for all participants in product development. As building blocks of PTC's industry-defining Creo design software solution, Creo Parametric, Creo® Direct, and Creo Simulate are now referred to as Creo. See Figure 1-1.

Creo Parametric—Designing Without Barriers

From http://www.ptc.com/products/creo/

PROBLEM

Customer requirements may change and time pressures may continue to mount, but your product design needs remain the same. Regardless of your project's scope, you need a powerful, easy-to-use, affordable solution.

SOLUTION

Creo Parametric 6.0 is the standard in 3D product design, featuring state-of-the-art productivity tools that promote best practices in design while ensuring compliance with your industry and company standards. Integrated, parametric, 3D CAD/CAM/ CAE solutions allow you to design faster than ever, while maximizing innovation and quality to ultimately create exceptional products.

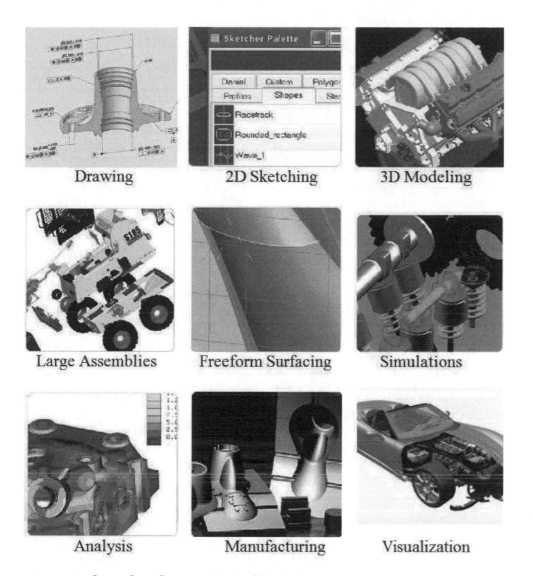

Figure 1-1 Some Creo Parametric Applications

Features and Benefits

▶ Powerful parametric design capabilities allow superior product differentiation and manufacturability.

▶ Fully integrated applications allow you to develop everything from concept to manufacturing within one application.

▶ Automatic propagation of design changes to all downstream deliverables allows you to design with confidence.

▶ Complete virtual simulation capabilities enable you to improve product performance and exceed product quality goals.

Benefits for New Designers or Students

> ▶ Easy to learn
>
> ▶ Easy to build parts from primitives, extrusions, and revolves
>
> ▶ Easy to make changes to your design
>
> ▶ Easy to explore variations in your design
>
> ▶ Easy to make 2D drawings for the shop
>
> ▶ Easy to assemble groups of components
>
> ▶ Easy to reuse existing data when creating new parts
>
> ▶ Easy to incorporate design intent into your design
>
> ▶ Easy to analyze your parts for stresses or possible failures

Help Center

You can access the Help Center by *clicking* **Help > Help Center** or using the context sensitivity menus. The following items are in the Help Center:

> ▶ Context-sensitive Help for Creo Parametric, including Creo Parametric Simulate
>
> ▶ Creo Parametric Installation and Administration Guide
>
> ▶ PTC Customer Service Guide
>
> ▶ Creo Parametric Resource Center

The default browser with its language setting is used for the Help Center. For the best results, view the Creo Parametric Help Center in the Creo Parametric browser.

Creo Parametric Resource Center—Web Tools and Tutorials

Web-based tools and tutorials can assist you with learning to use this release. The Creo Parametric Resource Center is accessible from the Creo Parametric browser, through the Creo Parametric Help Center (**Help > Help Center**), or on the Internet at:

> *http://www.ptc.com/community/creo/index.htm*

Feature Recognition Tool

The Feature Recognition Tool (FRT), a former plug-in application, is now integrated into Creo Parametric. FRT recognizes geometry representing features and replaces the geometry with Creo Parametric features.

   Note: Before starting Creo Parametric, you must set the **frt-enabled** configuration option to **yes** in the config.pro configuration file in the working directory.

Creo Interface—Data Exchange Support

For more information on data exchange formats and supported platforms, such as Windows 8 and 10, see the **Platform Support** page:

> *http://www.ptc.com/partners/hardware/current/support.htm*

Naming Conventions

Table 1-1 Naming Conventions Used in Creo Parametric

File Extension	File Type
.asm	Assembly file
.aux	Auxiliary parameter data file
.cel	Machine parameter data file
.dat	Data files created for editing, such as relations data
.drw	Drawing file
.edm	Contouring parameter data file
.gph	User-defined feature file (including work cells)
.inf	Information data file
.memb	Assembly member information file
.mfg	Manufacturing process file
.mtn	Tool motion parameter file
.ncd	CL syntax alias file
.nck	NC check image file
.ncl	CL data file (including pre- and post-machining files)
.plt	Plot file
.ppl	Route sheet data file
.ppr	PPRINT settings table file
.prt	Part file
.ptd	Part family table file
.sec	Section file
.shd	Shade display file
.sit	Site parameter data file
.smt	Parameter data file for all Punch Press NC sequences
.tph	Tool path storage file
.tpm	Tool parameter file

 ## Creo Parametric—Getting Started

Creo Parametric (Pro/ENGINEER Wildfire 5.0 renamed Creo Parametric in June 2011) is composed of four different applications: Modeling, Welding, Mold & Casting, and Simulate which share a common database. In this book, we will concentrate on the modeling application used to create solid models of parts, engineering drawings, and assemblies.

The tutorials in this book assume you are using the default settings of Creo Parametric in a university setting.

The startup screen should look similar to Figure 1-2. As with most Windows applications, it has pull-down menus, the ribbon (which is part of the Microsoft Office Fluent user interface), and a standard toolbar area. The start-up screen also has a navigator area, a web browser, and a message area. The Ribbon is designed to help you quickly find the commands that you need to complete a task. Commands are organized into logical groups, which are collected together under tabs. Each tab relates to a type of activity, such as doing file operations or modeling.

Creo uses context-sensitive menus. This means that additional menus and selections become available only when they are applicable. The layout of the main screen remains relatively constant throughout the program's execution. If a menu item is grayed out, it means that the listed option is not currently available.

The beginning user will typically refer to the tabs, the active ribbon, and the pull-down menus as they move through Creo Parametric. With experience, the standard toolbar and the keystroke shortcuts will probably become your standard mode of operation.

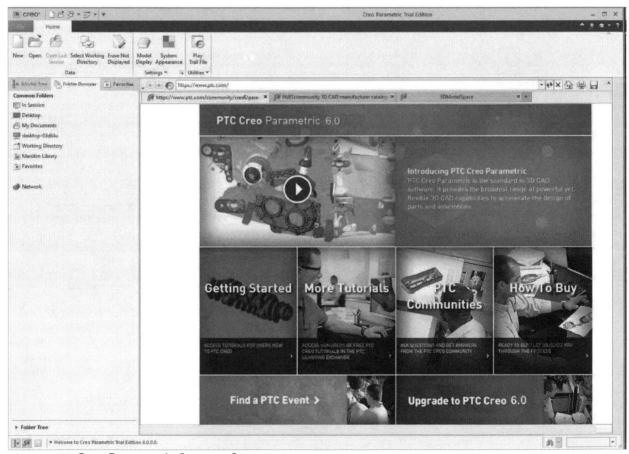

Figure 1-2 Creo Parametric Start-up Screen

The message area is a very important area. It tells you what Creo Parametric has done or what information it is currently looking for. Pay close attention to the message area.

The area behind the web browser is where solid models and drawings appear. This area is controlled by the environment settings.

The Navigator area allows the user to see the model tree, the available folders, and your favorite links.

The web browser allows the user to access model information or online documentation using its web browsing capabilities. The web browser closes when a model is opened. It can also be opened or closed by picking the web browser icon in the lower left corner of the window using the LMB.

Creo Parametric uses the mouse and its three buttons extensively. It is important to understand the mouse button's basic functions listed below. See Figure 1-3.

LMB—(LEFT MOUSE BUTTON)

Used for most operations such as selecting icons, picking graphic entities, and selecting menu items.

MMB—(MIDDLE MOUSE BUTTON)

Used to end a process or a command sequence or to accept the default setting or answer.

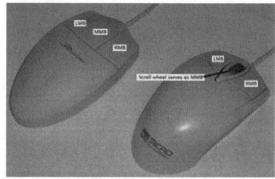

Figure 1-3 Three-button Mouse

RMB—(RIGHT MOUSE BUTTON)

Used to query selections. If held down, it typically brings up additional available options in a pop-up menu.

It is good practice to create a CAD files folder to store your data files in before starting Creo Parametric. Immediately after starting Creo Parametric you should set the working directory to your CAD folder.

When working with Creo Parametric it is strongly advised to _**save your work often**_ in case of a power failure or an unexpected program crash. However, if you happen to crash out of Creo Parametric without having saved your work, there is a way to let Creo Parametric recreate your work. We will demonstrate this procedure now by creating a block with a hole, then exiting from Creo Parametric without saving your work. Ready?

Step 1:   Start Creo Parametric by _double-clicking_ with the LMB on the CREO PARAMETRIC icon on the desktop, or from the Program list: Creo Parametric in Windows.

Step 2:   Set the working directory using **File>Manage Session>Select Working Directory** (Figure 1-4) or _pick_ the _**Select Working Directory**_ icon in the Home Ribbon (Figure 1-6), and then locate the folder you wish to use when storing your Creo Parametric files. _Pick_ **OK**. See Figure 1-5.

Figure 1-4 Select Working Directory

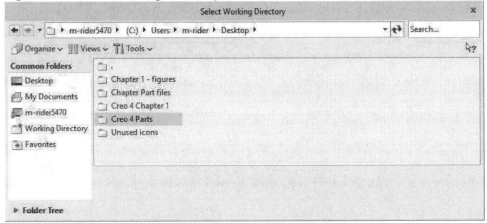

Figure 1-5 Select Your Working Directory

Step 3: Create a new part by *picking* the *New* icon at the top of the screen, or by *selecting* **File>New**. See Figure 1-6.

Figure 1-6 New Object

The following requirements apply when creating and performing operations on Creo files:

▶ File names have a 31 character limit. You cannot create or retrieve an object with more than a 31 character file name.

▶ You cannot use brackets, such as [ ], { }, or ( ), spaces and punctuation marks (.?!;) in file names.

▶ File names can contain hyphens (-) and underscores (_); however, the first character in a file name cannot be a hyphen.

▶ Use only alphanumeric characters (see hyphen and underscore exception in the previous bullet) in file names. Object files containing non-alphanumeric characters (such as @, #, and %) in their names do not display in, and are not retrievable from, dialog boxes. You cannot save new object files with non-alphanumeric characters in their names.

▶ Use only lowercase characters for file names. An object or file on disk always saves with the name in lowercase.

Step 4: *Select* **Part** and **Solid,** make sure the "Use default template" box is checked, then name the part "block_with_hole". *Pick* **OK**. See Figure 1-7.

Step 5: *Select* the **Extrude** icon (Figure 1-8) from the Model tab ribbon. Your tab may look different but can be customized accordingly.

Step 6: Pick the reddish Placement tab with the LMB, then pick Define...

Step 7: Pick the FRONT datum plane with the LMB. Verify that the RIGHT datum plane is oriented toward the right. Pick the Sketch button. See Figure 1-9.

Step 8: Select the Sketch View icon to orient the view so that the sketching plane is parallel with the screen. See Figure 1-10.

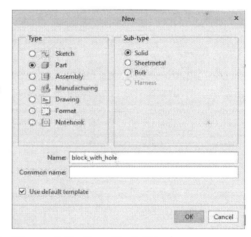

Figure 1-7 New Part Window

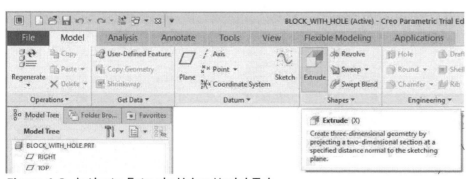

Figure 1-8 Activate Extrude Using Model Tab

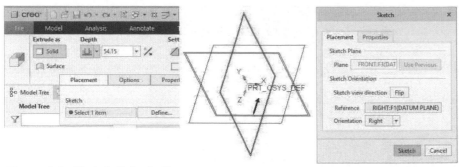

Figure 1-9 Sketch Orientation

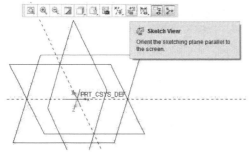

Figure 1-10 Sketch View

Step 9:   *Pick* the **Rectangle** tool from the ribbon. *Pick* the origin with the LMB, and then *pick* a point in the 1st quadrant to form a rectangle. See Figure 1-11.

Step 10:   *Press* the MMB to exit from this command.

Step 11:   Place the mouse cursor on top of the horizontal faint (weak) dimension, then *double-click* the LMB. Type "5" <Enter>.

Step 12:   Place the mouse cursor on top of the vertical faint (weak) dimension, then *double-click* the LMB. Type "4" <Enter>. You now have a rectangle that is oriented at the origin with sides of 4 units and 5 units. Pick the **Refit** icon to refit the rectangle to the screen.

Step 13:   *Pick* the **green checkmark** on your screen to exit from the sketcher and keep your sketch. See Figure 1-12.

Step 14:   *Type* 3.00 <Enter> for the thickness of the extrusion. See Figure 1-13.

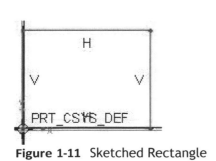

Figure 1-11 Sketched Rectangle

Figure 1-12 Exiting Sketcher

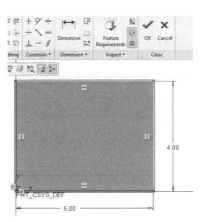

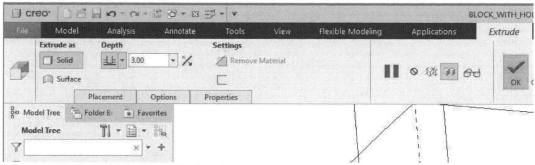

Figure 1-13 Set Thickness of Protrusion

Step 15: *Pick* the *glasses* icon to view your results.

Step 16: *Press down* the MMB, then move the mouse to see your 3D part rotate on the screen. If enabled, roll the scroll-wheel to zoom in or out on the part.

Step 17: *Pick* the **green checkmark** on your screen to accept the extrusion.

If you accidentally exit from this mode, highlight Extrude 1 on the left side of the window, *press and hold* down the RMB until a pop-up menu appears. *Pick* **Edit Definition** icon from this menu to place you back where you were.

Step 18: *Select* the **Named View List** using the LMB. Select FRONT, LEFT, etc., from the list and view the results. Select the **Standard Orientation** from the list. See Figure 1-14.

Step 19: *Select* the Hole tool from the ribbon.

Step 20: *Pick* the Placement tab.

Step 21: *Pick* the front surface of the part with the LMB.

Step 22: Change the hole depth to through all surfaces.

Step 23: Drag one of the broken diamonds to the bottom edge. Drag the other broken diamond to the right edge. They should change to closed diamonds if you placed them correctly. You can pick the edge of the block or its surface (Figure 1-15).

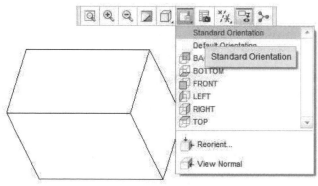

Figure 1-14 Named View List

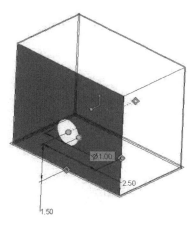

Figure 1-15 Adding a Hole and Setting References for Locating the Hole

Step 24: Change the vertical distance to 1.50 units. Change the horizontal distance to 2.50 units. Change the hole diameter to 1.00 units. Remember to change a value, double-click on it using the LMB, and then enter the new value followed by the <Enter> key.

Step 25: *Pick* the *glasses* icon to view your part.

Step 26: *Pick* the *green checkmark* to accept your hole creation. See Figure 1-16.

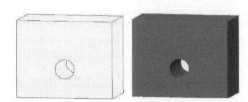

Figure 1-16 Completed Part

Now we are going to simulate a program crash by exiting from Creo Parametric without saving the part first.

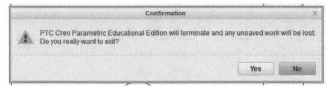

Figure 1-17 Terminate without Saving Part

Step 27: *Pick* the red **X** button in the upper right corner of the screen to terminate Creo Parametric.

Step 28: You get a chance to cancel your request before it is too late. However, we want to exit without saving the part so pick the **Yes** button. (Do not save the part.) See Figure 1-17.

▶ Recovering from a Program Crash

Step 1: In your working directory, or the default working directory set up by your system administrator, there will be a file called "trail.txt" that was created by Creo Parametric while you were working. If there is more than one such file, pick the latest one. See Figure 1-18.

Step 2: Open this "trail.txt" file in Notepad. If you cannot see the file, set the selection filter to **All Files**. This file contains all of the commands you just performed on your newly created part.

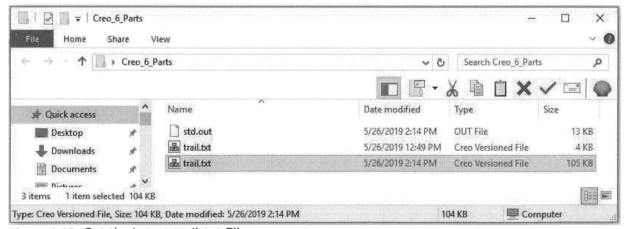

Figure 1-18 Get the Latest trail.txt File

```
~ Close `main_dlg_cur` `main_dlg_cur`
!Command ProCmdOSExit was pushed from the software.
! Message Dialog: Warning
!                    : Creo Parametric Trial Edition will terminate and any unsaved work will be lost.
!                    : Do you really want to exit?
~ FocusIn `UI Message Dialog` `yes`
~ Activate `UI Message Dialog` `no`
!End of Trail File
```

Figure 1-19 Modified trail.txt File using Notepad

Step 3: At the end of this file change the Activate "UI Message Dialog" from "yes" to "no" as shown above to cancel the request for exiting from the program. Save this file using **>File>Save As...** Rename the file something else like, "watch_this.txt". Pick the **Save** button. Exit from Notepad. See Figure 1-19.

Step 4: Restart Creo Parametric.

Step 5: Set your working directory.

Step 6: *Select* the ***Play Trail File*** icon in the Home tab ribbon. See Figure 1-20.

Step 7: *Select* "watch_this.txt" in the open window (Figure 1-21), then *pick* the **Open** button. Creo Parametric will repeat every move and command you made during your last session. Since we changed the confirmation for exiting from yes to no, Creo Parametric will not terminate so you can now save your part before terminating.

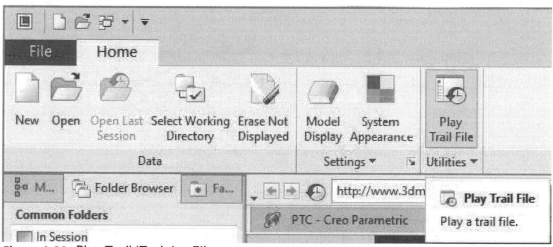

Figure 1-20 Play Trail/Training File

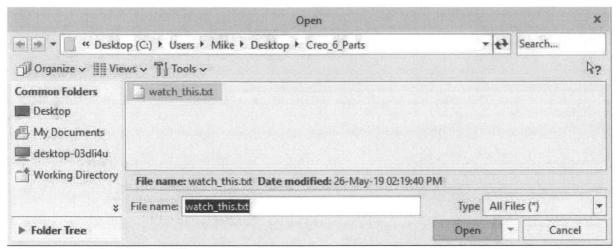

Figure 1-21 Select Replay File

When Creo Parametric pauses you should see your block with the hole on the screen. See Figure 1-22.

Step 8: Save your part by picking the *Save* icon from the Quick Access Toolbar or selecting File>Save from the pull-down menu. Pick the OK button.
Step 9: Exit from Creo Parametric as before.

Let's simulate another program crash. This time we had a power failure.

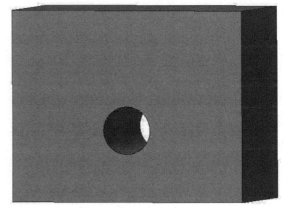

Figure 1-22 Block Recreated

Step 10: Using Notepad edit the "watch_this.txt" file as before. Locate the last line containing "!Dragger (Linear Dimension of Second Reference Dragger) start dragging." Delete everything below this line in the file. Save the file as "watch_this2.txt".
Step 11: Go into your working directory and delete the file "block_with_hole.prt" or rename it because the script file assumes that it is creating this part for the first time, thus it cannot exist.
Step 12: Delete the part "Block_with_hole.prt" from your working directory.
Step 13: Restart Creo Parametric.
Step 14: Set your working directory.
Step 15: *Select* the **Play Trail File** icon in the Home ribbon.
Step 16: *Select* "Watch_this2.txt" in the open window, then *pick* the **Open** button. Creo Parametric will repeat every move and command you made during your last session up to the point where you are creating the hole for the first time, and then it will pause.
Step 17: Finish the part as before, and then save it.
Step 18: **File>Close** to clear the graphics window or *pick* the ✖ in the upper left corner of the window.
Step 19: Pick the ***Erase Not Displayed*** icon, then *pick* **OK**.
Step 20: Exit from Creo Parametric.

If this had been an actual program crash, you would have recovered most of your work up to the point where the crash might have occurred.

If you have an actual program crash, simply remove the last set of commands from the file, and then repeat the above procedure. If this is the first time that you were creating the part and you had saved it in your working directory, you will need to delete all versions of this part before running the trail file. If the part already exists in your directory when you go to save it the first time, the trail file will exit without saving the part. If Creo Parametric still crashes, remove additional lines/commands until Creo Parametric remains running. Save your part, and then continue with your part creation.

Note: If you are creating a part for the first time when Creo Parametric crashes, you must remove any saved versions of this part file before running the trail.txt file under its new name.

Review Questions

1. Name three benefits of Creo Parametric over 2D drafting.

2. Name three different applications contained in Creo Parametric.

3. What file extensions are used for assembly files, part files, and drawing files?

4. What is the main function of the left mouse button (LMB)?

5. What are the main functions of the middle mouse button (MMB)?

6. What are the main functions of the right mouse button (RMB)?

7. What is the purpose of the working directory?

8. Why would a word like Placement be shown in red?

9. What is the name of the file that contains your step-by-step procedure when you are working in Creo Parametric and where is it located?

10. What is the first step in recovering from a system crash that occurred when you were in Creo Parametric?

Notes:

CHAPTER

2

INTRODUCTION

Objectives

▶ Introduction to design intent

▶ Introduction to the design process

▶ Introduction to brainstorming

▶ Go through the design process with an example

Introduction

This textbook is different from others on the market today because it combines design intent, design, and 3D modeling into one text. Design intent is mentioned at the beginning of the text and then referred to throughout the book. Design decisions for part sizes can be justified based on theories in statics, dynamics, strength of materials, and mechanical component design. A design solution can be obtained by designing a machine, mechanism, or part to perform a specified task using engineering theories. Creo Parametric, the 3D modeling/design software, can be used to create the model, lay out the possible solutions, determine part sizes, and verify resulting design solutions.

Design Process

New design concepts or ideas must exist in the mind of the designer before they can be put down on paper as a permanent record. If the ideas start as sketches, they will probably be converted to technical drawings later. At some point in the design process, you will need to convey your ideas to others, thus working as a group. Other people may be brought into the design process because they are more familiar with materials, production, assembly, or marketing. You must be able to express yourself clearly if you are to work effectively in a group. Do not underestimate the importance of your communication skills, verbally, symbolically, and graphically.

The design process is your ability to combine scientific principles, resources, and existing products into a problem solution. The design process can be subdivided into five steps as follows.

1. Identifying the problem
2. Brainstorming ideas
3. Evaluating the ideas, then selecting one
4. Modeling the solution
5. Working drawing and production

Ideally, the design moves through the steps sequentially, but at any particular step, it may be necessary to return to a previous step or begin the process anew since the problem definition may have changed or been ill-defined.

Step 1—Identifying the Problem

The design process begins with the determination of a need or want of a product, service, or system. Because the company needs to make a profit to stay in business, the solution must be economically feasible. The solution may be simple or quite complex. Even if the solution is simple, the tooling for production and/or the production process may be complex.

Information concerning the problem definition should take the form of a problem proposal which can be a paragraph or a multipage report. The proposal must consist of a plan for action that will be followed to solve the problem. The proposal, if approved, becomes the agreement to be followed. After approval of the proposal, further aspects of the problem are explored. Information relative to the problem is collected. A Gantt chart is a type of bar chart that illustrates the timeline for the proposed project solution. They illustrate the start and finish dates of the necessary tasks of a project. Other information in the proposal should include the following:

1. What is the product/design solution expected to do?
2. What is the market potential?
3. What is the maximum cost of the solution?
4. What is the production cost?
5. What can the product sell for?
6. When will initial design decisions be made?
7. When will a prototype be available for testing?
8. When will production drawings be ready?
9. When will tooling drawings be ready?
10. When will production begin?
11. When will the product be market-ready?

Nearly all design solutions are a compromise. The amount of time budgeted to each item above and to the project as a whole is no exception. Be conservative with your time estimates. Be aware that you have no control over how long it might take a supplier to furnish you with the needed parts or materials.

Step 2—Brainstorming Ideas

Brainstorming is a process where ideas about possible solutions to the defined problem are created. DO NOT evaluate the ideas at this time. Document all ideas using sketches and words that come up regardless of how silly or infeasible they may seem. Date and sign all notes and sketches, then retain these for possible patent proof. These silly ideas may spark a new idea in one of your group members that turns out to be the best solution. If you do not include all ideas at this step in the design process, you may not arrive at one of the better solutions which stemmed from it. The larger the collection of ideas, the better the chance of finding one that is a suitable solution. Sometimes the solution is a combination of several ideas, thus it is important to document all ideas.

Step 3—Evaluating the Ideas, then Selecting One

After careful consideration and possibly combining ideas from the previous step, select what appears to be the best solution thus far. Before selecting, add as much detail as possible to the solutions. Consider the following:

1. Will the design be manual, electric, or fuel powered?
2. What type of motion is required?
3. What kind of forces or torques may be present?
4. What kind of material should be used for construction?
5. How complex is the design solution? Use the KISS principle: **K**eep **I**t **S**imple **S**tupid. Simpler is typically better and costs less.
6. What might it cost to build? (The company needs to make a profit.)
7. Create layout drawings showing more details than the sketches, then estimate possible weak areas and necessary clearances between moving parts.
8. If the solution is an improvement of an existing product, is the solution an improvement or at least a production cost reduction?
9. If a part of the solution came from an existing product, how well did the existing product work? Should this solution work in a similar manner?

The ideal solution is the one that will perform the task at the lowest possible manufacturing cost.

Step 4—Modeling the Solution

A scale model should be made to study, analyze, and refine the design solution. In today's world, this model is often a computer-generated, 3D model found only inside a computer. With today's 3D graphics software, like Creo Parametric from PTC, the model can be constructed and analyzed without building a physical model. This drastically reduces the design cost of a possible solution.

The 3D computer model can be analyzed for stresses and clearances. Design changes can be made easily. If the solution proves to be unsatisfactory, the model can be modified or the designer can return to a previous step and begin again. If the next solution contains some of the same part characteristics, they can be taken from the previous solution at no additional cost.

If time permits, several of the design ideas can be modeled and analyzed to ensure that the best possible solution has been selected. Working drawings can be created from the full-scale computer model.

Step 5—Working Drawing and Production

A final set of working drawings need to be made, checked, and approved in order to manufacture a product. The design engineer is responsible for checking and approving the production drawings. Dimensions, tolerances, and manufacturing notes must be present on the working drawings.

Unaltered standard parts do not require a detailed drawing since they will not be manufactured locally. However, their 3D part creation is necessary since they will appear in the assembly and/or the subassembly drawings. Standard parts will be listed in the bill of materials as well.

To protect the manufacturer, a patent drawing, typically an assembly drawing, needs to be created and filed with the patent office.

▶ Design a New Product, an Example

Throughout this textbook a given design problem is taken through the steps necessary to design, model, simulate, and verify the solution. The necessary steps from problem definition and brainstorming to the final product are shown. This section may be skipped by the reader without loss of content.

Step 1—Identifying the Problem

Part of the assembly line requires that a 3.5-inch diameter piston be moved back and forth at a specified rate of 30 cycles per minute. This piston is used to compress and release a given mixture of particles. There is nearly a constant pressure of 100 psi pushing against the piston at all times. The piston must have a stroke of 3.0 inches and prevent leakage along its sides. The location of the piston is fixed along with the available space for the newly designed machine. The machine must be self-contained, attached at the base, and powered by either 220 VAC or 440 VAC. This power can be fed in from almost any point. The machine cannot be attached to either wall. See the available design space below (Figure 2-1).

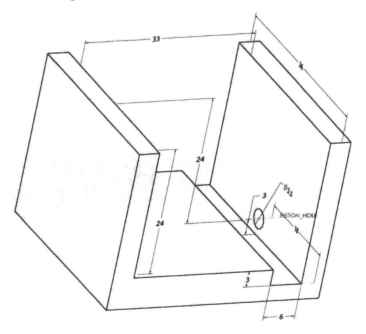

Figure 2-1   Design Space

Note: The left and right walls are not actually solid or thick. These walls represent the edges of the working space and could be any thickness for this design exercise.

1. What is the product/design solution expected to do?
   ▶ Move a 3.5-inch diameter piston back and forth using a 3-inch stroke at a specified rate of 30 cycles per minute with approximately 100 psi pushing against its head.

2. What is the market potential?
   ▶ Needed in a local assembly line.

3. What is the maximum cost of the solution?
   ▶ Maximum cost of materials and labor, $25K.

4. What is the production cost?
   ▶ Not applicable since only one system is needed.

5. What can the product sell for?
   ▶ Not applicable since only one system is needed.

6. When will initial design decisions be made?
   ▶ Six weeks from today.

7. When will a prototype be available for testing?
   ▶ Computer-generated 3D model ready in three months.

8. When will production drawings be ready?
   ▶ Three months from today.

9. When will tooling drawings be ready?
   ▶ Four months from today.

10. When will production begin?
    ▶ Five months from today.

11. When will the product be market-ready?
    ▶ Six months from today.

Figure 2-2 Hand Press Pump

Step 2—Brainstorming Ideas

The first step is to create/imagine many different possible solutions for the problem at hand. Look around you for ideas. Think about other machines that might contain a reciprocating, sliding piston. What comes to mind? Let's see. Do NOT evaluate your ideas at this time.

Idea 1: How about a hand pump? The handle moves back and forth while the piston moves up and down. See Figure 2-2.

Idea 2: How about an oil pumping rig? The weighted crank rotates through 360 degrees while the piston in the pipe moves up and down. See Figure 2-3.

Figure 2-3 Oil Pump Jack

Idea 3:   How about a hand press? The handle causes the piston in the cylinder to move up and down. See Figure 2-4.

Idea 4:   How about a hydraulic or air cylinder? In Figure 2-5 the air cylinder pushes and pulls on a link which in turn pushes and pulls on the connecting rod. As the connecting rod goes down, the piston goes up. As the connecting rod goes up, the piston goes down. There is a slot on the right end of the bell crank so the piston can move vertically while the end of the bell crank moves along an arc.

Figure 2-4   Hand Press

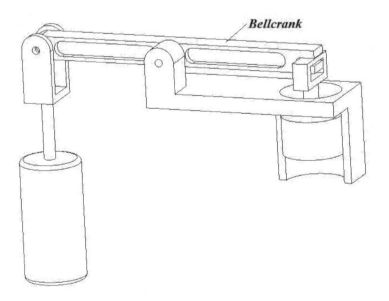

Figure 2-5   Bell Crank Mechanism

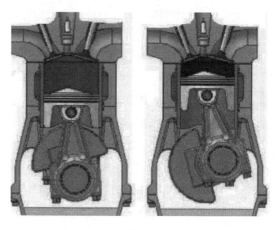

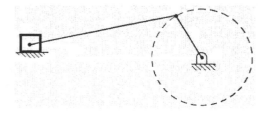

Figure 2-7 Offset Slider-Crank Mechanism

Figure 2-6 Internal Combustion Engine

Idea 5: How about an inline slider-crank mechanism? It works similar to the piston, connecting rod, and crankshaft of an engine as shown in Figure 2-6. With the engine, the piston drives the crankshaft. We want to do just the opposite. We want the crankshaft to drive the piston.

Idea 6: How about an offset slider-crank mechanism? The crank does not line up with the line of action of the slider (piston), thus the adjective "offset." This could be used since the center of the piston lines up with the area where the mechanism is to be mounted. See Figure 2-7.

Idea 7: How about a toggle mechanism? This would allow the driving crank to be above the line of action of the piston. See Figure 2-8.

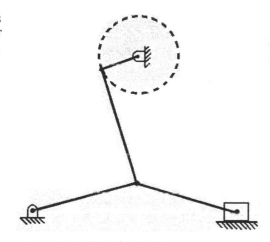

Figure 2-8 Toggle Mechanism

So far I have come up with seven possible solutions. Can you think of others?

Step 3—Evaluating the Ideas, then Selecting One

Now it is time to evaluate your ideas. To do this we need to determine some criteria in which to judge these ideas so we can pick the best one. If none of these ideas work, we will need to go back and do some more brainstorming. Let's see. What is important?

Criteria

1. The mechanism must be able to move the piston back and forth 3 inches without large perpendicular forces. (Large perpendicular forces would cause additional drag and energy loss.)

2. The portion of the mechanism that drives the piston must fit into the 6-inch wide gap.

Criteria 1: Motion without perpendicular forces (statics)

The proof of these comments is left up to the reader using ***engineering statics***.

Idea 1:   Hand pump—The hand pump on the left will produce a perpendicular force on the piston as the handle end moves through an arc. *(Not a good solution)* The hand pump on the right should produce no perpendicular force on the piston. *(Possible solution)*

Idea 2:   Oil rig—The oil rig will produce a very small perpendicular force on the piston if the connecting rod is very long. *(Possible solution)*

Idea 3:   Hand press—This functions much the way the hand pump on the right of idea 1 operates, thus it will not produce a perpendicular force on the piston. (*Possible solution*)

Idea 4:   Bell crank driven—The slot at the end of the bell crank is necessary to allow the piston to move vertically while the bell crank moves through an arc. When the bell crank is horizontal in this figure, the perpendicular force on the piston is zero. As the bell crank moves away from the horizontal position, the perpendicular force on the piston increases. For small swings of the bell crank the horizontal force on the piston is small. *(Possible solution)*

Idea 5:   Gasoline engine—When the piston is at the top or bottom dead center (crank at 90° or 270°), the horizontal force on the piston is zero. When the crank is at 0° or 180°, the perpendicular force on the piston is quite high. This helps to increase the torque produced by the engine but is not what we want for our design. *(Not a good solution)*

Idea 6:   Offset slider-crank—When link 3 in the figure is horizontal, the perpendicular force on the piston is zero. As the angle of link 3 increases, the perpendicular force on the piston increases. *(Not a good solution)*

Idea 7:   When link 5 in the figure is horizontal, the perpendicular force on the piston is zero. As the angle of link 5 increases, the perpendicular force on the piston increases. *(Not a good solution)*

Ideas 1, 2, 3, and 4 seem to be possible solutions using criteria 1.

Criteria 2: Driving mechanism fits in a 6-inch gap

We are going to evaluate ideas 1, 2, 3, and 4 since these are the only ideas that met criteria 1.

Idea 1:   Hand pump on the right—With a 3-inch stroke, at least 3 inches of the rod that slides through the top support bearing, and clearance at each end for the physical size of the parts, this idea will not fit into the 6-inch wide gap. *(Not a possible solution)*

Idea 2:   Oil rig—Because a long connecting rod is needed to reduce the perpendicular force on the piston and we have only 6 inches to work with, this idea will not work. Also, gravity is used to make the piston move downward in the well. This condition is not present in our design problem. *(Not a possible solution)*

Idea 3:   Hand press—This idea is similar to idea 1, thus it will not work. *(Not a possible solution)*

Idea 4:   Bell crank—The arc length of the right end of the bell crank will need to be about 3 inches to create a 3-inch stroke in the piston. If the distance between the ground pivot point and the bell crank's slot was 10 inches, the angle of rotation would be approximately 0.30 radians or 18 degrees. If the bell crank started 9 degrees above the horizontal and dipped to 9 degrees below the horizontal, then the perpendicular force on the piston will be approximately 15% the axial force on the piston. Making the bell crank radius longer would decrease the maximum perpendicular force on the piston. *(Possible solution)*

Of the seven ideas, only idea 4 above meets both criteria. It may not be the best solution, but it is a workable solution. A decision must be made to proceed with idea 4 or go back to brainstorming more ideas. Let's go with idea 4, the bell crank concept.

At this point, steps 3 and 4 of the design process are going to be combined since they are interrelated.

Final Preliminary Design

After some more brainstorming and sketching a decision was made to go with a mechanism similar to Figure 2-9 shown below. Since the motor's shaft may not be at the same level as the cam's shaft, we may need to put some type of belt, chain, or gear system between the two levels. The initial driving element is a cam, but a 4-bar linkage may be used after evaluating both choices.

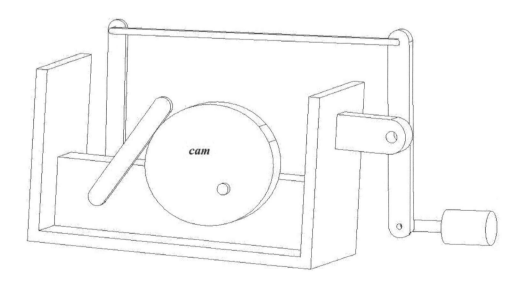

Figure 2-9   Preliminary Design

Next Step

We need to create the actual assembly line's working 3D space so as we design we can be assured that our mechanism fits in the constrained volume.

 ## Review Questions

1. What is the design process?

2. Name the five basic steps of the design process.

3. Why is the design process iterative?

4. What is brainstorming?

5. What is the design intent?

6. Brainstorm at least three ideas for a new way for a handicapped person to turn book pages.

SKETCHER

Objectives

▶ Introduction to 2D sketcher

▶ All basic tools of sketcher explored

▶ Using sketcher constraints

▶ Modifying all dimensions at once (scaling)

▶ Sketcher according to design intent

▶ Practice using the tools of sketcher

▶ Saving a copy of a 2D sketch

▶ Modifying an existing 2D sketch

▶ Sketcher exercise and the power of parametric model

▶ Sketcher design problems to reinforce concepts

▶ Sketcher Explored

Sketcher is the main creation tool of Creo Parametric. The sketcher toolbar ribbon is located at the top of the window. This section explores the many options of the sketcher. Its basic icons are shown in Figure 3-1. The LMB (Left Mouse Button) is used to select geometry or to select a location when creating geometry.

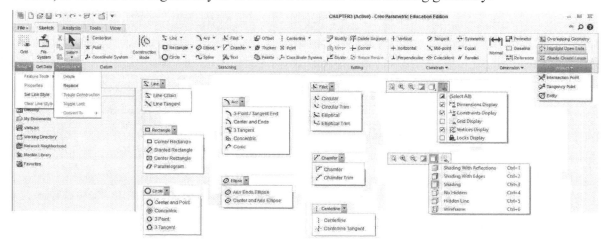

Figure 3-1   Sketcher Ribbon

Where multiple tools are present, the additional tools can be selected by moving the cursor on top of the small arrow icon (▾), then pressing the LMB. The row of options will appear. Move the cursor to the desired tool icon, then press the LMB again to select the tool.

As you create geometry, the sketcher will place enough dimensions on the sketch to completely define it. These original dimensions are referred to as weak dimensions. The value of a weak dimension can be changed, thus making it a strong dimension, or a weak dimension can be replaced by a strong dimension added by the user which matches design intent.

Reminder:

LMB = press the Left Mouse Button down; used to select points or features.

RMB = press the Right Mouse Button down; used to search through a series of features or used to bring up a pop-up menu.

MMB = press the Middle Mouse Button down; used to cancel a command, place a dimension, or accept the current value.

Entering Sketcher Directly from Creo Parametric

Select the NEW FILE icon from the ribbon at the top of the screen. Save as **>File>New**. When the New file window opens, *select* **Sketch** as the type, then enter a valid filename without the extension. *Pick* **OK**.

Figure 3-2 New Sketch Named

Entering Sketcher from PartMode within Creo Parametric

This section assumes you are in part creation mode already.

 and ![icon] To enter sketcher mode from part mode, *select* the SKETCHER icon at the top of the screen. *Pick* the plane that you wish to sketch on followed by orienting another plane toward the right, top, left, or bottom of the screen. *Pick* the **Sketch** button. If necessary in the toolbar at the top of the window, *select* the SKETCH VIEW icon to orient the sketching plane parallel to the display screen. The sketcher ribbon is slightly different at the left and right ends. These differences will be discussed later.

▶ Sketcher Tools Explained

Now let's explain most of the sketcher icons. It is important to note that the LMB activates each of the sketcher tools.

Define the grid settings. Grid type can be either Cartesian or Polar. Grid Spacing can be either Dynamic or Static. X Spacing and Y Spacing define the distance between grid points in the X and Y directions. Grid Orientation defines the origin (0, 0) and the angular position of the X-axis relative to the horizontal.

The File System icon allows you to import data into the active sketch.

Cut, Copy, and Paste icons perform the standard cut, copy, and paste functions. Cut is <Ctrl>x. Copy is <Ctrl>c. Paste is <Ctrl>v.

Select Items tool allows you to select features already on the screen by moving the cursor over the item, then pressing the LMB. Holding down <**CTRL**> and selecting a feature adds the feature to the list of selected features. Selecting a feature already in the list of selected features removes the feature from the list. Selecting a blank area of the screen removes all features from the list of previously selected features.

Datums

Geometry Centerline—Draw an infinitely long geometric centerline by selecting two points with the LMB. This series of center lines can be continued until a new sketcher tool is selected or the user presses the MMB once. These centerlines can reference existing geometry as well as be available outside of sketcher.

Geometry Point—Create a geometry point using the LMB. This point can reference existing geometry as well as be known and visible outside sketcher.

This series of geometry points can be continued until a new sketcher tool is selected or the user presses the MMB once.

Geometry Coordinate System—Create a geometric coordinate system at the specified point by pressing the LMB. This coordinate system is known and is visible outside sketcher. This series of coordinate systems can be continued until a new sketcher tool is selected or the user presses the MMB once.

Sketching Tools

There are numerous sketching tools with limited space for them at the top of the screen; therefore, many of these tools are stacked. If the tool is stacked, then a small down arrow icon appears to the right of the current icon tool. To see the stacked tools, pick the down arrow with the LMB. To select a stacked tool, move the cursor onto the desired tool, then press the LMB again.

Construction Mode—Toggles the creation of new geometry between construction line creation and solid geometry line creation. You can change the type of line segment created afterward by picking the line segments with the LMB, then pressing and holding the RMB until a pop-up menu appears. In the pop-up menu select Geometry to change construction lines to solid geometry lines, or pick Construction to change geometry lines into construction lines.

Line Chain—Draw a solid line from the first LMB pick location to the Second LMB location pick. After the first line has been drawn, the cursor can be moved to a new location, then the LMB pick will draw a solid line from the end of the previous line to the new location. To start a new series of solid lines, press the MMB once, then begin anew. When the cursor gets near the end of an existing line, a small circle will appear around its end. Pressing the LMB will cause the end point of the newly created line to be exactly the same as the endpoint of the existing line. This series of solid lines can be continued until a new sketcher tool is selected or the user presses the MMB twice. After the series is complete, enough weak dimensions will appear on the sketch to completely define the sketch. These dimensions can be changed or replaced according to design intent.

Line Tangent—Draw a solid line tangent between two arcs or circles, which are selected using the LMB. This series of solid lines can be continued until a new sketcher tool is selected or the user presses the MMB once.

Corner Rectangle—Sketch a rectangle by selecting two opposite corners of the rectangle using the LMB. This series of rectangles can be continued until a new sketcher tool is selected or the user presses the MMB once.

Slanted Rectangle—Sketch a slanted rectangle by sketching one side of the slanted rectangle using the LMB twice, then moving perpendicular to this side to create the slanted rectangle's size. This series of 3-point slanted rectangles can be continued until a new sketcher tool is selected or the user presses the MMB once.

Center Rectangle—Sketch a rectangle by selecting the center point of the rectangle, then one of its four corners using the LMB. This series of rectangles can be continued until a new sketcher tool is selected or the user presses the MMB once.

Parallelogram—Sketch a parallelogram by sketching one side of the parallelogram using the LMB twice, then moving away from this side to create the parallelogram's shape. This series of 3-point parallelograms can be continued until a new sketcher tool is selected or the user presses the MMB once.

Center and Point Circle—Draw a circle by selecting the location of the circle's center with the LMB, then moving away from that point to create its radius. When the desired radius is shown, press the LMB to set it. This series of circles can be continued until a new sketcher tool is selected or the user presses the MMB once. The symbol **R** will appear by the newly created circle and an existing circle when their radii match.

Concentric Circle—Draw a new circle using the same center point as an existing circle. First, the existing circle must be selected with the LMB, then a new circle appears. The new circle's radius follows the cursor and uses the center of the selected circle. The LMB sets the new circle's radius. Immediately another new circle appears using the same center point. The LMB will set this circle's radius or the MMB will cancel the creation of this new circle. After the MMB is pressed, then the user can repeat the entire sequence after picking a new reference circle. This series of concentric circles can be continued until a new sketcher tool is selected or the user presses the MMB once more.

3-Point Circle—Draw a circle through three points which are selected by pressing the LMB three times. This series of 3-point circles can be continued until a new sketcher tool is selected or the user presses the MMB once.

3-Tangent Circle—Draw a circle tangent to three features that are selected by pressing the LMB three times. This series of tangent circles can be continued until a new sketcher tool is selected or pressing the MMB once.

3-Point/Tangent End Arc—Draw a circular (constant radius) arc by selecting its two endpoints using the LMB, then moving the cursor to size the arc's radius or make one end of the arc tangent to an existing feature. A plus sign (+) will appear at the center of the arc. This series of 3-point arcs can be continued until a new sketcher tool is selected or the user presses the MMB once.

Center and Ends Arc—Draw a circular (constant radius) arc by first selecting its center point using the LMB, then moving the cursor to size the arc's radius. Pressing the LMB sets the arc's radius. A plus sign (+) will appear at the center of the arc. Now the cursor will only move in a circular fashion around the arc's center point. Press the LMB to select the other end of the arc. This series of 3-point arcs can be continued until a new sketcher tool is selected or the user presses the MMB once.

3-Tangent Arc—Draw an arc tangent to three other features. First select a feature where the arc begins and where the arc must be tangent to using LMB. Next, select a feature where the arc ends and where the arc must be tangent to using LMB. Third, select a feature that the arc must be tangent to using the LMB. If it is possible to create an arc tangent to all three features, one will be created. It is not necessary to select the exact tangent point on the features. Sketcher will adjust the endpoints of the arc to make the arc ends tangent to the features. This series of 3-point arcs can be continued until a new sketcher tool is selected or the user presses the MMB once.

Concentric Arc—Draw a concentric arc using the same center point as an existing circle or arc. First, the existing circle or arc must be selected using the LMB. Next, one end of the arc is selected using the LMB. The cursor will only move in a circular fashion around the selected center point. As the cursor moves, the new arc appears. Use the LMB to select the other end of this arc. Immediately the beginning of another new arc appears using the same center point. The LMB will set this arc's endpoint and radius or the MMB will cancel the creation of a new arc. After the MMB is pressed once, then the user can repeat the entire sequence after picking a new reference circle or arc. This series of concentric arcs can be continued until a new sketcher tool is selected or the user presses the MMB once more.

Conic Arc—Draw a conic (variable radius) arc by selecting its two end-points using the LMB, then moving the cursor to size the conic arc. The conic arc's shape can be changed by changing its tangent angle on either end or the value of the conic arc. This series of 3-point conic arcs can be continued until a new sketcher tool is selected or the user presses the MMB once.

Axis Ends Ellipse—Draw an ellipse by selecting the endpoints of the major or minor axis, then moving perpendicular to this axis to size the other axis. The ellipse is shown as the cursor is moved perpendicular to the axis drawn first. The second axis's size is set by pressing the LMB. This series of 3-point ellipses can be continued until a new sketcher tool is selected or the user presses the MMB once.

Center and Axis Ellipse—Draw an ellipse by selecting its center using LMB, one end of its major or minor axis using LMB, then moving perpendicular to this axis to size the other axis. The ellipse is shown as the cursor is moved perpendicular to the axis drawn first. The second axis's size is set by pressing the LMB. This series of 3-point ellipses can be continued until a new sketcher tool is selected or the user presses the MMB once.

Spline—Draw a free-hand spline curve by selecting spline points using the LMB. The spline curve must go through these spline points. If the curve is closed, then the curve's beginning slope will match the curve's ending slope. This series of spline curves can be continued until a new sketcher tool is selected or the user presses the MMB once.

Circular Fillet—Draw a circular fillet or round tangent to two features. The radius of this tangent arc is displayed after its creation. If the two features are intersecting lines, then the line segments at the intersection are trimmed back to the point where the arc is tangent to the lines. *The location of the intersecting lines is marked with a dot.* If neither feature is not a line, the features are not trimmed. This series of arcs tangent to selected features can be continued until a new sketcher tool is selected or the user presses the MMB once.

Circular Trim Fillet—Draw a circular fillet or round tangent to two features. The radius of this tangent arc is displayed after its creation. If the two features are intersecting lines, then the line segments at the intersection are trimmed back to the point where the arc is tangent to the lines. *The location of the intersecting lines is NOT marked with a dot.* If neither feature is not a line, the features are not trimmed. This series of arcs tangent to selected features can be continued until a new sketcher tool is selected or the user presses the MMB once.

Elliptical Fillet—Draw an elliptical fillet or round tangent to two features at the two points selected on the features using the LMB. If the two features are lines, then the line segments in the area of the elliptical arc are trimmed back to the selected point and the arc is tangent to the line at that point. *The location of the intersecting lines is marked with a dot.* If neither feature is not a line, the features are not trimmed. This series of elliptical arcs tangent to selected points on existing features can be continued until a new sketcher tool is selected or the user presses the MMB once.

Elliptical Trim Fillet—Draw an elliptical fillet or round tangent to two features at the two points selected on the features using the LMB. If the two features are lines, then the line segments in the area of the elliptical arc are trimmed back to the selected point and the arc is tangent to the line at that point. *The location of the intersecting lines is NOT marked with a dot.* If neither feature is not a line, the features are not trimmed. This series of elliptical arcs tangent to selected points on existing features can be continued until a new sketcher tool is selected or the user presses the MMB once.

Chamfer Fillet—Draw a chamfer between two intersecting lines starting at the points selected using the LMB, then remove the line segments in the area of the intersection of the two lines. *The location of the intersecting lines is marked with a dot.* Two weak dimensions are created that reflect the length of the two legs of the chamfer. This series of chamfers can be continued until a new sketcher tool is selected or the user presses the MMB once.

Chamfer Trim Fillet—Draw a chamfer between two intersecting lines starting at the points selected using the LMB, then remove the line segments in the area of the intersection of the two lines. *The location of the intersecting lines is NOT marked with a dot.* Two weak dimensions are created or modified that reflect the length of the two remaining line segments. This series of chamfers can be continued until a new sketcher tool is selected or the user presses the MMB once.

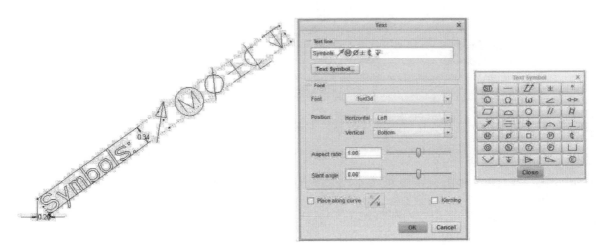

Figure 3-3   Text and Symbols

IA ***Text***—Create alpha characters and symbols on the sketch. Use the LMB to select a start point on the sketching plane and a second point which will set the text height and orientation. The length of the construction line determines the height of the text, while the angle of the line determines the text orientation. The two weak dimensions can be changed or replaced using the dimensioning tool. The Text dialog box opens. A red diamond appears at the start point of the text. Type a single line of up to 79 characters of text. Selecting the Text Symbol button brings up a new window containing many symbols that can be incorporated into the line of text. See Figure 3-3. This tool continues to operate until a new sketcher tool is selected or the user presses the MMB once.

The start point, by default, is the left bottom point of the text. The Left reference can be changed to Center or Right. The Bottom reference can be changed to Middle or Top. The aspect ratio affects the width to height ratio. Increasing this value spreads the characters out. The Slant angle tilts the characters making them appear more like italicized characters.

When you finish entering or modifying the text, *pick* the **OK** button. Double-clicking on the text using the LMB will bring this dialog box back so additional modifications can be made.

IA ***Text Along A Curve***—To place text along a curve create the start and height points as before, then *check* the "Place along curve" box (Figure 3-4) followed by selecting the curve to follow using the LMB. Type the text as before.

Use the [ ] button to change the direction of the text along the curve.

Offset—Create duplicate geometry an offset distance from the selected geometry. After selecting the Offset tool, the Type dialog box appears. In the type dialog box, select either single (one item), chain (series of interconnected items), or loop (series of interconnected items that make a closed loop). If you select single, then the feature selected with the LMB is offset a specified distance in the direction of the yellow arrow. If you select chain, then all features between the first feature picked with the LMB and the second feature picked are offset the specified distance. If it is not clear which line segments are selected, then a query box appears asking if this is the desired chain. You can accept this

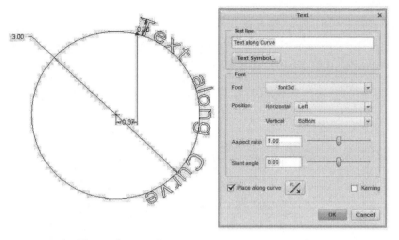

Figure 3-4 Text along a Curve

selection, go to the next selection, back up to the previous selection or quit. If you select loop, then all features in a loop that contains the first feature picked with the LMB are offset the specified distance. This tool continues to operate until a new sketcher tool is selected, the user closes the dialog window, or the user presses the MMB twice. If you want to change the type, leave the offset value blank and press the <Enter> key. Then select the desired type from the Type dialog box.

Thicken—Create entities by offsetting an edge or a sketched entity on two sides. The 2-line offset may have open, flat, or rounded ends. The user is prompted for the thickness, which is the distance between the two new lines, then the offset in the direction of the arrow for the further of the two new lines. The thickened edge can be a single line, a chain of lines, or a closed loop.

Palette—Provides you with a customizable library of predefined shapes that you can readily import onto the active sketch plane. These shapes are presented in a palette. You can resize, translate, and rotate the shape after placing it on the screen.

The sketcher palette has tabs representing categories of sections. Each tab has a unique name and contains at least one section of a certain category. The four tabs with predefined shapes are:

► Polygon—contains regular polygons such as Triangle (3), Square (4), Pentagon (5), Hexagon (6), Heptagon (7), Octagon (8), Nonagon (9), Decagon (10), Dodecagon (12), Hexdecagon (16), and Icosagon (20).

► Profiles—contains common profiles such as C-shape, I-shape, L-shape, and T-shape.

► Shapes—contains miscellaneous common shapes such as Arc racetrack, Cross, Oval, Racetrack, Rounded rectangle, Wave 1, and Wave 2.

► Stars—contains regular star shapes such as 3-tip, 4-tip, 5-tip, 6-tip, 7-tip, 8-tip, 9-tip, 10-tip, 12-tip, 16-tip, and 20-tip.

Using a shape from the palette is similar to importing the corresponding section from another sketch. Each shape in the palette appears as a thumbnail with its name. These cross-sections appear using the default line style and color.

After *selecting* the PALETTE tool, select the worded-tab corresponding to the desired shape (Figure 3-5), then *double-click* using the LMB on the specific shape. Move the cursor to the desired location in the sketcher window and press the LMB again. The shape can also be dragged onto the screen. To do this, *select* the desired shape with the LMB. While holding down the LMB move the cursor onto the screen. The selected shape will follow. To position the shape, release the LMB. In either case, the Move & Resize window will appear. At this point, the shape can be rotated or scaled in place by entering the corresponding value. On the screen the shape has two handles, one for rotation and one for scaling.

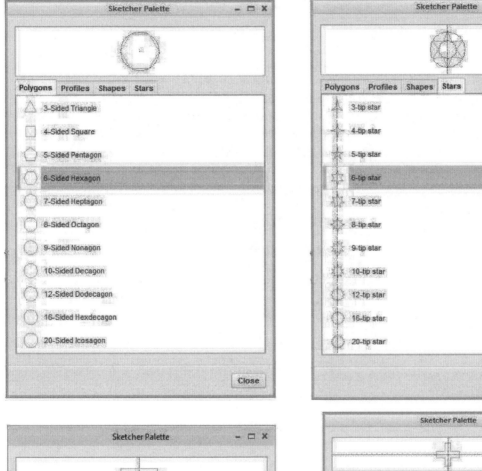

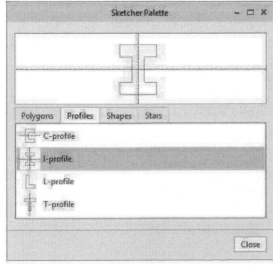

Figure 3-5   Sketcher Palette

Use the cursor and the LMB to free-hand rotate or scale the shape. Move onto the handle,　hold down the LMB, move the cursor, then release the LMB.

Construction Centerline—Draw an infinitely long sketcher centerline by selecting two points with the LMB. This series of centerlines can be continued until a new sketcher tool is selected or the user presses the MMB once. These sketcher centerlines are not known or visible outside of sketcher.

Construction Centerline Tangent—Draw an infinitely long sketcher centerline tangent to two circles or arcs using the LMB. This series of center lines can be continued until a new sketcher tool is selected or the user presses the MMB once. These sketcher centerlines are not known or visible outside of sketcher.

Construction Point—Create a construction point using the LMB. This point is known only inside sketcher and is not visible outside sketcher. This series of construction points can be continued until a new sketcher tool is selected or the user presses the MMB once.

Construction Coordinate System—Create a construction coordinate system at the specified point by pressing the LMB. This coordinate system is known only inside sketcher and is not visible outside sketcher. This series of coordinate systems can be continued until a new sketcher tool is selected or the user presses the MMB once.

Editing

Modify the selected dimension values in a separate window (Figure 3-6). This tool can be initiated in two different ways. The common way is to highlight the dimensions to be modified using the LMB for the first and <CNTL> and LMB for the others, then select the tool. A similar approach is to draw an imaginary box around the sketch and dimensions, then select this tool. The second way is to select the tool first, then select a dimension to be modified using the LMB. If additional dimensions are to be modified at the same time,　hold down

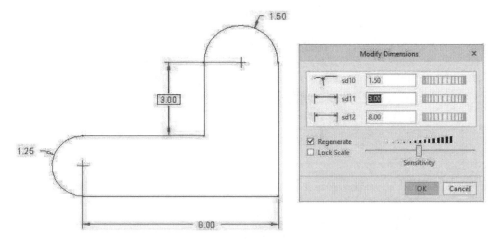

Figure 3-6 Modify Selected Dimensions

the <CTRL> key and select these dimensions with the LMB. Each dimension will be added to the list in the Modify Dimensions box as it is selected.

The Regenerate checkbox will cause the sketch to change shape each time a single dimension is changed. *Uncheck* this box if you want to change all dimensions without seeing their effect until the green checkmark is picked.

The Lock Scale checkbox, when checked first, allows the user to change just one dimension followed by picking the green checkmark and seeing all dimensions change by the same scale factor as the modified dimension. This will cause the original shape of the sketch to remain the same. This is a great way to get all dimensions close to their desired value at once if the original sketch is approximately the correct shape, but the wrong size.

Selecting the red X button cancels the request.

Mirror—Mirror a selection of features about a specified centerline, thus there must be a centerline present in the sketch. First, select the features to be mirrored either by selecting them one at a time using the LMB and <CTRL> LMB or drawing an imaginary box around them, then pick the Mirror tool using the LMB. Next, select the centerline the features are to be mirrored about. The mirrored copy will appear on the opposite side of the centerline. Note that you cannot pick this tool unless you have already selected at least one feature.

Divide Entity at This Point—Divide a feature at the point of selection located by pressing the LMB. This will break a straight or curved line segment into two parts. It will create a starting and ending point on a circle or an ellipse.

Delete Segment—Will remove any line segment that is drawn through while holding down the LMB. They will be erased when the LMB is released. All line segments drawn through with a red line will turn red and be erased when the LMB is released. Use <CTRL>Z or the undo tool to undo the deletions you just made if you accidentally remove a wanted line segment. Features can also be removed from the sketch by selecting the feature with the LMB, then pressing the <Delete> key.

Corner—This tool is used to trim the intersection of two line segments back to the intersection point. Select the two line segments using the LMB. The portion of the line segment where you selected the line will remain after the trim.

Another neat feature of this tool is its ability to connect at their intersection point two line segments (straight or circular) that do not touch but would touch if extended. Select each line segment with the LMB. Each line segment will be extended or trimmed as necessary so the two line segments meet and end at their intersection point.

This tool will not work on a circle, an ellipse, or a centerline since they are considered continuous, without ends.

Rotate Resize—The Rotate and Resize tool can be used to translate, rotate, or scale the selected features. First, select the features to be modified either by selecting them one at a time using the LMB and <CTRL> LMB, or drawing an imaginary box around them, then pick the Move & Resize tool using the LMB. The Move & Resize dashboard will appear. At this point, the features can be translated, rotated, or scaled by entering the corresponding value. On the screen the shape has three handles: one for translation, one for rotation, and one for scaling. Use the cursor and the LMB to free-hand move, rotate, or scale the features. Move the cursor onto the handle, hold down the LMB, and move the cursor. Release the LMB when the modification is complete.

If you want to move the selected features parallel to or perpendicular to a line segment, place the cursor in the first box in the dashboard, then select the line segment to be used as a reference.

If you want to rotate the selection relative to a point, place the cursor in the fourth box in the dashboard, then select the endpoint to be used as the rotational reference point.

Pick the green checkmark (or press the MMB) when all modifications are complete. Pick the red X to cancel the changes made.

Constraint

+ *Vertical Constraint*—Force the selected line to be vertical. After selecting this tool, select the desired line using the LMB. If it is possible to make this line vertical, the sketch will be modified accordingly. A "V" will appear near any line that is vertical. Instead of a line segment, two points can be forced to be vertically aligned by selecting the two points using the LMB. A small rectangular marker will appear near each of the two points indicating that they are vertically aligned. This tool continues to operate until a new sketcher tool is selected or the user presses the MMB once.

+ *Horizontal Constraint*—Force the selected line to be horizontal. After selecting this tool, select the desired line using the LMB. If it is possible to make this line horizontal, the sketch will be modified accordingly. An "H" will appear near any line that is horizontal. Instead of a line segment, two points can be forced to be horizontally aligned by selecting the two points using the LMB. A small rectangular marker will appear near each of the two points indicating that they are horizontally aligned. This tool continues to operate until a new sketcher tool is selected or the user presses the MMB once.

⊥ *Perpendicular Constraint*—Force two selected lines to be perpendicular to each other. After selecting the tool, select the two lines using the LMB. A perpendicular symbol ($\perp$) will appear near the intersection of these two lines indicating their status. If the two line segments do not intersect physically, then a perpendicular symbol with a subscript ($\perp_x$) is placed on each of the two lines. This tool is not limited to just line segments; any feature with a tangent endpoint can be made perpendicular to another feature. For example, the endpoint of an arc can be made perpendicular to a line segment. This tool continues to operate until a new sketcher tool is selected or the user presses the MMB once.

∝ *Tangent Constraint*—Force a line, an arc, or a circle to be tangent to an arc or a circle. After selecting the tool, select the two features using the LMB. The symbol "T" will appear near the intersection of these two features indicating their status. This tool continues to operate until a new sketcher tool is selected or the user presses the MMB once.

Mid-point Constraint—Force a point to locate itself at the midpoint of a line segment or arc. After selecting the tool, select the approximate location of the midpoint using the LMB, then select the line segment or arc using the LMB. The point will move to the midpoint of the selected line or arc. The symbol "M" will appear at this location. This tool continues to operate until a new sketcher tool is selected or the user presses the MMB once.

Coincident Constraint—Make two points coincident, that is, the exact same point. After selecting the tool, select the two points using the LMB. If the endpoints are governed by weak dimensions, then the common point is somewhere between the two original points. If the endpoints are constrained by strong dimensions, then a resolve sketch window will appear showing all possible necessary changes to accomplish the task.

In the example (Figure 3-7) there are seven possible modifications, four of them currently not visible in the scroll window. The possible solutions are highlighted. The possible changes to the sketch are:

1. Delete or make the 2.40 dimension a reference dimension.
2. Delete or make the 2.60 dimension a reference dimension.
3. Undo or delete the requested coincident point request.
4. Make the two vertical lines unequal by deleting left constraint L_1.
5. Allow the horizontal top line to become non-horizontal by deleting its horizontal constraint.
6. Allow the horizontal bottom line to become non-horizontal by deleting its horizontal constraint.
7. Allow the vertical left line to become non-vertical by deleting its vertical constraint.
8. Allow the vertical right line to become non-vertical by deleting its vertical constraint.
9. Make the two vertical lines unequal by deleting right constraint L_1.

Once a resolution has been reached, sketcher continues. This tool continues to operate until a new sketcher tool is selected or the user presses the MMB once.

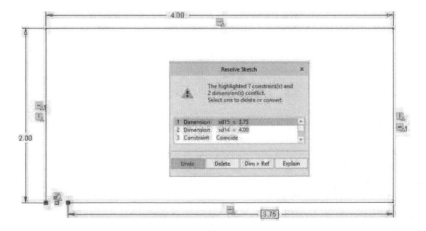

Figure 3-7   Resolve Sketch Constraints

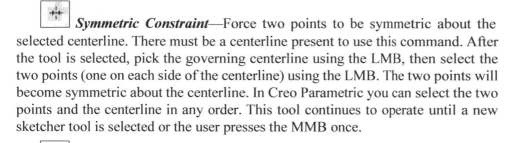 ***Symmetric Constraint***—Force two points to be symmetric about the selected centerline. There must be a centerline present to use this command. After the tool is selected, pick the governing centerline using the LMB, then select the two points (one on each side of the centerline) using the LMB. The two points will become symmetric about the centerline. In Creo Parametric you can select the two points and the centerline in any order. This tool continues to operate until a new sketcher tool is selected or the user presses the MMB once.

Equal Constraint—Force two or more features to be equal size. After the tool is selected, select the governing feature (line length or radius) using the LMB, then using the LMB select all features that you want to be the exact same size. The symbol "Lx" will appear next to line segments of the same length. The symbol "Rx" will appear next to radii of the same size. Line segments and radii cannot be made equal to each other using this constraint.

Press the MMB once to select a new governing feature (line length or radius). This tool continues to operate until a new sketcher tool is selected or the user presses the MMB twice.

Parallel Constraint—Force two lines to be parallel. After the tool is selected, select the governing line segment using the LMB, then using the LMB select all other line segments that you want to be parallel to the governing line segment. The symbol "//x" will appear next to line segments that are parallel.

Press the MMB once to select a new governing line segment. This tool continues to operate until a new sketcher tool is selected or the user presses the MMB twice.

Dimension

Normal Dimension—Add a strong dimension to the existing sketch. A weak dimension will disappear. If the dimension added causes a conflict with an existing dimension, then a conflict box will appear allowing you to resolve the conflict. The new dimension or an existing dimension can be made into a reference dimension or deleted at this time. Extension and dimension lines will be automatically created for the new dimension. This tool continues to operate until a new sketcher tool is selected or the user presses the MMB once.

To change the value of a dimension, double-click on it using the LMB, then type the new value followed by the <ENTER> key. The dimensioned feature will adjust its size to the new value if possible.

To dimension a single line segment, select the line using the LMB, then move perpendicular to the line segment. Position the cursor where you want the dimensional value to appear and press the MMB.

To dimension the distance between two points, use the LMB to pick each of the two points, then position the cursor where you want the dimensional value to appear and press the MMB. If you want the vertical distance between two offset points, move the cursor horizontally away from the two points before pressing the

MMB. If you want the horizontal distance between two offset points, move the cursor vertically away from the two points before pressing the MMB. If you want the length of the slanted line between the two points, move the cursor perpendicular to the slanted line before pressing the MMB.

To dimension the radius of an arc or circle, select the arc or circle by pressing the LMB, then position the cursor where you want the dimensional radius to appear and press the MMB.

To dimension the diameter of an arc or circle, select the arc or circle by pressing the LMB twice, then position the cursor where you want the dimensional diameter to appear and press the MMB.

To dimension the major or minor axis radius of an ellipse or an elliptical arc, select the ellipse by pressing the LMB, then press the MMB. A dialog box will appear asking if you want the major or minor axis. Select one of them, then pick the **Accept** button. Rx will appear in front of the first (major) axis's radius and Ry will appear in front of the second (minor) axis's radius. Note that the major axis is defined by the first axis line you drew when creating the ellipse; it is not necessarily the larger of the two axes nor is it necessarily horizontal or vertical.

To dimension the angle between two line segments, select the first line using the LMB, select the second line using the LMB, then position the cursor where you want the dimensional angular value to appear and press the MMB.

Figure 3-8 below shows four different ways to dimension an angle between two line segments. If you move the cursor between the line segments, then you get the angle between the line segments such as 80 degrees. If you move the cursor opposite to between the line segments, then you get the angle outside the line segments such as 280 degrees. If you move the cursor to the left of the line segments, then you get the angle between the line segments on the left side such as 100 degrees. If you move the cursor to the right of the line segments, then you get the angle between the line segments on the right side such as 100°.

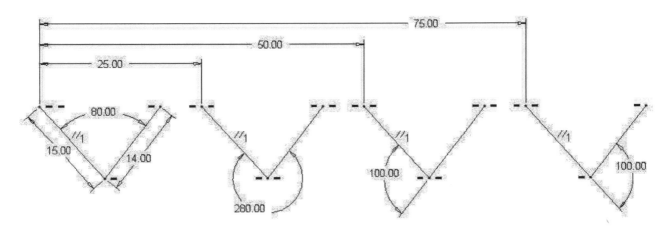

Figure 3-8   Dimensioning an Angle

To dimension the length of an arc, select one end of the arc using the LMB, select the arc using the LMB, select the other end of the arc using the LMB, then position the cursor where you want the dimensional arc length value to appear and press the MMB.

To convert an arc length dimension to an angular dimension, select the arc length dimension using the LMB, then press and hold the RMB to bring up a pop-up menu. Pick "Convert to Angle" from this list. To convert this angular dimension back to an arc length dimension, select the dimension with the LMB, press and hold the RMB to get the pop-up menu, then pick "Convert to Length" from the list.

To dimension a conic arc, select the conic arc by pressing the LMB, then position the cursor in the area where you want the dimension to appear and press the MMB.

Perimeter—Add a perimeter dimension to the sketch after selecting a dimension which can vary when the perimeter dimension is modified. First, select the geometry that makes up the perimeter using the LMB. Hold down the <CTRL> key to select multiple features. Press the MMB once when finished. Using the LMB, select the dimension which will vary to keep the perimeter dimension correct. Type the desired dimension followed by the <Enter> key.

A perimeter dimension is used to dimension the total length of a chain or loop of entities. You must select a dimension that the system can adjust to obtain the desired perimeter. This dimension is called the varying dimension and will have the letters "var" after its value. When you modify the perimeter dimension, the system modifies the varying dimension accordingly. You cannot modify the varying dimension because it is a driven dimension. If you delete the varying dimension, the system deletes the perimeter dimension as well. You cannot create perimeter dimensions for parallel blends and variable section sweeps.

Baseline—Create an ordinate dimension baseline, either vertical or horizontal, representing 0.00. The baseline dimension is typically placed on a drawing to locate the origin of a part when N/C machining is involved in its production. Add a baseline dimension of 0.00 to your sketch by selecting the feature using the LMB, then moving the cursor to the desired location for the baseline value and pressing the MMB. Select Horizontal or Vertical in the pop-up window. This tool continues to operate until a new sketcher tool is selected or the user presses the MMB.

Reference Dimension—Add a reference (driven by other values) dimension to the existing sketch. Reference dimensions are added the same way as regular dimensions. (See adding dimensions above.) These dimensions cannot be changed by selecting them, then typing a new value. This tool continues to operate until a new sketcher tool is selected, or the user presses the MMB once.

Reference dimensions appear on sketches for information purposes only. Therefore, they are read-only and cannot be used to modify the shape. They are automatically updated during regeneration if changes are made to the sketch.

Inspect

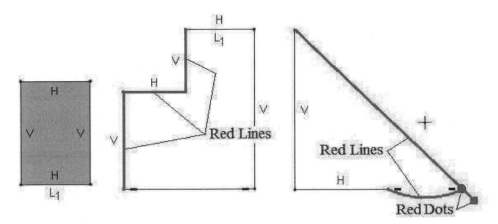 **Overlapping Geometry**—Pick the Overlapping Geometry icon to highlight the sketcher geometry that overlaps so that you can correct the problem.

Highlight Open Ends—Pick the Highlight Open Ends icon to highlight using green dots, the line segments that are not connected to anything, thus they are open ends. (Should be active.)

Shade Closed Loops—Pick the Shade Closed Loops icon to fill in all closed figures so you can see which sections of your sketch are not closed for one reason or another. (Should be active.)

Figure 3-9 below shows a rectangle that is properly closed on the left. The middle figure has red highlighted lines because of overlapping line segments. This could be caused by duplicate line segments where at least one of the red lines is duplicated. The triangular figure on the right has an open-ended line segment (green dots) and overlapping line segments in its lower right corner.

Save Sketch

File>Save saves the sketch under the name you selected prior. *Pick* **OK** to save the file.

Close Sketcher from Part Mode

If you entered sketcher from part mode, then the following two icons appear at the far right side of the ribbon.

✔ Accept the changes made in the sketcher and exit sketcher.

✗ Cancel the changes made in sketcher and exit sketcher.

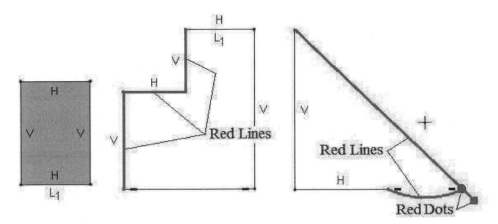

Figure 3-9   Sketched Sections

Set up Sketcher from Part Mode

After selecting the sketching plane and the orientation plane, the user typically orients the sketching plane parallel to the display screen using the ***Sketch View*** icon in the upper left corner of the window. See Figure 3-10.

Sketch Setup—Allows you to redefine the sketch plane and the sketch orientation. *Pick* the **Sketch** button to continue with changes or **Cancel** to continue without changes.

Select References—From previous work to be used when defining the current sketch. These references can be used for constraints or dimensions. *Select* the references using the LMB. *Pick* the **Close** button when done.

Sketch View—Orients the sketching plane parallel to the display screen. This icon is also found in the graphics toolbar just below the ribbon.

Sketcher Graphics Toolbar

The sketcher graphics toolbar is located just below the ribbon in the graphics area. It contains many of the commands that are used regularly. See Figure 3-11.

Refit—Adjust the zoom level to fully display the object on the screen.

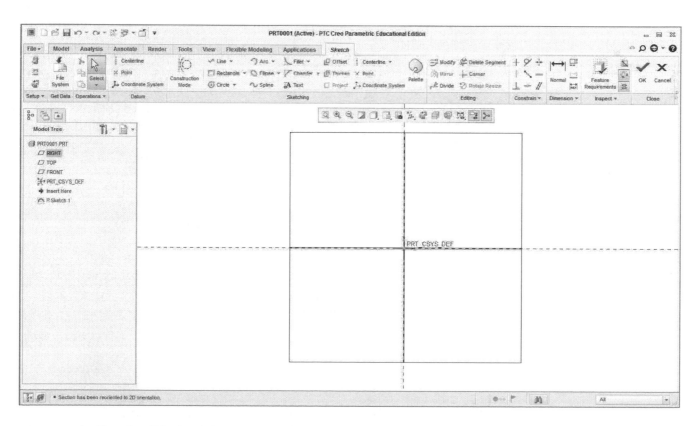

Figure 3-10 Sketcher Window when Entered from Part Mode

Figure 3-11   Sketcher Graphics Toolbar for Part Creation

Zoom In—Zoom in on the target geometry to view it in greater detail.

Zoom Out—Zoom out to gain a wider perspective on the geometry.

Repaint—Redraw the current view (refresh).

Display Style—Determines how the object is to be viewed. See Figure 3-12.

Shading With Reflections	Ctrl+1
Shading With Edges	Ctrl+2
Shading	Ctrl+3
No Hidden	Ctrl+4
Hidden Line	Ctrl+5
Wireframe	Ctrl+6

Figure 3-12 Object Viewing Options

Saved Orientations—Selects the view orientation including user-defined orientations. See Figure 3-13. The *reorient…* tool at the bottom of the pull-down menu lets the user create new user-defined orientations.

View Manager—Lets the user create, set, or delete simplified representations; create cross-sections; create, redefine, or set layers; create, set, or delete views; and create, redefine, or set combined states.

Datum Display Filter—Allows the user to control the display of datum axes, datum points, datum coordinate systems, and datum planes. See Figure 3-14.

Sketch View—Orient the sketching plane parallel with the screen.

Clip Model—Hide the model geometry in front of the sketching plane.

Perspective View—Toggle perspective view.

Sketcher Display Filter—Allows the user to control the display of sketcher dimensions, sketcher constraints, the sketcher grid, and sketcher vertices (endpoints). This tool only shows up when you are in sketcher. See Figure 3-15.

Figure 3-13 Named View List

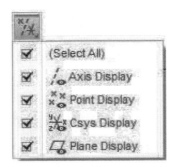

Figure 3-14 Datum DisplayControl

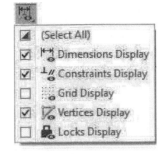

Figure 3-15 Sketcher DisplayControl

Annotation Display—This allows the user to turn on or turn off 3-D annotations and annotation elements.

Spin Center—Allows the user to show or hide the spin center. The object will rotate about the spin center if the spin center is visible. The object will rotate about the current position of the mouse cursor if the spin center is hidden.

▶ Sketcher Practice

This section allows you to practice the necessary skills needed to use sketcher in Creo Parametric. Do each of the exercises below as quickly as possible. Go back and read about the sketcher command if you have trouble using the tool or forgot how to use it.

Step 1: Start Creo Parametric by *double-clicking* with the LMB on the CREO PARAMETRIC icon on the desktop, or from the Program list: Creo Parametric.

Step 2: Set your working directory, by *selecting* the **Select Working Directory** icon in the Home ribbon. Locate your working directory, and then *pick* **OK**. See Figure 3-16.

Step 3: Create a new sketch, named "sketcher_exercises" (Figure 3-17).

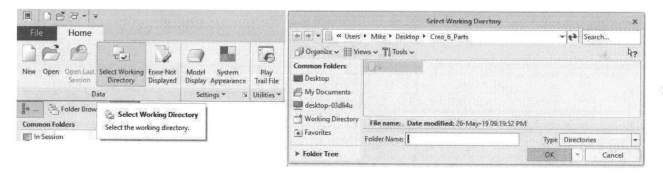

Figure 3-16 Set Working Directory

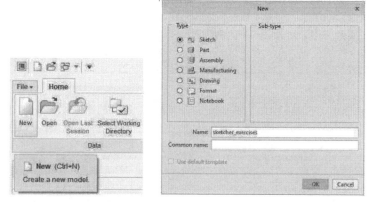

Figure 3-17 New Sketcher File

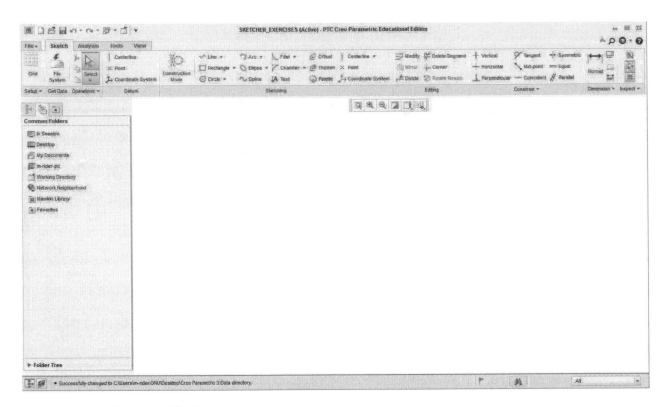

Figure 3-18   Sketcher Window

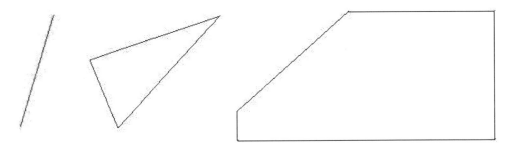

Figure 3-19   Sketch Lines

Your sketcher window's drawing area may be colored. For purposes of this book, the sketcher window will be white. (This can be done by *selecting* **File>Options**. *Select* System Appearances. *Set* System Colors to "Black on White.") See Figure 3-18. Sketcher commands and frequently used icons are located at the top of the window. For now, let's concentrate on the sketcher tools located in the sketch tab.

Step 4:   Use ⌄ to draw Figure 3-19. Do not concern yourself with the size or dimensions of the sketch.

Did you use the MMB to end one figure before drawing the next figure? If not, repeat the exercise. Draw the figure on the right, press MMB, draw the triangle in the middle, press MMB, then draw the line on the left, and press MMB twice to terminate this tool.

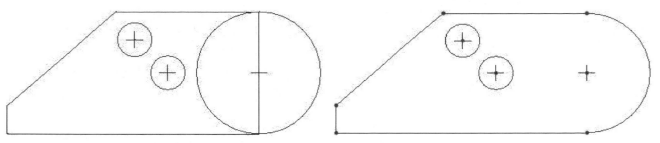

Figure 3-20 Delete Lines and Add Circles   **Figure 3-21** Delete Lines

Step 5: Use ⬚ to erase the line on the left and the middle triangle. With the cursor near the line, *press and hold* the LMB. *Drag* the cursor through the line and the three sides of the triangle. They should turn red. *Release* the LMB, and they are erased. *Press* MMB to terminate.

Step 6: Use ⬚ to draw a circle on the right end as shown in Figure 3-20. Also, draw two circles in the interior of the figure. Draw the first circle at the same horizontal position as the big circle, then draw a second circle the same size as the previous circle centered up and to the left of it as shown. *Press* MMB to terminate this command.

Step 7: Use ⬚ to erase the line on the right end of the original figure and the left half of the circle. See Figure 3-21. *Press* MMB to terminate.

Step 8: Use ⬚ to dimension the sketch according to design intent. The first step is to dimension the figure per design intent as shown in Figure 3-22. Don't worry about the value of the dimensions yet. *Press* MMB to terminate.

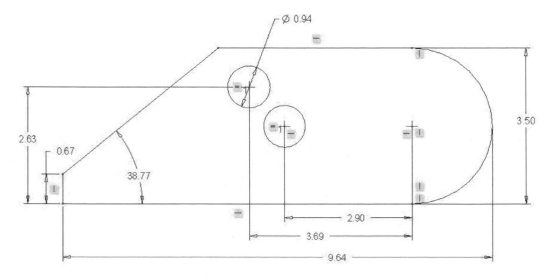

Figure 3-22 Add Dimensions

Design Intent

The height of the figure must be 3.00 inches. The overall length must be 7.50 inches. The slanted side must be at a 40-degree angle. The two small holes must be 0.75 inches in diameter. The center of the first small hole must be 3.00 inches to the left of the big radius's center. The center of the second small hole must be 1.00 inches from the bottom of the figure and 4.00 inches from the center of the big radius. The short vertical leg on the left side must be 0.50 inches long.

Now modify each dimension to reflect the design intent. See Figure 3-23.

Step 9:   Use [REF] to add a reference dimension showing the diameter of the other small hole, the value of the big radius, and length of the slanted line. These dimensions are not needed to define the figure, thus they are for reference only (Figure 3-24). Normally you would not do this, but we are practicing here.

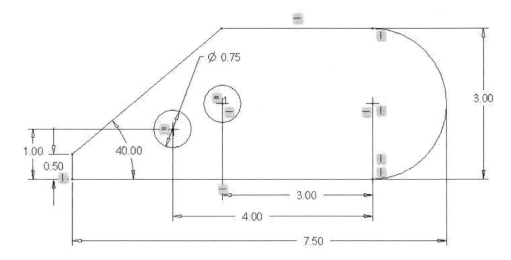

Figure 3-23   Design Intent Dimensions

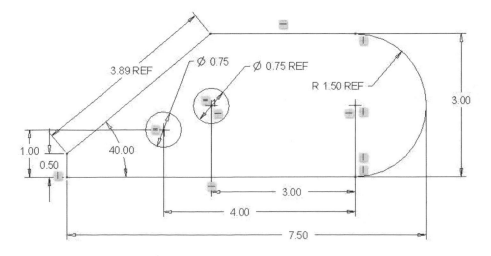

Figure 3-24   Design Intent Dimensions

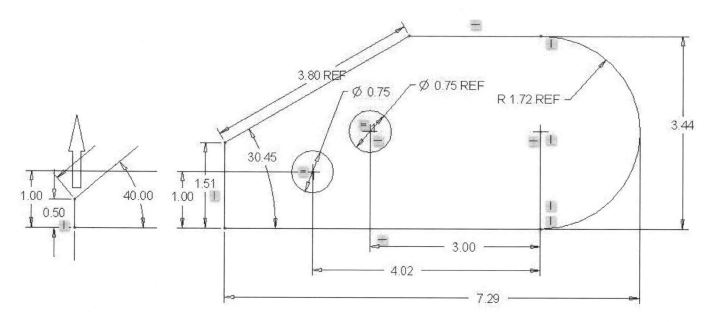

Figure 3-25 Design Intent, 0.50 Changed to 1.51 and diameter changed to 3.44

Step 10: Move the cursor onto the point where the left vertical line meets the slanted line, then *press* the LMB. The point will turn red. *Press down* the LMB, then move the cursor upward. The point will follow the cursor. Move the point to the left and right. Pick the point where the top of the arc meets the horizontal line. Move it upward as well. As you move the point observe the dimensions on the figure. Note that the left small hole and its size do not change since it references the horizontal distance to the center of the big radius and the vertical distance from the bottom of the figure. See Figure 3-25. What dimensions change or do not change, and why?

Step 11: Draw an imaginary box around the entire figure, then press the <Delete> key. What happens? (The screen should be completely empty again.)

Step 12: Use 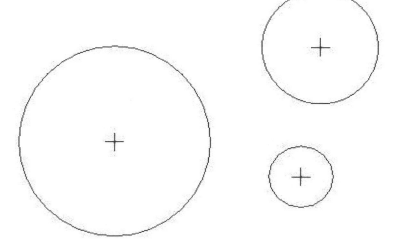 to draw three circles somewhere on the screen as shown in Figure 3-26. Do not let the circles intersect. Do not make the circle the same diameter.

Figure 3-26 Three Circles

Step 13: Use 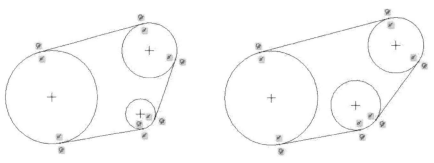 to draw three tangent lines as shown in Figure 3-27.

Step 14: Use the LMB to select any of the tangent points or the center of any circle. *Press down and hold* the LMB while moving the cursor.

Design Intent

Note that the lines remain tangent to the circles because you specified a design intent of tangent lines. See Figure 3-28.

Step 15: Use ⎍ to draw a vertical centerline to the left of the figure and a horizontal centerline above the figure.

Step 16: Use ⧉ to create a mirror image of the entire figure from step 13 or 14 above the centerline. Then mirror both figures about the vertical centerline as seen in Figure 3-29.

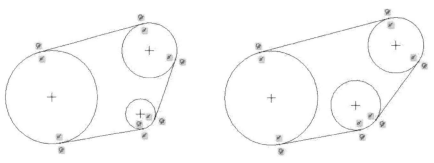

Figure 3-27   Lines Tangent to Circles   **Figure 3-28**   Design Intent with Tangents

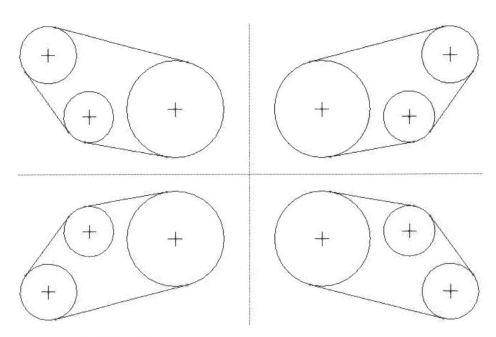

Figure 3-29   Mirror Images

Step 17: Draw an imaginary box around all the figures including the two centerlines, then press the <Delete> key to clear the screen. If the centerlines remain, *select* them individually with the LMB, then press the <Delete> key.

Step 18: Use ⬜ to draw two rectangles as shown in Figure 3-30. (R)

Step 19: Use ◇ to draw two angled rectangles as shown in Figure 3-30. (AR)

Step 20: Use ⬚ to draw two center-located rectangles as shown in Figure 3-30. (CR)

Step 21: Use ▱ to draw two parallelepipeds as shown in Figure 3-30. (PP)

Step 22: Use 🔧 to remove the bottom eight figures or draw an imaginary around the bottom eight figures, then press the <Delete> key to clear the screen.

Step 23: Use ◯ to draw a circle that goes through three points as seen in Figure 3-31.

Step 24: Draw an imaginary box around all of the figures, then press the <Delete> key to clear the screen.

Step 25: Draw the figure shown below on the right. Use ◎ to create a circle that is tangent to the three circles and a circle that is tangent to the two bigger circles and the straight line segment. See Figure 3-32.

Step 26: Draw an imaginary box around all the figures, then press the <Delete> key to clear the screen.

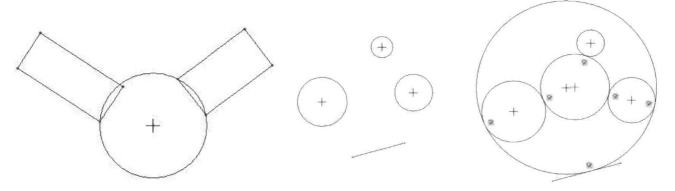

Figure 3-30   Drawing Rectangles

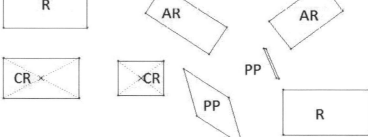

Figure 3-31 Circle through 3 Points   **Figure 3-32**   Circles Tangent to Others

Step 27: Use 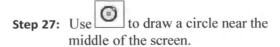 to draw a circle near the middle of the screen.

Step 28: Use 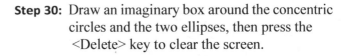 to draw five concentric circles. Note that you must first select the desired circle, then create the new concentric circles. See Figure 3-33.

Step 29: Use and to create two ellipses similar to the ones shown in Figure 3-34.

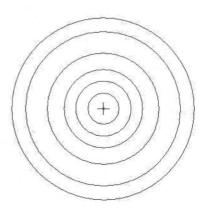

Figure 3-33   Concentric Circles

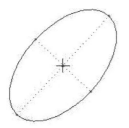

Figure 3-34   Ellipses

Step 30: Draw an imaginary box around the concentric circles and the two ellipses, then press the <Delete> key to clear the screen.

Step 31: Use ⌒ to create an arc, then use ⌇ to create six concentric arcs as shown on the left in Figure 3-35.

Step 32: Use ⌐ to create the upper and lower right most arcs in Figure 3-35.

Step 33: Use ⌕ to draw the arc that is tangent to three different arcs as shown in Figure 3-35.

Step 34: Draw an imaginary box around the arcs, then press the <Delete> key to clear the screen.

Step 35: Draw an angled rectangle. Use ⌁ to draw a conic curve on the upper left side of the rectangle. See the left side of Figure 3-36.

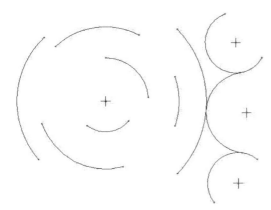

Figure 3-35   Arcs

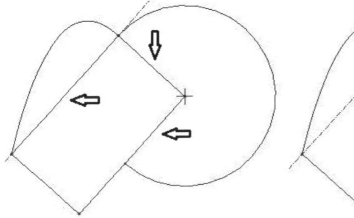
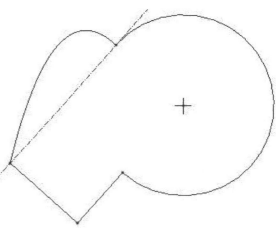

Figure 3-36   Review

Step 36: Use ![arc icon] to locate the center of an arc at the corner of the angled rectangle, then draw the arc shown.

Step 37: Use ![icon] to remove the interior lines. See the right side of Figure 3-36. How did you do? Review the tools if necessary.

Step 38: Add a concentric circle just inside the big arc. Use ![icon] to create an arc tangent to the concentric circle and the two straight lines. Draw a line between the inner circle and the big arc where the construction line crosses both. The dashed lines represent the object's shape before the tangent arc was added.

Step 39: Use ![icon] to divide the big arc at the point where you drew the previous short line. Use this tool to break the line where the small arc intersects the bottom line.

Step 40: Remove the unwanted lines as shown in Figure 3-37. Use the **_undo_** tool, ![undo icon] , if you accidentally erase the wrong line segment.

Step 41: Figure 3-37 is a bit weird looking, but it does demonstrate the use of some of sketcher's tools. Draw an imaginary box around the figures, then press the <Delete> key to clear the screen.

Step 42: Create a rectangle, then use ![icon] to create four rounded corners. Use ![= icon] to make all four rounded corners the same radius. Adjust the dimensions so the rectangle is 4.00 inches long and 2.00 inches high with 0.50-inch radius corners. There should be only three dimensions present on the sketch. See Figure 3-38. Remember that ![icons] stand for horizontal, vertical, and tangent.

Step 43: Use ![icon] to draw a vertical construction centerline near the middle of the rectangle. Add three lines that represent the cutout, then remove the horizontal line at the top of the cutout.

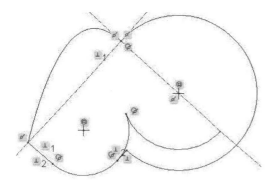

Figure 3-37 Tangent and Divide

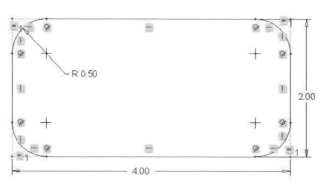

Figure 3-38 Rounded Corners

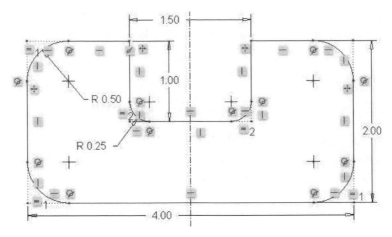

Figure 3-39 Symmetry and Rounds

Step 44: Use 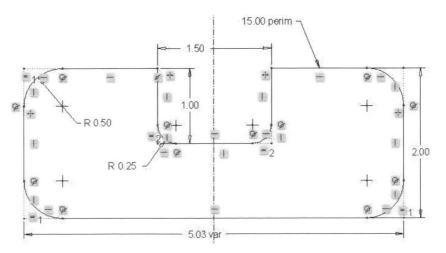 to make the cutout symmetric about the vertical centerline. Use it again to make the rectangle symmetric about the centerline. Make the cutout 1.50 inches wide and 1.00 inches deep. The ⊹ symbols are positioned on the sketch to show two points that are symmetric about the centerline.

Step 45: Use ⌞ to round the inside corners of the cutout to a radius of 0.25 inches (Figure 3-39). There should be only three additional dimensions on the sketch, 1.50, 1.00, and 0.25. Note that this tool leaves a dot where the corner used to be.

Step 46: Determine the circumference of this figure using ⊡ Remember to *pick* on the line segment with the LMB, then hold down <CTRL> and pick the rest of the segments around the perimeter of the figure. *Pick* **OK**. *Pick* the 4.00-inch dimension to be the adjustable dimension. You should get a perimeter of 12.93 inches. What happens to the overall length of the figure if the perimeter is changed to 15 inches? The overall length becomes 5.03 inches. Can you verify this? See Figure 3-40.

Figure 3-40 Perimeter Dimension Added

Step 47: Draw an imaginary box around the entire figure, then select 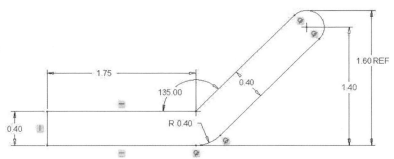. Note that you cannot modify the overall length of the figure since its value is driven by the perimeter dimension. Pick the red X to cancel the request without making any changes.

Step 48: At this point, we have covered most of the commands found in sketcher. Use your knowledge to draw and dimension the following figures, Figures 3-41, Figure 3-42, and Figure 3-44.

Figure 3-41 Control Bracket

Now, practice on your own without detailed instructions to see how you do. Refer back to the needed instruction if you forgot how to implement it.

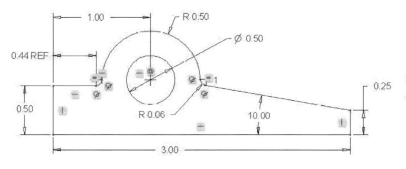

Step 49: Draw control bracket side view (Figure 3-41).

Step 50: Draw angle bearing side view (Figure 3-42).

Figure 3-42 Angle Bearing

Step 51: Draw sheet metal pattern (Figure 3-43). Note that <u>there are only five dimensions on this sketch</u>. If you fold the sheet metal pattern at the edges, can you guess what it forms? (Answer: Right prism.)

The phantom lines shown below at the folds were created using the LINE tool, then *selecting* the line, holding down RMB until **Menu** pops up, *selecting* **Properties** from the pop-up menu, changing SOLIDFONT to PHANTOMFONT, *picking* **Apply**, then *picking* **Close**. Is it easier to determine what shape is created with the phantom lines added? See Figure 3-44.

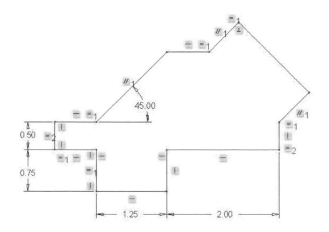

Figure 3-43 Sheet Metal Pattern

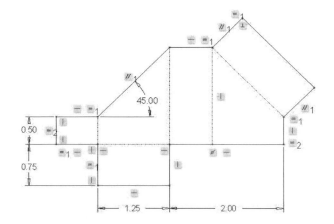

Figure 3-44 Pattern with Phantom Lines

▶ **Sketcher Exercise**

Design Intent

A pipe needs to be pushed up and over to the right using a simple, inexpensive jig. There is a small protruding tab mounted to a base plate just to the left of the pipe. After brainstorming and analysis, it is decided that a simple tapered block will solve the problem. The slanted face of the block will push the pipe over and up as dictated by the problem statement. If the block is sized correctly, then it can be held in place by the protruding tab. See the layout below, Figure 3-45.

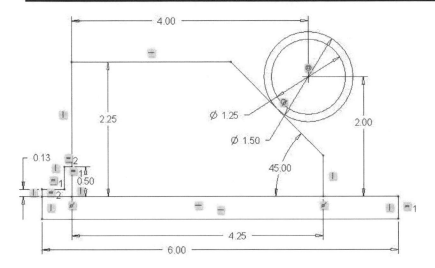

Figure 3-45 Layout 1

Design the tapered block (shown above) by properly sizing it according to the data given in Table 3-1.

Table 3-1 Design Data			
Dimension	**Design_1a** Size (inches)	**Design_1b** Size (inches)	**Design_1c** Size (mm)
Pipe_Locator	4.00	5.00	100
Pipe_Height	2.00	2.50	50
OD	1.50	2.00	42
ID	1.25	1.50	32
Push_Angle	45	60	45
Height	2.25	2.50	60
Base	?	?	?
Top	?	?	?

Let's use Creo Parametric's sketcher to draw the sketch above (Figure 3-45) to scale, and then determine the appropriate sizes for the tapered block.

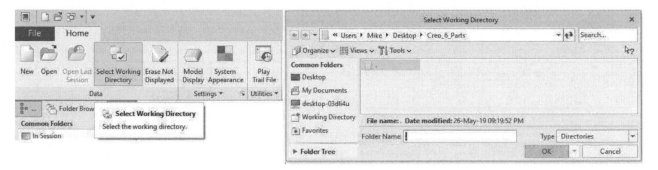

Figure 3-46 Set Working Directory

Using Creo Parametric to Sketch the Jig, Design_1a

Figure 3-47 New Sketcher File

Step 1: Start Creo Parametric by *double-clicking* with the LMB on the ***Creo Parametric*** icon on the desktop, or from the Program list: Creo Parametric.

Step 2: Set your working directory by *selecting* the ***Select Working Directory*** icon in the Home ribbon. Locate your working directory, and then *pick* **OK**. See Figure 3-46.

Step 3: Create a new sketch by *picking* the ***New*** icon at the top of the screen (Figure 3-47), or by *selecting* **File>New**.

Step 4: *Select **Sketch*** and name the sketch, "exercise1a". *Pick* **OK**. See Figure 3-48.

Step 5: 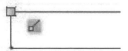 *Select* the ***Rectangle*** tool, and then draw a rectangle as shown below by *selecting* the location of two of its opposite diagonal corners using the LMB. Don't worry about its size right now. See Figure 3-49. The ▬ icon lets you know this line is horizontal. The ▮ icon indicates that the line is vertical.

Step 6: *Select* the ***Line*** tool and draw the backward "L-shaped" tab on the left end of the base. Be sure to return to the starting point by drawing the last horizontal line. Place the cursor near the upper left corner of the base. Creo Parametric will use the same point as previously defined when an X and a box appear over the top of the point as shown in Figure 3-50. Draw a short vertical line (left end of the backward L shape). When drawing the upper horizontal line and the left vertical line of the tab, extend the lines until the = symbol appears. When drawing the top horizontal line of the tab, again extend the line until the = symbol appears.

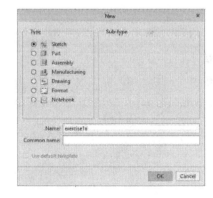

Figure 3-48 Name Sketcher File

Figure 3-49 Rectangle

Figure 3-50 Same Point

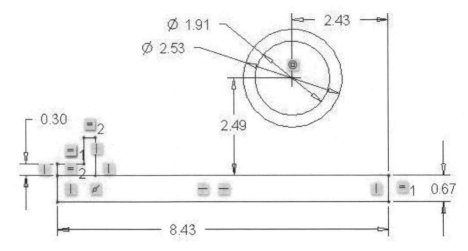

Figure 3-51   Initial Layout

Step 7: *Select* the ***Center and Point Circle*** tool. Move the cursor to the approximate center of the circles, then *press* the LMB. Move the cursor away from this point, thus increasing the size of the circle. When the circle appears approximately the correct size, *push* the LMB again. Draw two concentric circles near the right end of the base rectangle as shown in Figure 3-51. Don't worry about the circles' diameters.

Step 8: Your sketch should look similar to the one pictured in Figure 3-51, although the dimension locations and values may be different. The two shorter lines of the "L-shape" should be labeled with =2. The two longer lines of the angle tab and the thickness of the plate should be labeled with =1. The numbers 1 and 2 may be reversed.

Step 9: Because the dimensions shown above do not match our design intent, we will use the ***Normal dimension*** tool to add the appropriate dimensions. Note that as you add dimensions, others disappear. Creo Parametric will show only enough dimensions to completely define the shape. Dim dimensions are referred to as weak dimensions. Normal text dimensions are referred to as strong dimensions. All dimensions should be strong before you complete any drawing or sketch.

To place a dimension between two points, *pick* the first point with the LMB, *pick* the second point with the LMB, then move the cursor to where you want the dimension to show up and press the MMB. After placing the dimension, type the desired value in the box and press the <Enter> key.

To dimension a line length, *pick* the line with the LMB, then move the cursor to where you want the dimension to show up and press the MMB. After placing the dimension, type the desired value in the box and press the <Enter> key.

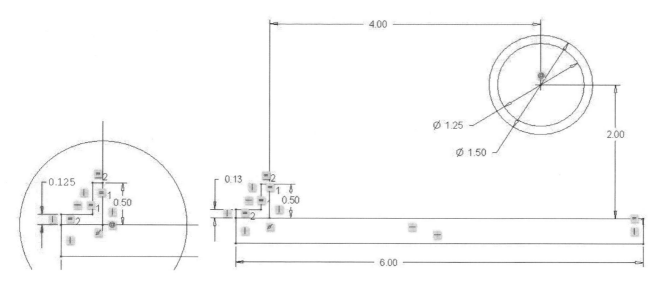

Figure 3-52 Reflects Design Intent

Modify your sketch so that it is similar to the one shown in Figure 3-52. Note that 0.125 rounds to 0.13 for 2-place decimals. To change a dimension's value, *double-click* using LMB, then type its new value followed by the <Enter> key. To move a dimension, *select* it with the LMB, then *press and hold* the LMB while moving the cursor. The dimension will follow. *Release* the LMB when the dimension has been properly placed.

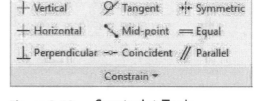

Figure 3-53 Constraint Tools

Step 10: If any 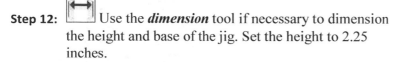's are missing from your sketch, they can be added using the ***constraint*** tools found on top of the screen (Figure 3-53). If there are extra symbols, they can be erased by *selecting* them with the LMB, then pressing the <Delete> key.

Step 11: Use the ***Line*** tool to draw the slanted-surface of the jig. Start at the base of the tab, move upward a couple of inches, then horizontally toward the two circles. Do not draw horizontally through the center of the circles. Stop before reaching the circles. Draw a slanted line that touches the outer circle tangent (Figure 3-54), then draw a vertical line back down to the base.

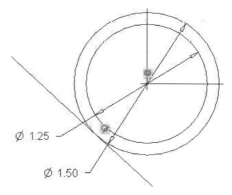

Figure 3-54 Slanted Surface Touches Outer Circle

Step 12: Use the ***dimension*** tool if necessary to dimension the height and base of the jig. Set the height to 2.25 inches.

Step 13: Use the ***dimension*** tool to define the angle between the base of the jig and the slanted edge. Set this value to 45 degrees. To dimension an angle, *pick* the slanted line with the LMB, *pick* the baseline with the LMB, then move the cursor between the two and press the MMB to locate the dimension. Type 45 <Enter> to set the angle to 45

Step 14: The base dimension can be any value; however, a value near 4.25 inches looks reasonable. See Figure 3-55. To change a dimension, *double-click* LMB on it, then type in the new value followed by the <Enter> key. Will a base size of 4.5, 4.0, or 3.25 work?

Step 15: It may appear that all necessary dimensions are shown for the jig; however, this is not true. The slanted side of the jig is controlled by the line being tangent to the outer diameter of the pipe. Without the pipe present, this slanted line could move. We need to add another dimension in this sketch to completely define the jig. *Select* the ***Dimension*** tool again.

Step 16: *Select* the horizontal line at the top of the jig with the LMB. Move upward a bit, then press the MMB to locate the dimension. Note that a "Resolve Sketch" window appears. See Figure 3-56. This appears because the sketch as defined is completely defined and we are trying to add another dimension. Several constraints and dimensions need to be modified in order to add this dimension. A reference dimension is a dimension shown on a sketch that is not necessary to define the sketch. When the window appears, be sure the new dimension is highlighted, then *pick* the **Dim>Ref** button. *Pick* **OK**.

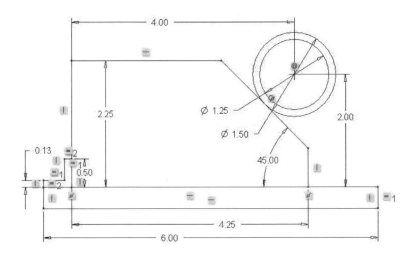

Figure 3-55 Sizing the Block

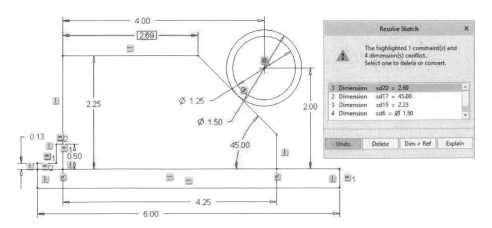

Figure 3-56 Resolve Sketch

Step 17: The letters REF will appear after the reference dimension. At this point, your sketch should look similar to the one shown in Figure 3-57. The top horizontal line of the jig needs to be 2.69 inches. This will cause the slanted side of the jig to just touch the outside diameter of the pipe.

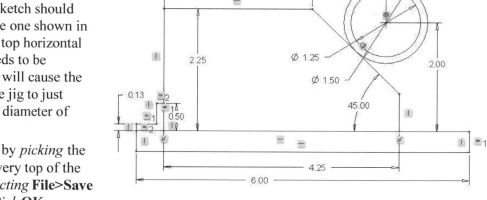

Figure 3-57 Reference Dimension Added

Step 18: Save your sketch by *picking* the **Save** icon at the very top of the screen, or by *selecting* **File>Save** from the menu. *Pick* **OK.**

Step 19: What if the pipe had an outside diameter of 1.75 inches, then the horizontal top surface of the jig would be_____inches. (Answer: Top = 2.51 inches.)

Step 20: Set the pipe's outside diameter back to 1.50 inches. What if the height of the jig was 2.00 inches, then the horizontal top surface of the jig would be____inches. (Answer: Top = 2.94 inches.)

Step 21: With the jig height at 2.00 inches, the base of the jig could be reduced to_____inches so that the pipe hits near the middle of the slanted surface. (Answer: Base = 4.00 inches.)

Do you see some of the power of parametric dimensioning? You can quickly see the effects of changing a dimension or parameter after the initial design is set up.

Using Creo Parametric to Design a Jig, Design_1b

Step 22: With sketch "exercise1a" visible on the screen, *select* **File> Save As>Save a Copy.** See Figure 3-58.

Step 23: In the New Name area, type "exercise1b", then *pick* **OK.** Do not change the model name. See Figure 3-59.

Figure 3-58 Save a Copy

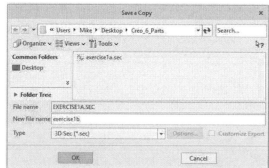

Figure 3-59 Save As "exercise1b (*.sec)"

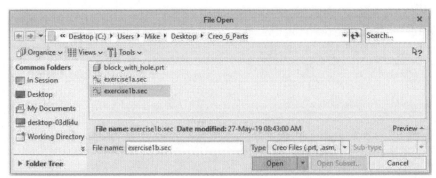

Figure 3-60   Close Window

Figure 3-61   Open "exercise1b.sec"

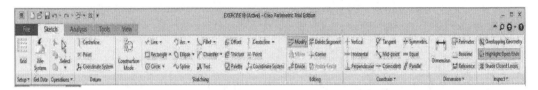

Figure 3-62   Filename at Top of Screen

Step 24:   **File>Close** to close the window and clear the screen (Figure 3-60).

Step 25:   **File>Open** or pick the ***Open*** icon.

Step 26:   Locate "exercise1b.sec" in your working directory. *Pick* the **Open** button as shown in Figure 3-61.

Note that "EXERCISE1B" appears at the top of the Creo Parametric window (Figure 3-62).

Step 27:   Modify the dimensions according to Design_1b's parameters shown in Table 3-2 and Figure 3-63. To modify a dimension, *double-click* on its value using the LMB, and then type a new value followed by the <Enter> key.

Table 3-2 Design 1b's Data	
Dimension	**Design_1b Size (inches)**
Pipe_Locator	5.00
Pipe_Height	2.50
OD	2.00
ID	1.50
Push_Angle	60
Height	2.50
Base	?
Top	?

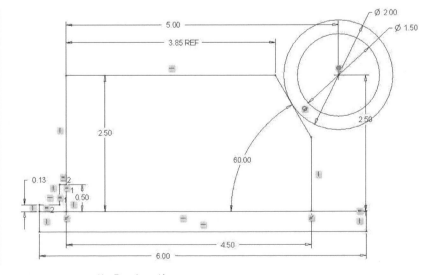

Figure 3-63   Jig Design 1b

Step 28: Within a few minutes the six defined parameters should be changed. Setting the jig's base to 4.50 inches allows the pipe to touch near the center of the slanted surface. The jig's new top dimension is _____ inches. (Answer: Top = 3.85 inches.)

The power of parametric dimensioning is the ability to change dimensional values without redrawing the object.

Step 29: 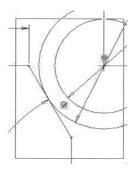 *Select* the **Zoom In** icon from the graphics toolbar. Draw an imaginary box around the area where the outer diameter of the pipe touches the slanted surface using the LMB. See Figure 3-64. Press the MMB to exit from this mode.

Figure 3-64 Zoom In

Step 30: 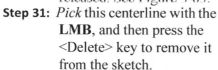 *Select* the **centerline** tool. Using the LMB *pick* the center of the pipe, then the point where the line touches the jig's slanted surface. The normal force produced by the pipe pushing on the slanted surface will lie along this line. If this line crosses the left edge of the jig above the fixed tab, then the jig will not stay in place by itself. In this case, this line crosses below the fixed tab, thus this jig will stay in place when released. See Figure 3-65.

Step 31: *Pick* this centerline with the **LMB**, and then press the <Delete> key to remove it from the sketch.

Step 32: **File>Save** or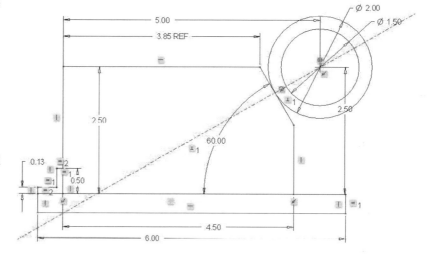

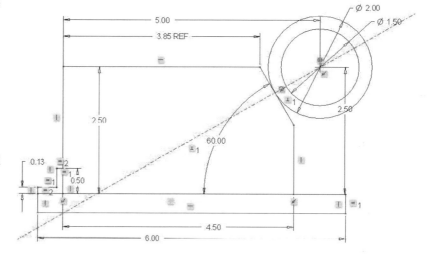

Figure 3-65 Line of Action for Pipe Force

Step 33: **File>Manage File>Delete Old Versions.** See Figure 3-66.

Step 34: *Pick Yes* to remove all previously saved versions of this file in your working directory (Figure 3-67).

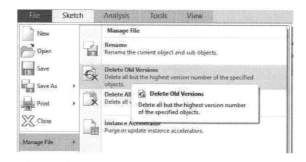

Figure 3-66 Delete Old Versions

Figure 3-67 Accept Command Actions

Using Creo Parametric to Design a Jig, Design_1c

Step 35: With sketch "exercise1b" visible on the screen, *select* **File>Save As> Save a Copy**.

Step 36: In the New Name area, type "exercise1c", then *pick* **OK**. Do not change the model name.

Step 37: **File>Close** to close the window and clear the screen or *pick* ⊠

Step 38: **File>Open** or *pick* the **Open** icon.

Step 39: Locate "exercise1c.sec" in your working directory. *Pick* the **Open** button. Note that "EXERCISE1C" appears at the top of the Creo Parametric window.

Step 40: Draw an imaginary box around the entire sketched jig including all dimensions. *Pick* the first corner with the LMB, then *press and hold* the LMB while moving to the opposite corner of the figure. *Release* the LMB.

Step 41: ⊅ *Select* the **Modify** tool.

Step 42: *Check* the "Lock Scale" box, then type a value of 100 where there was previously a value of 5.00. See Figure 3-68. *Pick* **OK** to update the entire sketch. (Be sure to *check* the **Lock Scale** box before changing the dimension to 100.) Use the **Refit** tool to refit the jig to the screen.

Step 43: Change the tab's height dimension from 10 to 12, and the tab thickness dimension from 2.5 to 3. Change the rectangular base's length from 120 to 140. See Figure 3-69.

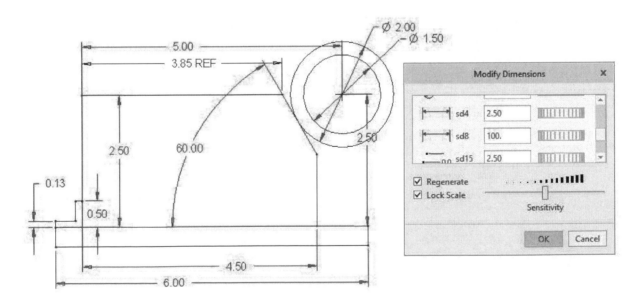

Figure 3-68   Scale All Dimensions

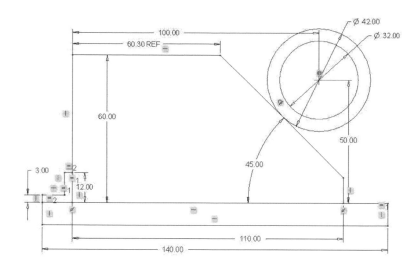

Figure 3-69 Jig Design 1c

Table 3-3	Design 1c's Data
Dimension	**Design_1c Size (mm)**
Pipe_Locator	100
Pipe_Height	50
OD	42
ID	32
Push_Angle	45
Height	60
Base	?
Top	?

Step 44: Continue to modify the dimensions according to Design_1c's parameters shown in Table 3-3. To modify a dimension, *double-click* on its value using the LMB, and then type a new value followed by the <Enter> key.

Step 45: Within a few minutes the six defined parameters should be changed. Setting the jig's base to 110 millimeters allows the pipe to touch near the center of the slanted surface. The jig's new top dimension is _____ mm. (Answer: Top = 60.3 mm.) See Figure 3-69.

The power of parametric dimensioning is the ability to change dimensional values without redrawing the object.

Step 46: File>Save.

Step 47: File>Manage File>Delete Old Versions. *Pick Yes* to remove all previously saved versions of this file in your working directory.

Step 48: File>Close.

Step 49: If you are done at this time, then exit from Creo Parametric by picking the ✖ in the upper right corner of the window. When you move the cursor over the icon, it changes to red. When asked, "Do you want to exit?" *Pick* **Yes**.

End of Sketcher Exercise.

▶ Review Questions

1. What is the main function of the left mouse button (LMB) in sketcher?

2. What are the main functions of the middle mouse button (MMB) in sketcher?

3. What are the main functions of the right mouse button (RMB) in sketcher?

4. What might happen to your sketch if you delete a sketcher constraint such as ▬ │ ⊘ ?

5. What is the difference between a light blue dimension and a blue dimension?

6. How do you create an explicit dimension for a line?

7. How do you convert a weak dimension into a strong dimension?

8. How do you create a diameter dimension?

9. How can you verify that a given section is closed?

10. Can you trim a construction line?

11. What is a reference dimension?

12. Why do weak dimensions show up on your sketch after creating a series of constructions?

13. How can you find where two lines overlap?

14. How do you scale all dimensions in a sketch at one time?

15. How do you change the length of a line?

16. How do you change the diameter of a circle?

17. How do you force a line to be tangent to a circle?

18. How do you find the name of a sketcher variable?

19. How do you get rid of a weak dimension placed on the sketch by the intent manager?

20. How do you zoom in on a selected area of a sketch?

21. How do you zoom out?

22. If you are sketching on the FRONT datum plane and the sketch gets rotated in 3D, how do you reposition the sketch so the FRONT datum plane is parallel with the screen again?

23. Should you set your sketcher constraints or your sketcher dimensions up first? Why?

24. How do you create a symmetrical sketch?

25. What is the main advantage of parametric design?

Sketcher Problems

3.1 Draw "**shear_plate**" (Figure 3-70). **Design Intent**—The left hole must be directly 15/16 inches above the lower left corner at point A. The right hole must be 2.875 inches from the left hole. The slot must be oriented 30 degrees from the indicated centerline. What are the length of line KA, height at point L, and the angle at K? (>**File**>**Options**>**Sketcher**, change to 3 decimal places. *Pick* **OK**.)

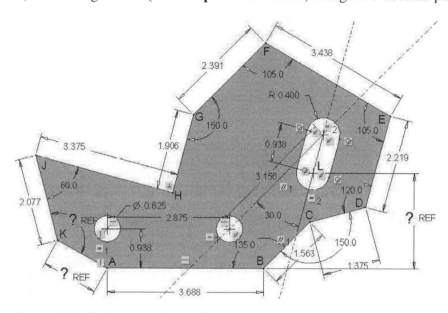

Figure 3-70 Problem 1—Shear Plate

3.2 Draw "**latch_plate**" (Figure 3-71). **Design Intent**—The plate must be symmetrical about a horizontal centerline (draw 1st). The four holes must be at the corners of a 3.000-inch by 3.500-inch imaginary box. Note that the beginning of the 15° line is directly below or above the center of the circles of radius R1. What is the length of the far right vertical line L_2 and the horizontal line L_1? Note that 2.093 must be a reference dimension if the sketch is drawn correctly. (>**File**>**Options**>**Sketcher**, change to 3 decimal places before sketching. *Pick* **OK**.)

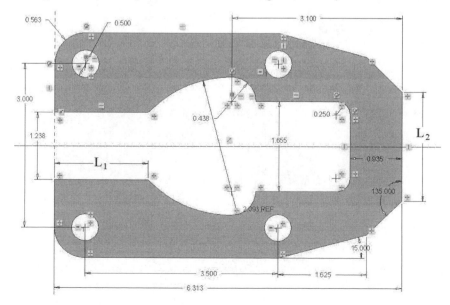

Figure 3-71 Problem 2—Latch Plate

3.3 Draw the front view of "**special_cam**" (Figure 3-72). **Design Intent**—The cam must be a symmetrical ellipse (draw centerlines). The hexagonal shaped hole must be 1.500 inches from the geometric center of the cam. The cutout and small hole are for mass reduction only. Mating arcs are tangent. What is the maximum height of the circular arc cutout? The 3.625 radius arcs are centered at the top and bottom of the ellipse. (>**File**>**Options**>**Sketcher**, change to 3 decimal places before sketching. *Pick* **OK**.)

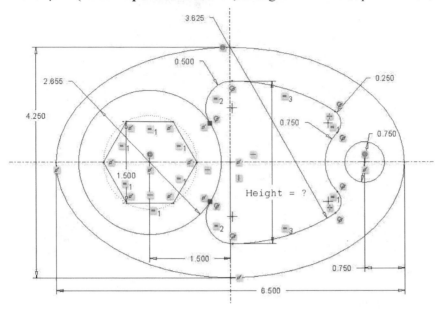

Figure 3-72   Problem 3—Special Cam

3.4 Draw the front view of "**rocker_arm**" (Figure 3-73). **Design Intent**—The hexagonal shaped hole of the rocker must be offset by 15 degrees. The right 0.65-inch diameter hole must be 3.00 inches from the center hex hole. **Design change:** The left 0.65-inch diameter hole must be 5.62 inches from the right hole instead of 5.82 inches as shown. What is the overall size of the rocker arm? (>**File**>**Options**>**Sketcher**, change to 2 decimal places before sketching. *Pick* **OK**.)

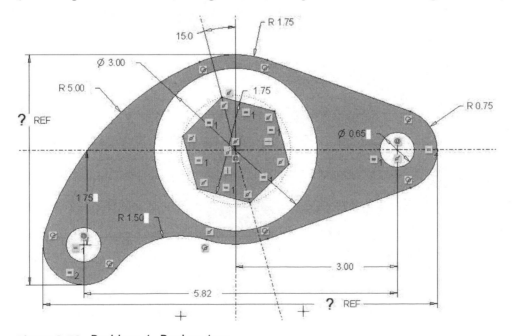

Figure 3-73 Problem 4—Rocker Arm

3.5 Draw the front view of the "**rocker**" (Figure 3-74). **Design Intent**—The rocker is symmetric about the center hole. The right 1.00-inch diameter hole must be 4.00 inches right of the center hole and 1.25 inches below it. The left 1.00-inch diameter hole must be 4.00 inches left of the center hole and 1.25 inches above it. The arcs that form the perimeter are tangent to each other. **Hint**: draw circles tangent to each other, then trim them to form the arcs. Note that there are four different radius arcs 1.00, 1.60, 2.00, and 5.00. (>**File>Options>Sketcher**, change to 2 decimal places before sketching. *Pick* **OK**.)

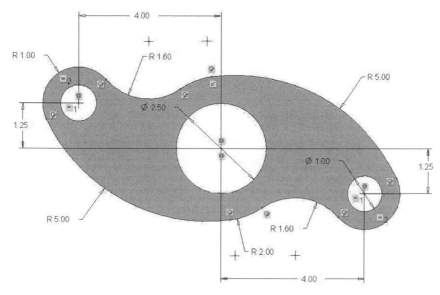

Figure 3-74 Problem 5—Rocker

3.6 Draw the front view of "**cover_plate**" (Figure 3-75). **Design Intent**—The plate is symmetric about its center in both directions. Why does the 3.00 dimension have the letters REF after it? (>**File>Options>Sketcher**, change to 2 decimal places before sketching. *Pick* **OK**.)

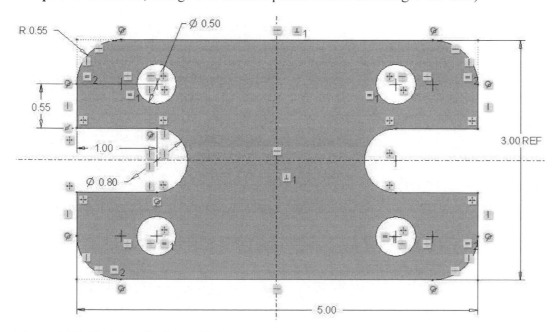

Figure 3-75 Problem 6—Cover Plate

3.7 Draw the "**front_cover_plate**" (Figure 3-76). **Design Intent**—All features are located from the outside edge of the cover plate, and not relative to each other. (>**File**>**Options**>**Sketcher**, change to 2 decimal places before sketching. *Pick* **OK**.)

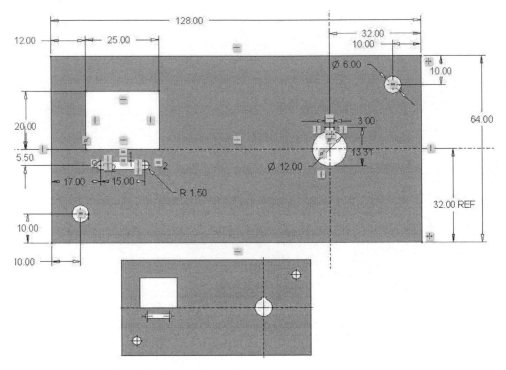

Figure 3-76 Problem 7—Front Cover Plate

3.8 Draw the "**support_frame**" (Figure 3-77). **Design Intent**—The thickness of the frame is ¾ of an inch. What is the volume of the shaded area if the support frame was 10 inches deep? Can you determine the volume underneath the support frame (white space under) if it is 10 inches deep? (>**File**>**Options**>**Sketcher**, change to 2 decimal places before sketching. *Pick* **OK**.)

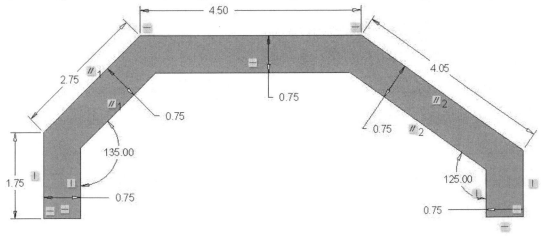

Figure 3-77 Problem 8—Support Frame

3.9 Draw the "**adjustable_sector**" (Figure 3-78). **Design Intent**—The two holes need to be 4.00 inches apart. The angular slot needs to be 25 degrees clockwise from vertical to 10 degrees counterclockwise below from horizontal. To reduce stresses round all sharp corners using a ¼-inch round or fillet. (Fillets and rounds are not shown in the figure.) (>**File**>**Options**>**Sketcher**, change to 2 decimal places before sketching. *Pick* **OK**.)

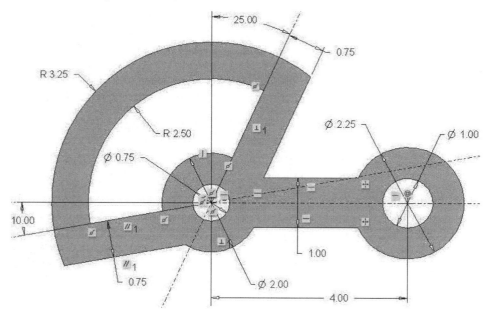

Figure 3-78 Problem 9—Adjustable Sector

3.10 Draw the "**cork_gasket**" (Figure 3-79). **Design Intent**—The three holes need to be on a 5.50-inch bolt circle spaced 120° apart. All boundary arcs are tangent to each other. (Note (=3) on numerous arcs equal 2.75 inches or half of the 5.50 bolt circle.) (>**File**>**Options**>**Sketcher**, change to 2 decimal places before sketching. *Pick* **OK**.)

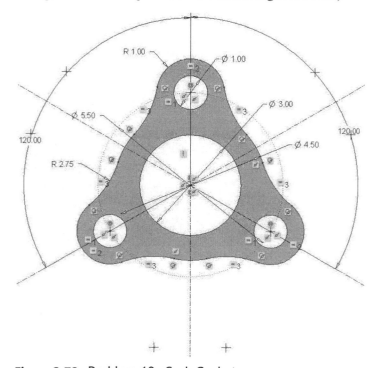

Figure 3-79 Problem 10—Cork Gasket

3.11 Challenge problem - Draw the **"Special_S-wrench"** (Figure 3-80). **Design Intent** — All arcs are tangent to each other and the edges of the two end shapes.   The diamond shape is 1.30 inches per side. The outer pentagon fits inside a circle of 1.75 inches in diameter. The inner pentagon uses a 1.20-inch diameter circle.   The 3-inch radius centerline is tangent at both ends and at its midpoint. The tangent points for the outer arcs are directly above and below the tangent point for the centerline arc. (**>File>Options>Sketcher**, change to 2 decimal places before sketching. *Pick* **OK**.)

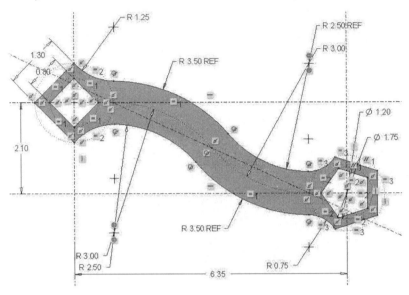

Figure 3-80 Problem 11—Special S Wrench

3.12 Challenge problem - Draw the **"Ratchet_Pawl"** (Figure 3-81). **Design Intent**—All arcs are tangent to each other or the straight edges. The R 12.00 arcs are centered along the horizontal lines. Pay close attention to the constraints shown. (**>File>Options>Sketcher**, change to 2 decimal places before sketching. *Pick* **OK**.)

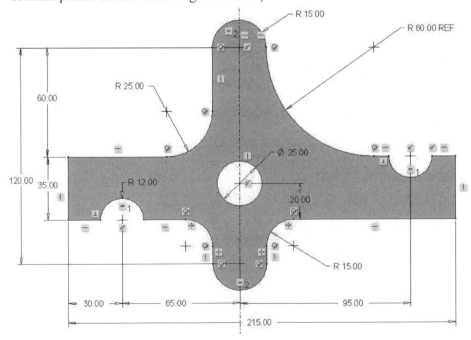

Figure 3-81 Problem 12—Ratchet Pawl

CHAPTER
4

EXTRUSIONS

Objectives

- ▶ 3D parts from extrusions explored
- ▶ Changing units of measure
- ▶ Practice creating extruded parts
- ▶ Saving a new copy of a modified 3D extruded part
- ▶ Dynamic view
- ▶ Modifying an existing 3D extruded part

- ▶ Extrusion exercise according to design intent
- ▶ Creating additional references in sketcher
- ▶ Adding rounds to a 3D part
- ▶ Extrusion design problems to reinforce concepts

▶ Extrusions Explored

At this point, you should be familiar with sketcher. If not, go back and work through the previous chapter because sketcher is the main tool of part creation. This section introduces extrusions, which are 2D area sketches extended into a third dimension, thus creating a three-dimensional volume. These extrusions can be positive, which adds material, or negative, which subtracts material. The first extrusion is always positive and is referred to as the parent. The additional extrusions will be child features of the first extrusion or previously defined extrusions.

A child feature is dependent upon the parent feature. To create a block with a hole, first, create the block (parent), and then add the hole (child). You cannot create the hole, then add the block because the child cannot come before the parent.

Before we begin creating extrusions in Creo Parametric let's practice visualizing how to create a part by drawing one of its cross-sections and extending it in the third dimension to create a volume. For example, let's visualize the creation of a block with two holes as shown in Figure 4-1.

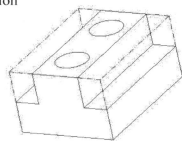

Figure 4-1 Block

75

We could create the upside-down tee-shaped cross-section, extend it into the third dimension, and then add the two holes. See Figure 4-2.

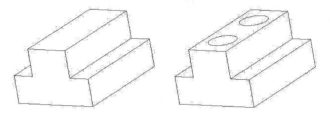

Figure 4-2   Block Creation

We could create the wide bottom section, and then add the top section and the two holes. See Figure 4-3.

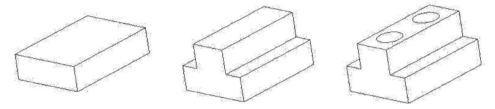

Figure 4-3   Block Creation 2

We could create the middle section with two holes and then add the lower extensions. See Figure 4-4.

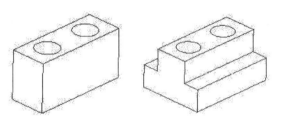

Figure 4-4   Block Creation 3   **Figure 4-5**   Block15703

As you can see there are several ways to create this part. Based on design intent, one way may be better than the others, but each way did create the same part. Don't be too concerned if you are creating a part in a different order than your classmates. The important thing is: "Can you create the part?"

Let's look at block15703, a bushing support, (Figure 4-5) that can be created from extrusions. How would you create this part?

Assuming Creo Parametric's top plane is the bottom of this part, sketch the footprint of this part. The sketched section must be closed to extrude it. Extrude it upward the specified height. See Figure 4-6.

Assuming the front plane is the back of this part, sketch the protrusion above the base, then extrude it forward the specified thickness (Figure 4-7).

Figure 4-6   Extrude Base

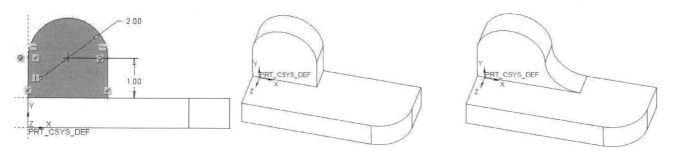

Figure 4-7 Add Top Portion

Figure 4-9   Add Rounds

Add the round where the upper protrusion meets the base protrusion. See Figure 4-8.

The holes can be part of the two extrusions or they can be added afterward (preferred method). In this case, the holes were added afterward. See Figure 4-9.

Let's look at block15701, an adjustable sliding axle support (Figure 4-10). How would you create this part? The paper scale is laid out in inches.

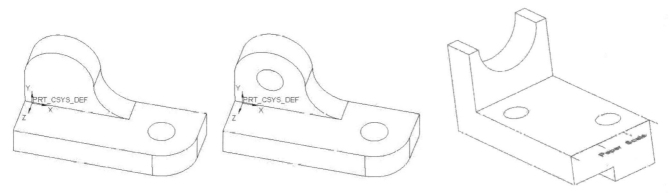

Figure 4-9 Add Holes

Figure 4-10   Block15701

Assuming the paper scale is on the right end of the part, sketch its cross-section, then extrude it toward the right the entire length of the part. See Figure 4-11.

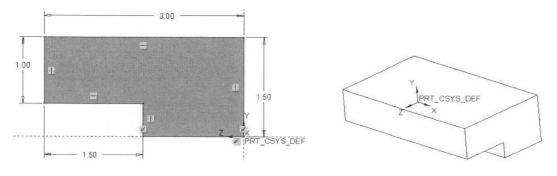

Figure 4-11 Extrude Base

Use the left end of the part as the sketching plane, sketch the protrusion above the base section, and then extrude it to the appropriate width (Figure 4-12).

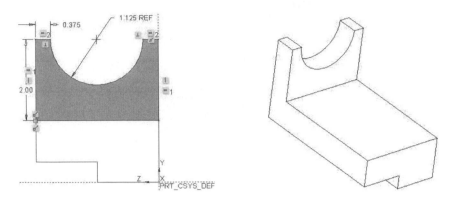

Figure 4-12   Extrude Upper Section

Using the top of the base as a reference, add the two holes in the base section. See Figure 4-13.

Let's look at block15713, a support bracket (Figure 4-14). Its construction should be obvious.

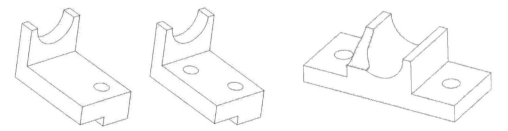

Figure 4-13   Add Holes   **Figure 4-14** Block15713

Sketch the front view, then extrude it the appropriate depth. See Figure 4-15. Let's assume the origin is in the exact middle of the bottom plane of the part. (The cutaway shows the second hole.)

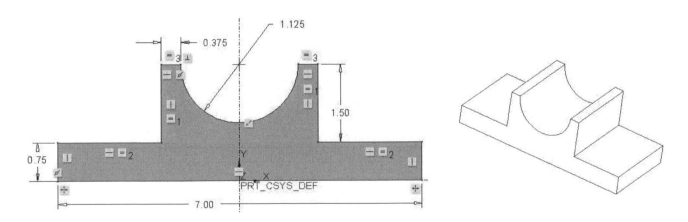

Figure 4-15   Extrude Base

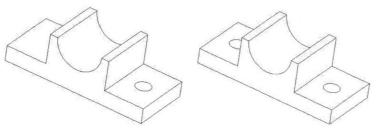

Figure 4-16 Add Holes

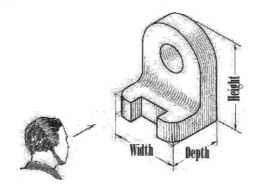

Add the two holes in the top of the base (Figure 4-16).

The man pictured in Figure 4-17 is looking at the front of this part. The origin of this part is directly below the hole on the back side of the part. How might you create this part? Are you seeing a pattern yet?

Figure 4-17   Front View

One approach might be to sketch the footprint of the bracket which will define the width and depth of the part along with the front cutout and rounded corners. See Figure 4-18. Note that the origin is located in the middle of the back edge of the part.

Add the upper protrusion by sketching on the front plane and extruding it forward the appropriate thickness. See Figure 4-19.

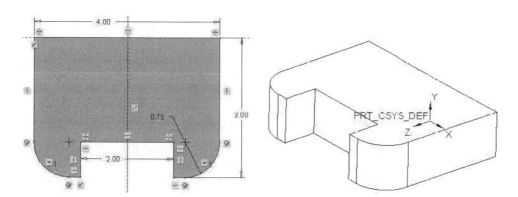

Figure 4-18 Extrude Base

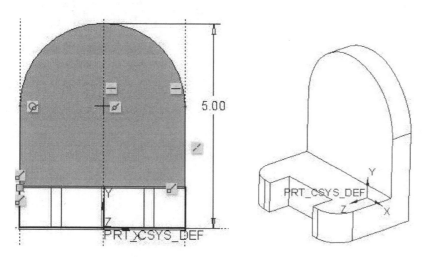

Figure 4-19 Extrude Upper Section

Add the hole to the top protrusion centered at the radius center. Finally, add a rounded edge between the two protrusions. See Figure 4-20. Note that these two operations could be done in either order.

Do the constructions make sense? Describe how you might construct the following part (Figure 4-21) in Creo Parametric. The lower rectangular slot goes all the way through the part.

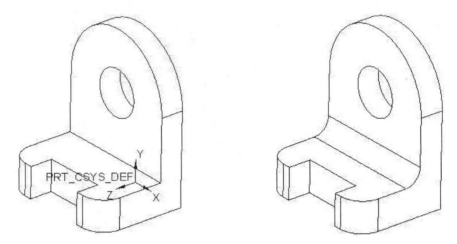

Figure 4-20   Add Holes

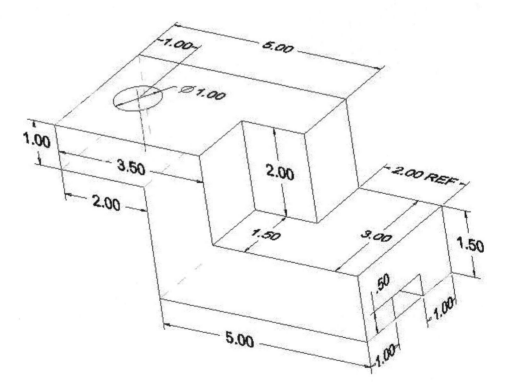

Figure 4-21   Slotted Block Showing Design Intent

Extrusions Practice

New part creation follows a standard procedure in Creo Parametric. This procedure is used for most thin-walled and solid extrusions. This text will make several assumptions when you are creating a new part. We will use the default template file when creating a new part. This template contains the FRONT, TOP, and RIGHT side datum planes along with an XYZ coordinate system locating the origin (0, 0, 0). We will use the default system of units, IPMS, inch-lbm-second. When we are doing stress analysis, we will use the IPS, inch-pound-second, system of units because stresses make more sense in these units. New parts will not be assigned a material property, so if an analysis is done requiring a material property, a material must be assigned by the user.

There are several ways to extrude a feature. Let's begin by starting Creo Parametric and setting the working directory.

Step 1: Start Creo Parametric by *double-clicking* with the LMB on the **Creo Parametric** icon on the desktop, or from the Program list: Creo Parametric.

Step 2: Set your working directory by *selecting* the **Select Working Directory** icon in the Home ribbon. Locate your working directory, and then *pick* **OK.** See Figure 4-22.

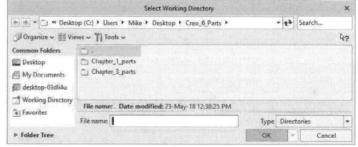

Step 3: Create a new part by *picking* the

Step 4: *Select* **Part** from the window and name it "my_first_part." The sub-type should be **Solid.** Make sure the "Use default template" box is *checked*, and then *pick* **OK.** See Figure 4-24.

Figure 4-22 Set Working Directory

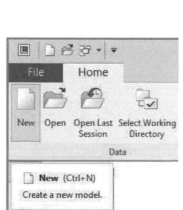

Figure 4-23 New Sketcher File   **Figure 4-24**   New Part

An XYZ coordinate system will appear in the middle of the screen along with FRONT, TOP, and RIGHT side planes (Figure 4-25). (**File>Options** → Entity Display → Show Datum Plane Tags must be checked to see the words FRONT, TOP, and RIGHT.)

The red-green-blue lines in the middle of the screen are in the X-Y-Z directions. See Figure 4-26. This symbol is the spin center. It can be turned ON or OFF by clicking on its icon at the top of the window.

We will create a 3D part by sketching a 2D shape, then extruding it in the third direction. In this case, the pipe support jig can be created by sketching the front view (Figure 4-27) and extending it 2.0 inches in the third direction. First, we need to set the units and properties.

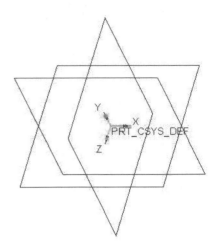

Figure 4-25 XYZ Coordinate System and Spin Center

Figure 4-26   Spin Center

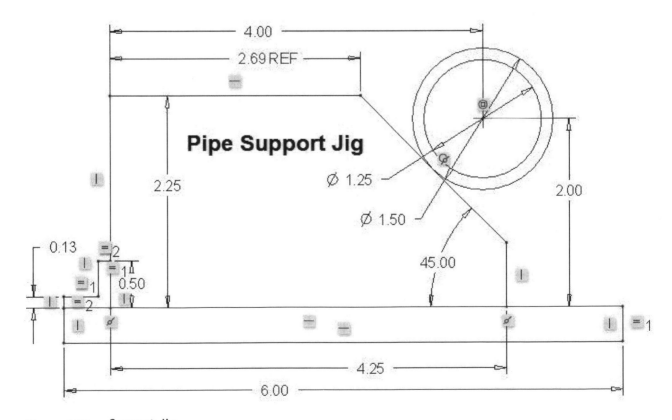

Figure 4-27   Support Jig

Step 5: We need to set the proper units for our pipe support jig. Let's use the IPS (Inch-Pound-Second) system of units for practice. Use **File>Prepare> Model Properties**. See Figure 4-28.

Step 6: The following Model Properties window (Figure 4-29) appears. *Pick* the word **change** on the right end of the Units row. Note that a material property is not assigned.

Figure 4-28 Model Properties

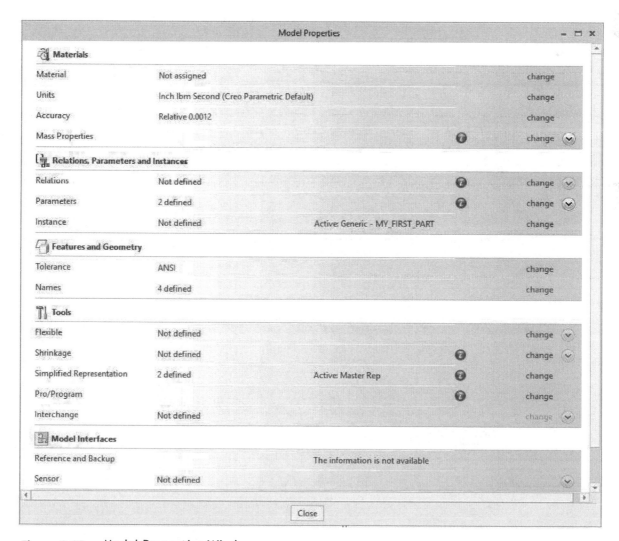

Figure 4-29 Model Properties Window

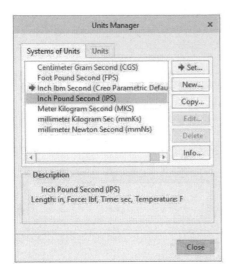

Figure 4-30 Window Units Manager

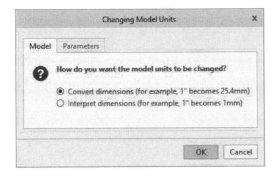

Figure 4-31   Changing Model Units

Step 7: The Units Manager window (Figure 4-30) appears. If the system of units, **IPS,** is not selected as shown in Figure 4-30, *pick* it from the list, then *pick* the [→ Set...] button.

Step 8: After *picking* the [→ Set...] button, the following window appears. *Select* the Convert dimensions (for example 1″ becomes 25.4mm) choice, then *pick* **OK.** See Figure 4-31.

Step 9: *Pick* **Close** on the Units Manager window. *Pick* **Close** on the Model Properties window.

Step 10: [Extrude icon] *Select* the **Extrude** tool icon located at the top of the screen under the Model tab.

Step 11: The upper left corner of the screen should look similar to the figure below with the Extrude tab selected. The two icons above the word Placement are for creating a **Solid** or a **Surface.** Be sure the Solid icon is selected. Note that the word Placement is red. This means that this section needs additional information to continue. *Pick* the word Placement. See Figure 4-32.

After you pick the word Placement, the following appears. Note that "• **Select 1 item**" has a red dot on your screen (Figure 4-33), thus it needs additional information before you can continue. *Pick* the **Define...** button.

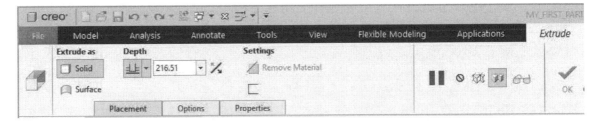

Figure 4-32 Placement Needs More Information

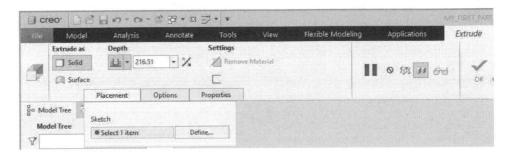

Figure 4-33   Define Sketch

Step 12: Move the cursor onto the FRONT plane so that it turns color. *Select* it by clicking the left mouse button (LMB). The RIGHT plane turns green and the area in the upper right corner of the screen shows the following sketch window. See Figure 4-34.

The FRONT plane will be the sketching plane. This is the plane we will use to create the 2D sketch. The sketch orientation indicates that the RIGHT plane will be toward the right side of the screen. This is what we want. *Pick* **Sketch.**

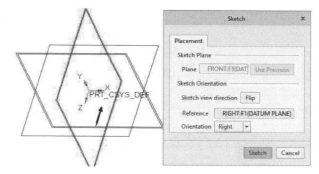

Figure 4-34 Sketch Plane and Orientation

Step 13: If needed, *pick* the ***Sketch View*** icon to orient the sketch plane so that it is parallel with the display screen.

Step 14: A new orientation (Figure 4-35) appears. In the middle of the screen, you can see the XYZ coordinate system. This is (0, 0, 0), our reference point in 3D space. We are sketching on the FRONT plane. The RIGHT side plane is perpendicular to the screen and facing toward the right side on the screen. The TOP plane is perpendicular to the screen and facing upward.

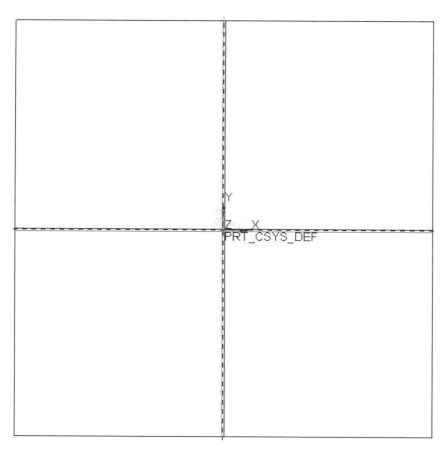

Figure 4-35 Sketch Window

The sketcher screen may be colored. For the purposes of this textbook, the sketcher screen will be shown as dark lines on a white background as shown in Figure 4-35. Note that a new set of tools is available at top of the screen under the Sketch tab. These are the sketcher tools described previously. Looking at the pipe support jig we note that it can be drawn with a series of straight lines so we will *select* the **line** icon. A small X appears at the end of the mouse cursor.

Step 15: Sketch the pipe support jig with its lower left corner being at the origin. Begin sketching the front view as shown below starting at the origin and drawing the vertical line first. To draw a line, use the LMB and *select* the origin. Move the mouse cursor vertically. When the line looks long enough, *select* the location on the screen using the LMB. A new line will begin at this point. Move the mouse cursor to the right. *Select* the location, then continue.

Step 16: When finished drawing the lines, click the MMB twice to exit drawing a line mode or move the cursor to the **arrow** icon at top of the screen and *pick* it.

Step 17: If the **Shade Closed Loops** icon is active, then the closed section should be colored in. If it is not selected, then *select* it and verify that you have created a closed section.

Sketcher places enough dimensions on your sketch to completely define it. Your sketch may look similar to the one pictured in Figure 4-36. Don't worry about the physical size of the jig at this time. The actual values of the dimensions are not important at this time. Note that the default dimensions do not match our design intent so we must use the dimension tool to add the appropriate dimensions. In my case, the 142.56-degree angle and the 38.97-inch vertical height for the short side do not match design intent. We need to add the overall length of the jig and the slanted side's angle relative to the base. (Note that: | = this line is vertical, ═ = this line is horizontal, and ✗ = this point is coincident.)

The dimmed dimensions on your sketch are not the correct lengths and are referred to as weak dimensions. Here is where the power of parameter modeling shows up. The lines in the sketch will change the size to match the dimensions you enter.

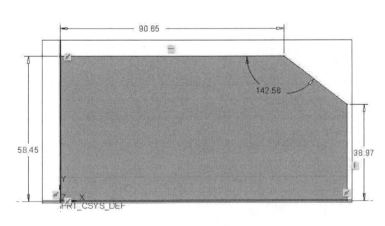

Figure 4-36   Pipe Support Jig Sketch

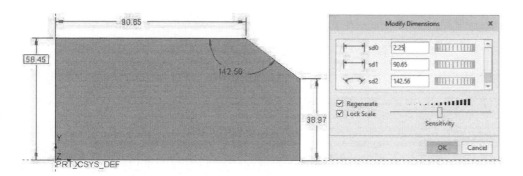

Figure 4-37 Lock Scale and Modify **all Dimensions at Once**

Step 18: To get all the dimmed dimensions drawn to approximately the correct value, highlight the entire sketch by drawing an imaginary box around it and the dimensions. *Pick* a point above and to the left of the sketch, *hold* the LMB (left mouse button) *down,* and then drag the mouse to a point below and to the right of the sketch. *Release* the LMB. All selected entities turn color.

Step 19: ⌐ Modify *Select* the *Modify* icon from the top of the screen. This brings up the following window (Figure 4-37). Your dimensions will be different because no two people can sketch a view using exactly the same dimensions. Clicking on any of the dimensions in this window highlights the appropriate dimension on your sketch. *Select* the **Lock Scale** checkbox first, and then locate the height dimension of the jig. Type 2.25 <Enter> since this is the desired height. *Pick* **OK**. Use the *Refit* icon to resize the sketch according to the new dimensions entered.

Step 20: ↔ *Select* the *normal dimension* tool and add the overall length of the jig and the angle between the base of the jig and the slanted side. Note that two weak dimensions will disappear when these dimensions are added. To change an existing dimension, *double-click* on it with LMB, type the new value, then press <Enter>. Your sketch should appear as shown in Figure 4-38. (Datum Planes have been turned off.)

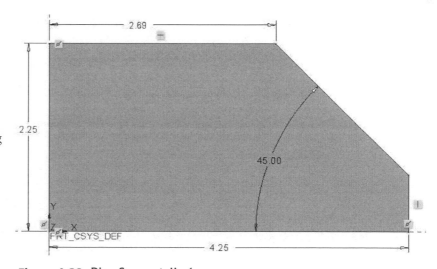

Figure 4-38 Pipe Support Jig 1a

Step 21: *Select* the ***Feature Requirements*** icon (Figure 4-39) in the Sketch ribbon. This will verify that your sketch is complete and ready to be extruded.

Step 22: *Pick* **Close** in the Feature Requirements window.

Step 23: *Pick* the ***green checkmark*** at top of the window to exit from sketcher and extrude the sketch.

Step 24: In the toolbar type 2.00 <Enter> for the depth of the extrusion (Figure 4-40), then *pick* the ***green checkmark*** to complete the extrusion.

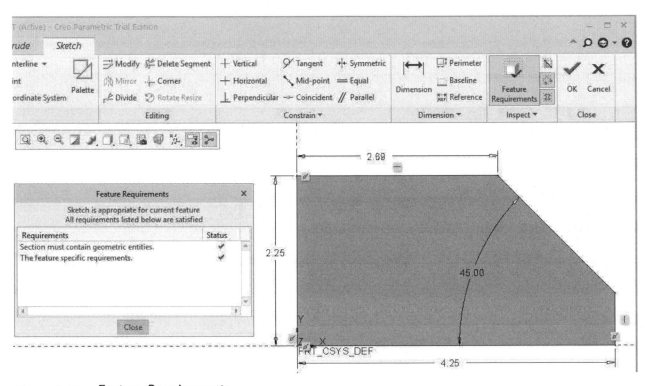

Figure 4-39   Feature Requirements

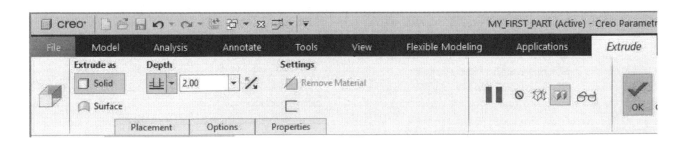

Figure 4-40   Set Depth of Extrusion

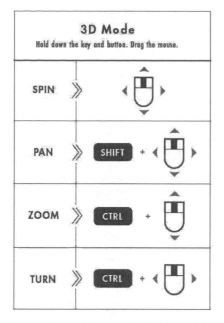

Figure 4-41 Mouse Button Functions

Note: The zoom when spinning the mouse wheel works only if the "Scroll inactive windows when I hover over them" feature is turned OFF in Windows 10.

Step 25: Use the mouse to move the jig around. See Figure 4-41. Practice using all of these options.

Step 26: 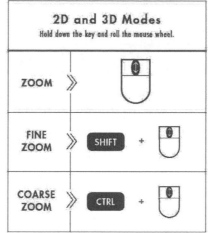 *Pick* the **Refit** icon at the top of the window.

Step 27: Turn off **Plane Display** to hide the FRONT, TOP, and RIGHT datum planes. See Figure 4-42.

Let's look at the different model display options. Become familiar with each. See Figure 4-43.

Figure 4-42 Turn Off Plane Display

Step 28: *Pick* **No Hidden** to remove all hidden lines and show only visible lines. Note that there is a Ctrl+4 shortcut.

Step 29: *Pick* **Hidden Line** to show the hidden lines as faint lines.

Step 30: *Pick* **Wireframe** to show all hidden lines as solid lines.

Step 31: *Pick* **Shading with Edges** to see the edges of the object drawn with black lines.

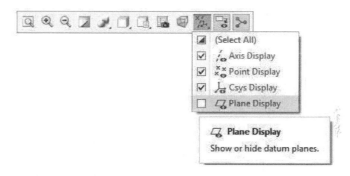

Step 32: *Pick* **Shading** to show all surfaces as shaded planes.

Figure 4-43 Part Viewing Mode

Step 33: *Pick* **Shading with Reflections** to add light sources and shadows. The number and location of the light sources can be changed. We will not do that in this text.

We will continue with the design.

Step 34: Set the model back to **No Hidden** mode. Fit to screen.

Step 35: **Save** your part. See Figure 4-44. *Pick* **OK.**

Figure 4-44 Save Part

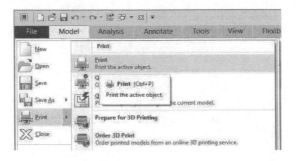

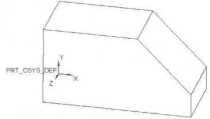

Figure 4-45 Print Part   **Figure 4-46** Printed Part

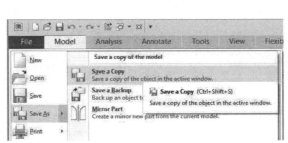

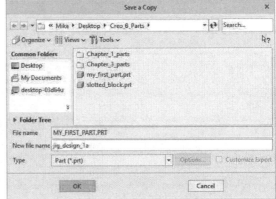

Figure 4-47 Save a Copy

Step 36: **File>Print>Print** to make a hard copy of your jig as proof you completed this practice. See Figure 4-45 and orient the part similar to Figure 4-46 before printing.

Step 37: Use **File>Save As>Save a Copy** to save a copy of your jig, naming it "jig_design_1a" (Figure 4-47). Type the new name in the blank area beside New Name, then *pick* **OK.** This is the part file for our 1st design.

Step 38: Keep working on the original file. Using the LMB *pick* the arrow in front of Extrude 1 in the model tree. Use LMB to *pick* the sketch, Section 1, underneath. *Select* **Edit Definition** icon from this pop-up menu. This will place you back in sketcher mode.

Step 39: If needed, *pick* the **Sketch View** icon to orient the sketch plane so that it is parallel with the display screen.

Step 40: Change the value on the sketch to match the values shown in Figure 4-48 for Design 1b, then exit from the sketcher by *picking* the **green checkmark**.

Step 41: Verify your design. See Figure 4-49. Use **File>Save As>Save a Copy** to save a copy of your jig, naming it "jig_design_1b". Type the new name in the blank area beside New Name, then *pick* the **OK** button. This is the part file for our 2nd design.

Isn't parametric dimensioning nifty? Now let's do design 1c. Since design 1c is in metric units, we need to switch to metric units before continuing.

Step 42: **File>Prepare>Model Properties.**

Step 43: *Pick* **change** on the Units—Inch Pound Second (IPS) line.

Step 44: *Select* "**millimeter Newton Second (mmNs),**" then *pick* the 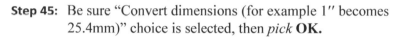 button.

Step 45: Be sure "Convert dimensions (for example 1″ becomes 25.4mm)" choice is selected, then *pick* **OK.**

Step 46: **Close** the Units Manager window.

Step 47: **Close** the Model Properties window.

Step 48: 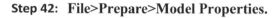 Using the LMB *pick* the arrow in front of Extrude 1 in the model tree. Use LMB to *pick* the sketch, Section 1, underneath. *Select* the **Edit Definition** icon from this pop-up menu. This will place you back in sketcher mode.

Step 49: 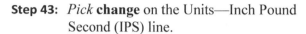 If needed, *pick* the ***Sketch View*** icon to orient the sketch plane so that it is parallel with the display screen.

Step 50: 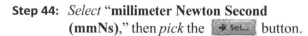 *Pick* the ***Refit*** icon at the top of the window.

Step 51: Change the value on the sketch to match the values shown in Figure 4-50 for Design 1c, then exit from the sketcher by *picking* the ***green checkmark***.

Step 52: 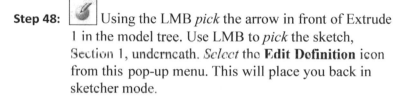 Using the LMB *pick* Extrude 1 in the model tree. *Select* **Edit Definition** icon from this pop-up menu. The Extrude tab will reappear. Change the depth of the extrusion from 50.8 to 50 mm. *Pick* the ***green checkmark***.

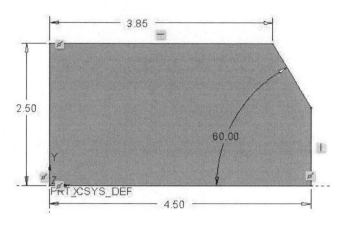

Figure 4-48   Jig Design 1b

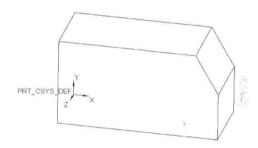

Figure 4-49   Pipe Support Jig 1b

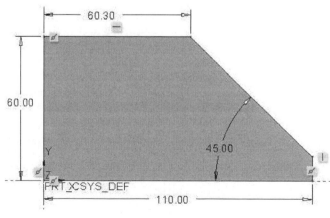

Figure 4-50   Jig Design 1c (mm)

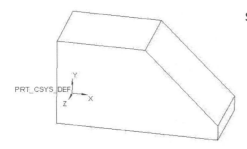

Figure 4-51   Pipe Support Jig 1c

Step 53: Verify your design. See Figure 4-51. Use **File>Save As>Save a Copy** to save a copy of your jig, naming it "jig_design_1c". Type the new name in the blank area beside New Name, then *pick* the **OK** button. This is the part file for our 3rd design.

Step 54: **File>Close** to clear the window or pick in the upper left corner of the window.

Step 55: *Pick* Erase Not Displayed icon or **File>Manage Session>Erase Not Displayed.** *Pick* **OK.** See Figure 4-52.

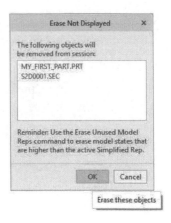

Figure 4-52 Erase Not Displayed

End of Extrusion Practice

▶ Extrusions Exercise

As an introduction to Creo Parametric extrusions, let's create the **stocktail clamp 138** shown on the next page with some modifications.

Design Intent

Some dimensions on Figure 4-53 do not reflect design intent. For example, the 0.43-inch wall thickness at the top of the clamp is not important. Instead, the flat cutout needs to be 1.38 inches wide. The 0.44-inch depth of the top cutout is not important. Instead, the thickness of the clamp needs to be 1.06 inches. The two slots need to be located along the geometric center with the origin of the clamp on the bottom surface at its geometric center.

Step 1: Start Creo Parametric by *double-clicking* with the LMB on the **Creo Parametric** icon on the desktop, or from the Program list: Creo Parametric.

Step 2: Set your working directory by *selecting* the **Select Working Directory** icon in the Home ribbon. Locate your working directory, and then *pick* **OK.** See Figure 4-54.

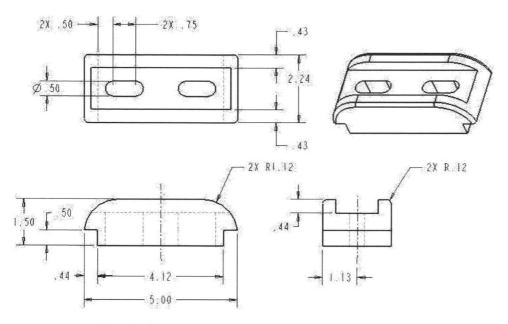

Figure 4-53 Tailstock Clamp 138

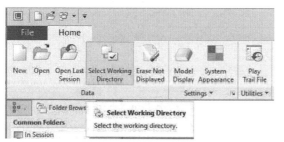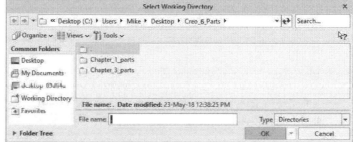

Figure 4-54 Set Working Directory

Step 3: Create a new part by *picking* the *New* icon at the top of the screen, or by *selecting* **File>New.**

Step 4: *Select* **Part** and sub-type **Solid,** then name the part, "tailstock_clamp_138". Be sure that "Use default template" box is *checked*. *Pick* **OK.**

Step 5: Select the **sketch** tool in the Model tab at the top of the screen. This time we are going to create the sketch first, then extrude it just to show you another way of creating an extrusion. Normally, we *pick* the **Extrude** tool, then create the sketch.

Step 6: With the sketch window present, *select* the FRONT plane as the sketch plane and the RIGHT plane oriented toward the right. *Pick* the **Sketch** button. See Figure 4-55.

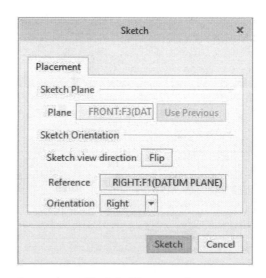

Figure 4-55 Sketch Plane and Orientation

Step 7: *Pick* the ***Sketch View*** icon to orient the sketch plane so that it is parallel with the display screen.

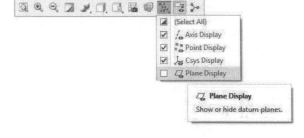

Step 8: Turn off ***Plane Display***. See Figure 4-56.

Figure 4-56 Plane Display Off

Step 9: Draw a vertical construction centerline down the middle of the screen and through the origin.

Step 10: Draw a rectangle symmetric about the centerline and sitting on the horizontal dashed line axis. The symbol ⊹ will appear when the left and the right side of the rectangle are equal or symmetrical about the centerline.

Step 11: Press MMB to exit from rectangle drawing mode.

Step 12: *Double-click* with LMB the vertical height of the rectangle. Type 1.50<Enter>. *Double-click* with LMB the horizontal length of the rectangle. Type 5<Enter>. *Pick* the ***Refit*** icon. See Figure 4-57.

Step 13: Draw two lines in both lower corners of the rectangle that will represent the two end cutouts.

Step 14: *Hold down* the LMB and drag the cursor through the two lower corners of the rectangle, thus removing both lower corners. See Figure 4-58.

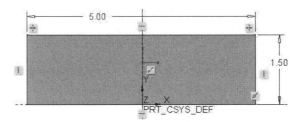

Figure 4-57 Symmetric Rectangle

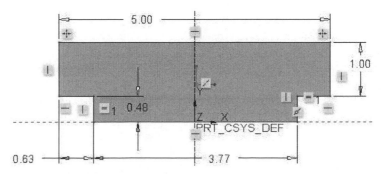

Figure 4-58 Corner Cutouts

Step 15: 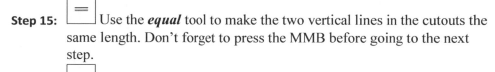 Use the *equal* tool to make the two vertical lines in the cutouts the same length. Don't forget to press the MMB before going to the next step.

Step 16: Use the *equal* tool to make the two horizontal lines in the cutouts the same length.

Step 17: Use the *normal dimension* tool to add dimensions that reflect the design intent as shown in Figure 4-59.

Step 18: Use the *circular trim* tool to round the upper left and right corners. *Pick* the top line with the LMB, then *pick* the left vertical edge line. Repeat picking the top line and the right vertical line.

Step 19: Use the *equal* tool to make the radii equal to each other by *selecting* both with the LMB. Then press the MMB twice to exit from this mode.

Step 20: *Double-click* on the value for the radius using the LMB, then type: 1.12<Enter>. If done correctly, the overall length of 5.00 should disappear. A new weak dimension will appear somewhere on the sketch. This weak dimension does not reflect our design intent so replace it with the overall dimension. Section not shaded because it is not a closed section yet.

Step 21: Use the *delete segment* tool to remove the two short vertical lines and extending into the end cutouts. See Figure 4-60.

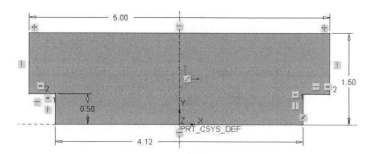

Figure 4-59 Design Intent Dimensions

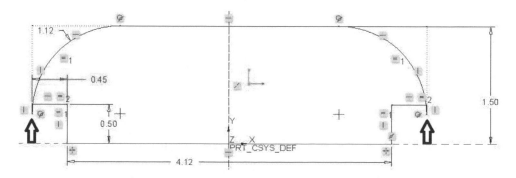

Figure 4-60 Remove Extra Line Segments

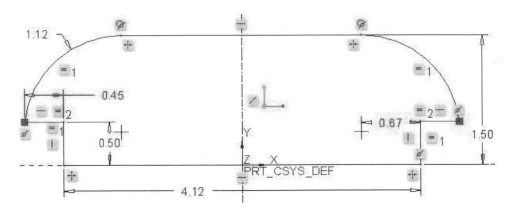

Figure 4-61   Highlight Open Ends

Step 22: [icon] *Select* the ***Highlight Open Ends*** tool, if not already selected, from the Sketch ribbon. You should see red dots at each end of the sketch as shown in Figure 4-61. Section not shaded because it is not closed yet.

Step 23: [icon] Use the ***Zoom In*** tool from the graphics toolbar and draw a box around the left red dots (Figure 4-62), thus zooming in on it.

Step 24: [icon] Use the ***delete segment*** tool to remove the short horizontal line.

Figure 4-62 Remove Extra Horizontal Line

Step 25: [icon] Use the ***Refit*** tool to zoom back out.

Step 26: Repeats the above three steps on the right red dots.

Step 27: [icon] Use the ***Normal Dimension*** tool to add a dimension which represents the overall length of the clamp. Using the LMB *pick* the furthest leftmost point, then the furthest rightmost point on the clamp. Move below the sketch and press the MMB to place the dimension.

Step 28: Press the MMB twice.

Step 29: Change the overall length to 5.00 inches. Remember to press <Enter> after typing the value.

Step 30: Verify that the sketch is a closed loop and is ready to be extruded. Your sketch should look similar to Figure 4-63.

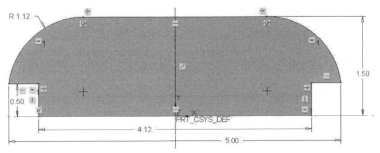

Figure 4-63   Closed Section with Design Intent

Figure 4-64 Highlight
Sketch 1 in Model Tree

Figure 4-65 Extrude Options

Figure 4-66 Solid Extrusion

Step 31: 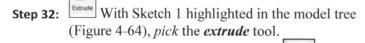 *Pick* the **green checkmark** to exit sketcher and
keep the sketch.

Step 32: *With* Sketch 1 highlighted in the model tree
(Figure 4-64), *pick* the **extrude** tool.

Step 33: *Select* extrude as a **solid**, **symmetric** , and enter
a width of 2.24 inches (Figure 4-65).

Step 34: *Select* the **glasses** to view your solid.
See Figure 4-66.

Step 35: *Pick* the **green checkmark** to accept
the extrusion. See Figure 4-67.

Step 36: If you need to modify the extrusion, *pick*
Extrude 1 in the model tree, press and hold
the RMB, then use the LMB to *pick* **Edit
Definition** icon from the pop-up menu.

Step 37: **File>Save.** *Pick* **OK.**

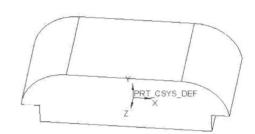

Figure 4-67   Clamp Extrusion

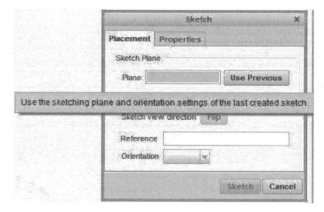

Figure 4-68   Use Previous

Design Intent

Add the top 1.38-inch wide cutout with the thickness of the clamp below
the cutout being 1.06 inches.

Step 38:  *Pick* the **sketch** tool again. *Pick* the **Use Previous** button. See
Figure 4-68.

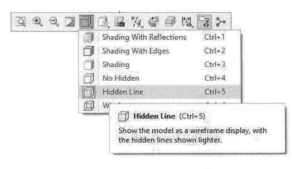

Figure 4-69　Hidden Line Mode

Step 39: If needed, *pick* the **Sketch View** icon to orient the sketch plane so that it is parallel with the display screen.

Step 40: *Select* **Hidden Line** mode from the graphics toolbar. See Figure 4-69.

Step 41: Use the **Project** sketcher tool to *select* the horizontal, top line, and the two end curves as shown in Figure 4-70.

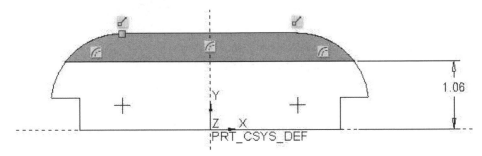

Figure 4-70　Sketch Top Cutout

Step 42: Use the **Line** tool to create a horizontal line 1.06 inches above the base of the part.

Step 43: Use the **Delete Segment** tool to remove the curved lines outside the enclosed area. See Figure 4-70.

Step 44: *Pick* the **green checkmark** to exit sketcher and keep the sketch.

Step 45: With Sketch 2 highlighted in the model tree, *pick* the **extrude** tool.

Step 46: *Select* extrude as a **solid**, **symmetric**, and **remove material**. Enter a width of 1.38 inches, then *select* the **glasses** to view your solid with the top cutout. See Figure 4-71.

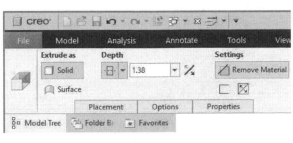

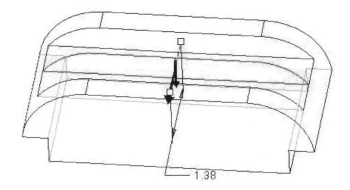

Figure 4-71　Top Cutout Extruded

Step 47: 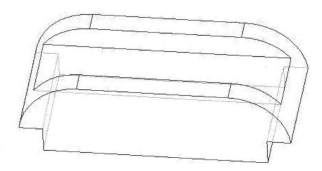 If all looks correct, *pick* the **green checkmark**. See Figure 4-72.

Step 48: If you need to modify this extrusion, *pick* **Extrude 2** in the model tree, press and hold the RMB, then use the LMB to *pick* **Edit Definition** icon from the pop-up menu. If you need to modify the sketch for Extrude 2, *pick* the down arrow to the left of Extrude 2 with the LMB, *select* **Sketch 2,** then press the LMB. *Select* the **Edit Definition** icon from this pop-up menu.

Figure 4-72   Clamp with Top Cutout

Step 49: File>Save.

Design Intent

Add the two symmetric, ½-inch wide through slots in the top cutout area. The straight portion of these slots needs to be 0.75 inches.

Step 50: **Sketch** *Pick* the **sketch** tool again. *Select* the top of the cutout as the sketching plane and the cutout's back surface wall (shaded in figure) oriented toward the bottom. *Pick* **Sketch.** See Figure 4-73.

Step 51: If needed, *pick* the **Sketch View** icon to orient the sketch plane so that it is parallel with the display screen.

Step 52: Turn Plane Display back on, if it is off.

Step 53: Draw a vertical construction centerline through the middle of the part. (Also, through the origin.)

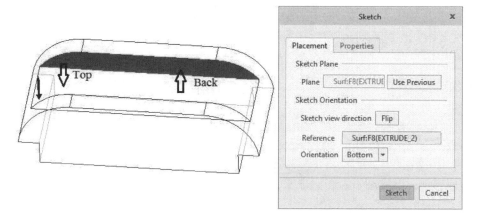

Figure 4-73 Sketch Plane and Orientation

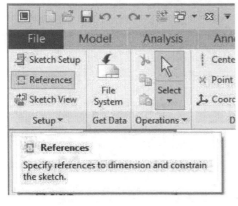

Figure 4-74   References for Sketcher

Step 54: *Select* the ***References*** icon in the Sketch tab. See Figure 4-74.

Step 55: *Pick* the FRONT plane and the two hidden lines that represent the outside undercuts. Dashed lines will appear on the selected references. *Pick* **Close.** See Figure 4-75.

Step 56: Draw four equal radius circles centered on the FRONT plane. See Figure 4-76.

Step 57: Draw horizontal lines at the very top and bottom of the left two circles and the right two circles. See Figure 4-77. Did you remember to press the MMB after drawing each line, and twice after drawing the fourth line?

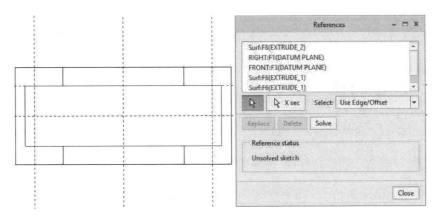

Figure 4-75   Add References

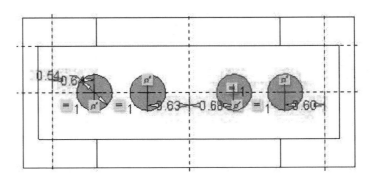

Figure 4-76   Four Equal Size Circles

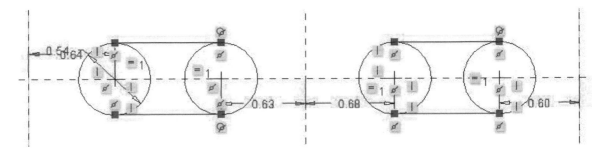

Figure 4-77   Connect Circles

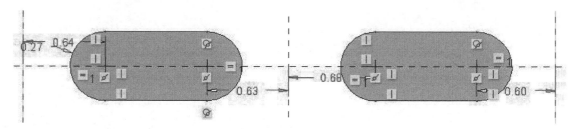

Figure 4-78   Two Slots

Step 58: 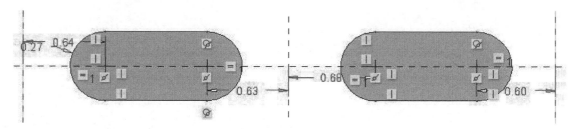 Use the *delete segment* tool to remove the inner parts of the circles, thus leaving the 2 slots. See Figure 4-78.

Step 59: Use the *normal dimension* tool to add dimensions per our design intent. Each slot is located from the edge of the undercut. Remember to create a diameter dimension, *pick* the curve twice with the LMB, then use the MMB to locate the dimensional value. Set the values as shown in Figure 4-79.

Step 60: *Select* the *Shade Closed Loops* icon at the top of the screen if not already selected to verify that both slots are closed shapes. See Figure 4-80.

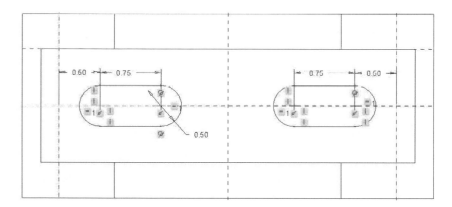

Figure 4-79   Design Intent Dimensions

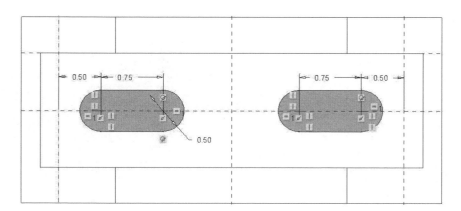

Figure 4-80   Closed Sections

Step 61: ✅ *Pick* the ***green checkmark*** to exit from sketcher and keep the sketch.

Step 62: *Extrude* With Sketch 3 highlighted in the model tree, *pick* the ***extrude*** tool.

Step 63: *Select* extrude as a ***solid***, ***though*** and ***remove material*** ▨ .

Step 64: ▧ Use the leftmost ***change depth direction*** icon to change the direction of the yellow arrow so it points downward into the part. (If necessary, use the rightmost ***change material direction*** icon to make the yellow arrow point inside the slot.) See Figure 4-81.

Step 65: ✅ If all is correct, *pick* the ***green checkmark***. See Figure 4-82.

Step 66: If you need to modify this extrusion, *pick* **Extrude 3** in the model tree, press and hold the RMB, then use the LMB to *pick* **Edit Definition** icon from the pop-up menu.

Step 67: File>Save.

Design Intent

To avoid cuts when handling the part, the two outside edges need to be rounded using a 0.12-inch radius. Leave all other edges unrounded.

Figure 4-81 Extruded Slot Cut

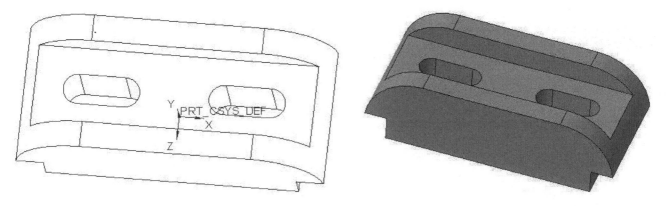

Figure 4-82   Clamp Design

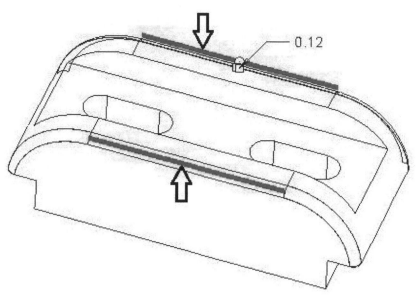

Figure 4-83 Round Upper Edges

Step 68: 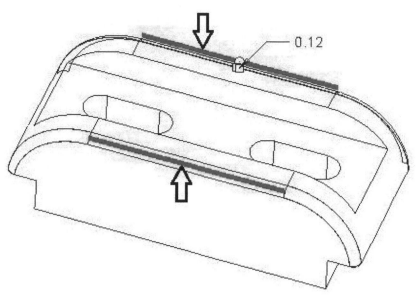 *Select* the **Round** tool under the Model tab. *Pick* the top front edge of the part, then hold down the <CTRL> key and *pick* the back top edge of the part. Type 0.12<Enter> for the radius. See Figure 4-83.

Step 69: *Pick* the **green checkmark** to accept the rounded edges.

Step 70: **File>Save.**

Step 71: **File>Manage File>Delete Old Versions.** *Pick Yes*.

Step 72: *Pick* **Refit** tool from the graphics toolbar to reposition the tailstock clamp 197.

Step 73: Be sure *No Hidden* is *selected* at the top of the screen. Turn off **Plane Display**.

Step 74: Orient the part similar to Figure 4-84, then print the part to prove you have completed this exercise. **File>Print>Print.** *Pick* **OK.**

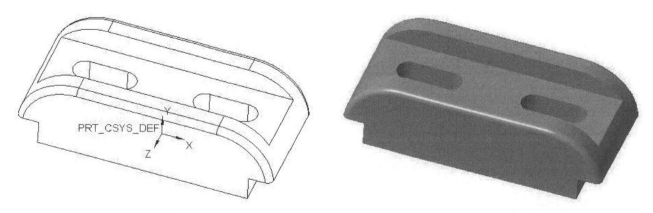

Figure 4-84 Final Design

Step 75: File>Close.

Step 76: **File>Manage Session> Erase Not Displayed.** *Pick* **OK.** See Figure 4-85.

End of Extrusion Exercise.

Figure 4-85 Erase Not Displayed

▶ **Review Questions**

1. What is an extrusion?

2. How do you name a part?

3. Where will a part file be saved?

4. Can you change the radius of a round on a 3D part without going back into sketcher? If so, how?

5. What is the difference between Edit and Edit Definition for an extrusion?

6. What happens if you pick the glasses icon while editing an extrusion?

7. What is a blind extrusion width?

8. Is a symmetric extrusion the specified width or two times the specified width?

9. How do you find the name of an extrusion variable in Edit mode?

10. When a word like Placement shows up in red, what does this mean?

11. If an extruded part is not visible on the screen because it is to the far right or left, what is the easiest way to center it on the screen?

12. What does THRU NEXT depth do?

13. What is the difference between a round and a fillet?

14. How do you change the units of measure for a part that already exists?

Extrusions Problems

4.1 Design the symmetrical "**holder_block**" (Figure 4-86) with the origin located in the middle of the bottom 1.40 inches from each edge. The 1.50-inch diameter through hole is horizontally centered on the slanted plane, thus a symmetrical extrusion is required about the FRONT plane. The holder_block is 2.44 inches high. All dimensions are in inches.

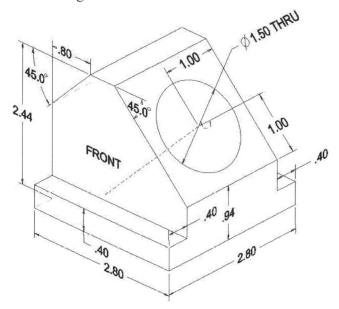

Figure 4-86 Problem 1—Holder Block

4.2 Design the symmetrical "**Operating_Arm**" (Figure 4-87) with the origin at the bottom of the large 1.25-inch diameter through hole. The keyway is ¼ inch wide. The distance from the top of the keyway to the opposite side of the hole is 1.363 inches. The through slot is ¾ inch wide. The straight portion of the slot is 1.75 inches long. The portion of the part containing the slot is 2.00 inches wide and 1.00 inches thick. The center of the slot aligns with the center of the hole. All dimensions are in inches.

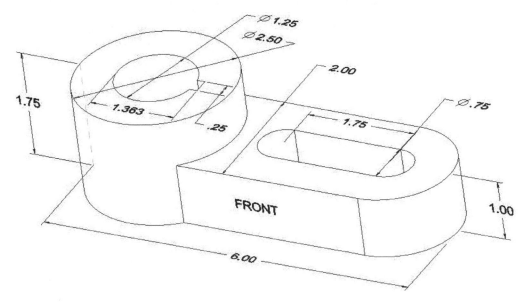

Figure 4-87 Problem 2—Operating Arm

4.3   Design the "**anchor_bracket**" (Figure 4-88) with the origin located at the back center of the 1.250-inch diameter through hole. The straight portion of the through slot must be 2.00 inches long with the furthest center point 5.50 inches from the 1.250-inch reamed hole's center. Add the .06-inch saw cut, fillets, and rounds last during the creation of this part. The small rightmost hole is 1.25 inches from the centerline of the 1.250 diameter reamed hole and centered on its 1.75-inch width. All dimensions are in inches.

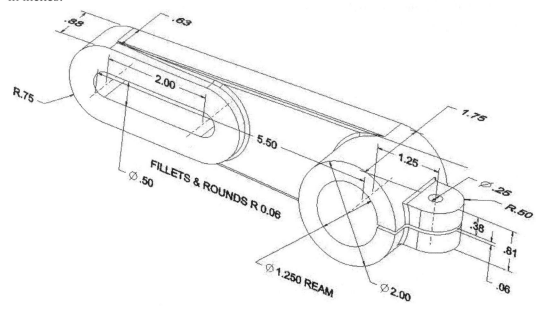

Figure 4-88 Problem 3—Anchor Bracket

4.4   Design the symmetrical "**boiler_stay**" (Figure 4-89) using the FRONT and TOP views shown. Place the origin at the bottom of the leftmost hole. The three through holes must be located on 2.00-inch centers. The ¾-inch through hole must be 4.625 inches from the leftmost hole's center. All dimensions are in inches. **Design change:** increase the .75-inch diameter to .88 inches.

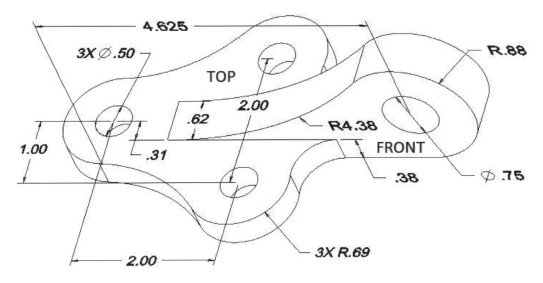

Figure 4-89 Problem 4—Boiler Stay

4.5   Design the "**clutch_lever**" (Figure 4-90) with the origin located at the front center of the large 1.250-inch reamed through hole. The top 3/8-inch hole is centered between the front and back planes of the part and goes through the entire part. The counterbore is located at the bottom of the part. The two 3/8-inch holes go through their respective walls. Add the ⅛-inch wide saw cut last during the creation of the part. All dimensions are in inches.

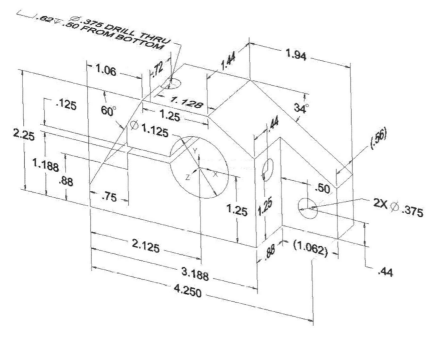

Figure 4-90 Problem 5—Clutch Lever

4.6   Design the symmetrical "**cross-feed_stop**" (Figure 4-91) with the origin located in the middle of the back, bottom edge. The two through holes must be 2.64 ± 0.01 inches apart. The ⅞-inch by ⅛-inch bottom slot runs completely through the part. Add the fillets and rounds last. All dimensions are in inches. **Design change:** increase the 0.44-inch diameter holes to 0.50 inches.

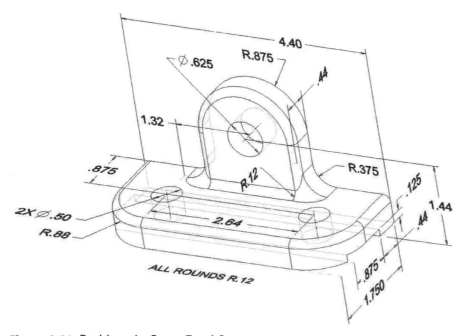

Figure 4-91 Problem 6—Cross Feed Stop

4.7   Design the symmetrical **"cutoff_holder"** (Figure 4-92) with the origin located in the middle of the bottom leftmost edge. The vertical hole must be 1.50 inches from the right end and 1.06 inches from the front surface. The horizontal hole must be 2.094 inches above the top of the flat cutout (not the bottom of the part) and in line with the vertical hole. The front cutout needs to be .31 inch by .16 inch. The back cutout needs to be located 1.50 inches from the edge of the front cutout and .16-inches high. All dimensions are in inches.

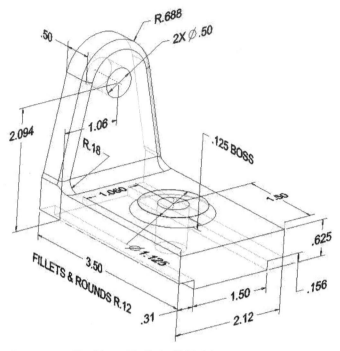

Figure 4-92 Problem 7—Cutoff Holder

4.8   Design the symmetrical **"bearing_holder"** (Figure 4-93) with the origin located in the back center of the part. The ⅛ inch drill goes through the top wall of the part. Add the ⅛-inch fillets and rounds last. **Design change:** The manufacturing engineer wants a .03x45° chamfer (not shown) added to both ends of the 1.625-inch diameter hole. All dimensions are in inches.

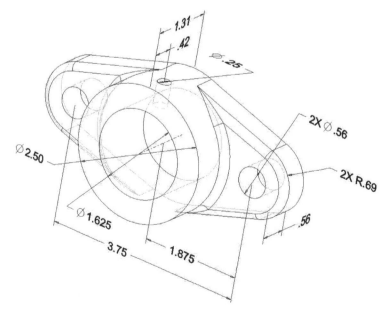

Figure 4-93 Problem 8—Bearing Holder

4.9 Design the symmetrical "**frame_guide**" (Figure 4-94) with the origin located on the centerline of the 1.50-inch diameter through hole directly below the center of the 0.19-inch diameter hole. The two 3/8-inch through holes need to be 1.69 inches from the centerline of the large hole. All dimensions are in inches. **Design change:** increase the 0.19-inch diameter hole to 0.25 inches.

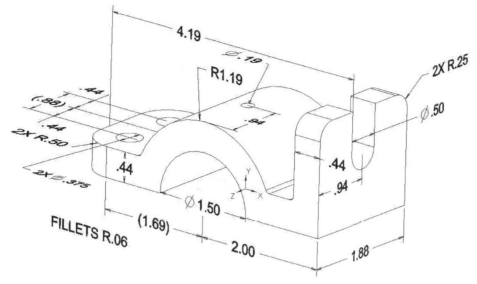

Figure 4-94 Problem 9—Frame Guide

4.10 Design one of the blocks shown in Figure 4-95 with the origin located in the back left corner of the part. All holes and cutouts go through the parts. All dimensions are in inches.

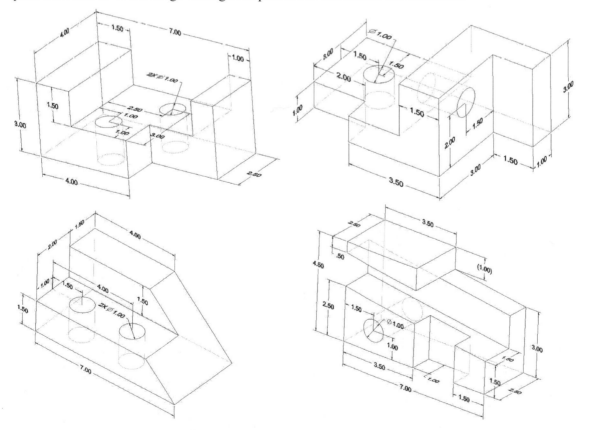

Figure 4-95 Problem 10—Assorted Blocks

4.11 Design the symmetrical "**shifter_fork**" (Figure 4-96) with the origin located on the centerline of the 1.423-inch diameter through hole directly above the center of the 0.68-inch diameter holes. The upper yoke is centered about the 0.68-inch diameter through hole. The support under the U-shape is 0.50 inches thick. Add fillets and rounds last during the creation of this part. All dimensions are in inches.

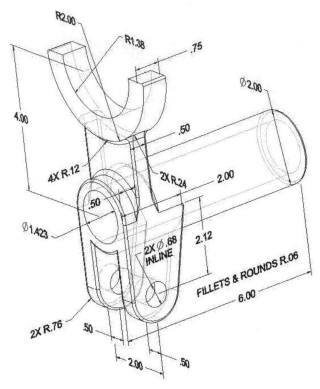

Figure 4-96 Problem 11—Shifter Fork

4.12 Design the "**index_feed**" (Figure 4-97) with the origin located in the back leftmost corner of the part. The 0.19-inch deep tab will be below the TOP plane while everything else is above the TOP plane. The 40-degree and 50-degree angles are important. All dimensions are in inches.

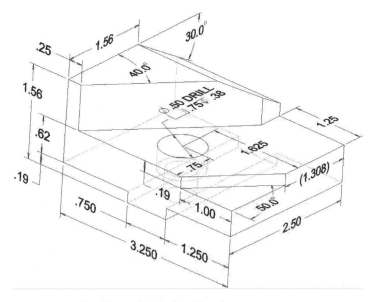

Figure 4-97 Problem 12—Index Feed

4.13 Design the symmetrical "**support_bracket**" (Figure 4-98). The tab portion is 0.75 inches thick. The through hole in the tab is 0.50 inches in diameter with a 1.00-inch counterbore that is ¼ inch deep. The top flat portion of the part is ¼ inch by 1.00 inch. The slanted surfaces are 45 degrees. The middle slot is ¼-inch wide by ⅛ inch deep. The origin is centered below the slot on the left edge. All dimensions are in inches.

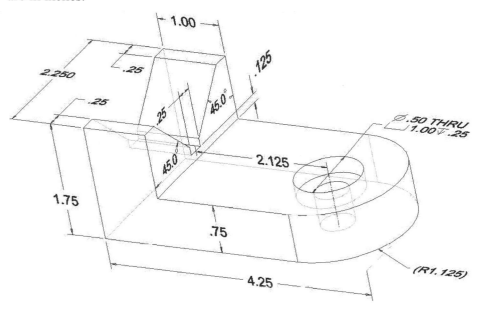

Figure 4-98 Problem 13—Support Bracket

4.14 Design "**vibrator_arm**" (Figure 4-99). Make the reamed through hole 0.625 inches in diameter. The bottom slot goes through the bottom portion of the part. The two vertical through holes must be 1.88 inches apart. The origin is in the bottom, front corner of the left rounded protrusion. All dimensions are in inches. **Design change:** increase the 0.625-inch diameter holes to 0.750 inches.

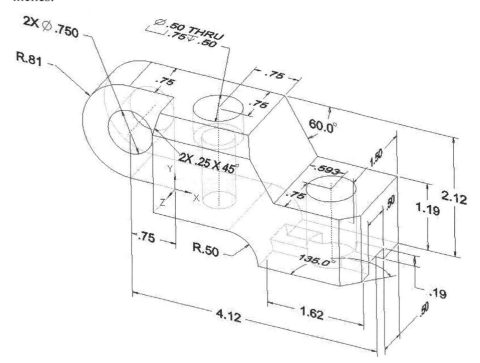

Figure 4-99 Problem 14—Vibrator Arm

4.15 Design the symmetrical **"holder_block_mm"** (Figure 4-100) with the origin located in the middle of the bottom 35 millimeters from each edge. The 40-millimeter through diameter hole is horizontally centered on the slanted plane, thus a symmetrical extrusion is required about the FRONT plane. The holder_block_mm is 62 mm high. All dimensions are in millimeters.

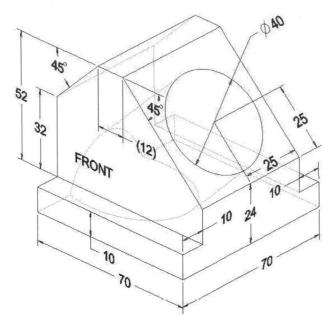

Figure 4-100 Problem 15—Holder Block (metric)

4.16 Design the symmetrical **"Operating_Arm_mm"** (Figure 4-101) with the origin at the bottom of the large 32-millimeter diameter through hole. The keyway is 7 millimeters wide. The distance from the top of the keyway to the opposite side of the hole is 35.1 millimeters. The through slot is 20 millimeters wide. The straight portion of the slot is 45 millimeters long. The portion of the part containing the slot is 50 millimeters wide and 25 millimeters thick. All dimensions are in millimeters.

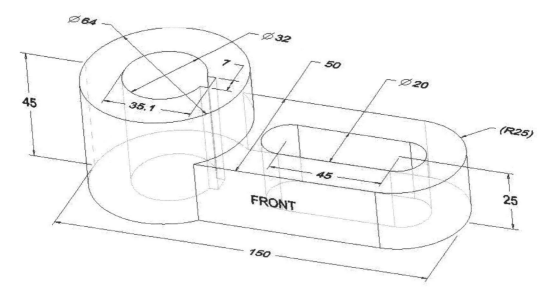

Figure 4-101 Problem 16—Operating Arm (metric)

4.17 Design the symmetrical **"boiler_stay_mm"** (Figure 4-102) using the FRONT and TOP views shown. Place the origin at the bottom of the leftmost hole. The three through holes must be located on 50-millimeter centers. The 20-millimeter through-hole must be 116 millimeters from the leftmost hole center. All dimensions are in millimeters. **Design change:** increase the three 12-mm diameter holes to 13 mm.

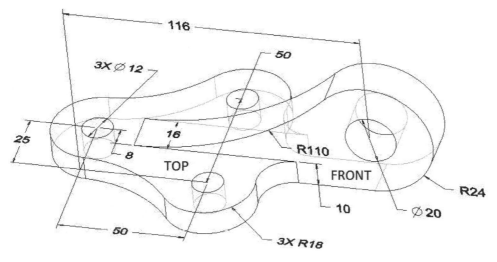

Figure 4-102 Problem 17—Boiler Stay (metric)

4.18 Design the symmetrical **"bearing_holder_mm"** (Figure 4-103) with the origin located in the back center of the part. The 6-millimeter drill goes through the top wall of the part. Add the 3-millimeter fillets and rounds last. All dimensions are in millimeters. **Design change:** The design engineer wants a 1x45° chamfer (not shown) added to both ends of the 40-mm diameter hole.

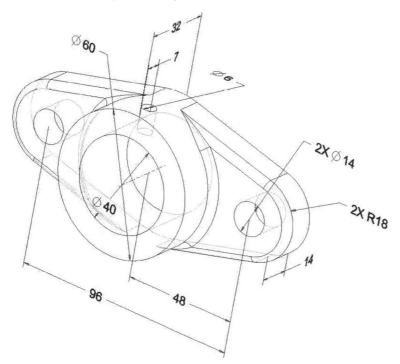

Figure 4-103 Problem 18—Bearing Holder (metric)

4.19 Design the "**index_feed_mm**" (Figure 4-104) with the origin located in the back leftmost corner of the part. The 5-millimeter deep tab will be below the TOP plane while everything else is above the TOP plane. The 40-degree and 50-degree angles are important. All dimensions are in millimeters.

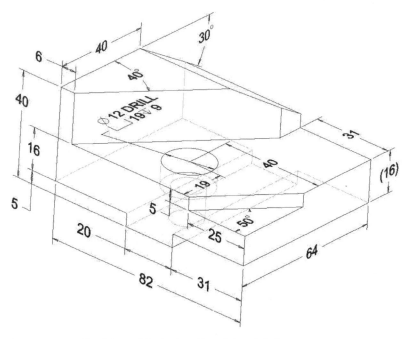

Figure 4-104 Problem 19—Index Feed (metric)

4.20 Design the symmetrical "**support_bracket_mm**" (Figure 4-105). The tab portion is 20 millimeters thick. The through hole in the tab is 12 millimeters in diameter with a 25-millimeters counterbore that is 6 millimeters deep. The top flat portion of the part is 6 millimeters by 25 millimeters. The slanted surfaces are 45 degrees. The middle slot is 6 millimeters wide by 3 millimeters deep. The origin is centered below the slot on the left edge. All dimensions are in millimeters.

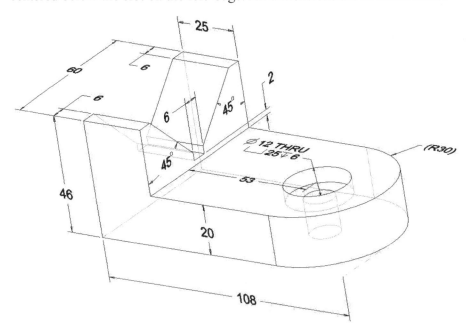

Figure 4-105 Problem 20—Support Bracket (metric)

REVOLVES

Objectives

▶ Designing with symmetrical features

▶ Creating 3D parts from revolved sections is explored

▶ Creating a keyway properly

▶ Converting an engineering sketch into a 3D revolved part

▶ Learn about different hole options and placement

▶ Creating a set of common features

▶ Adding rounds to reduce stresses

▶ Creating a slot for a retaining ring on a shaft

▶ Adding chamfers to a shaft

▶ Creating a tapered hole

▶ Practice creating revolved parts

▶ Revolved part exercise according to design intent

▶ Designing based upon engineering calculations

▶ Revolved part design problems to reinforce concepts

▶ Revolves Explored

Symmetrical Features in Designs

Symmetry is an important characteristic seen in many designs. Symmetrical features can be created by an assortment of tools available in Creo Parametric. You can create multiple copies of a symmetrical feature with the Feature Pattern command, or you can create a mirrored image of the model or a 2D section using the Mirror command.

It is important to identify the features that exist in a design when doing parametric modeling. Feature-based parametric modeling enables you to build complex designs by creating a series of simple features. This approach simplifies the modeling process and allows you to concentrate on the features of the design.

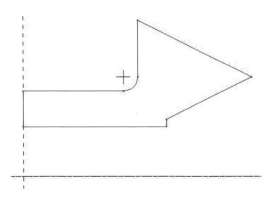

Figure 5-1   Sketched Cross-section

Figure 5-2   Revolved Feature

The modeling technique of extruding two-dimensional sketches perpendicular to the sketch to create three-dimensional features was discussed in the previous chapter. For cylindrical parts with tapered surfaces, the protrusion procedure will not work. For designs that involve cylindrical shapes revolving a 2D sketch (Figure 5-1) about an axis will form the needed 3D feature. In solid modeling, this is called a revolved feature. See Figure 5-2.

The tapered clutch was created using the revolve command in Creo Parametric. Because the clutch has a taper, it cannot be created using the extrude command. The keyway was created using an extrusion cut through the clutch.

A revolve can be created by picking the revolve tool, then sketching the cross-section to revolve, or you can create the cross-section in sketcher, highlight it, then pick the revolve tool. The first approach (revolve tool, then sketch) is preferred. For this example, we are going to draw the upper cross-section of the clutch in sketcher along with a centerline that will be visible outside of sketcher, then revolve the sketch. Since the section will be revolved, the radial dimensions must be diameter dimensions. This is done by picking the diameter point, picking the centerline, then picking the diameter point again followed by pressing the MMB to place the diameter dimension at the current cursor position. See Figure 5-3. When you have finished dimensioning the section, pick the checkmark to exit from sketcher and keep the sketch.

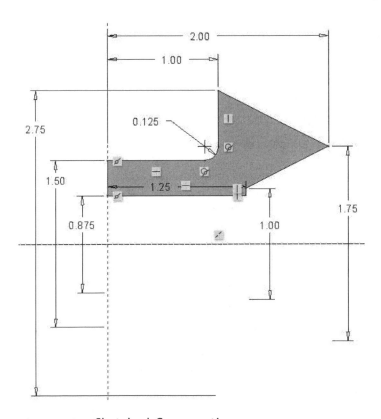

Figure 5-3   Sketched Cross-section

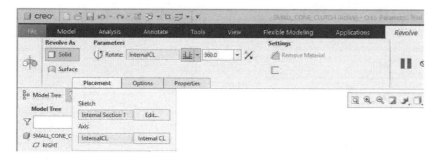

Figure 5-4 Revolve Section Tab

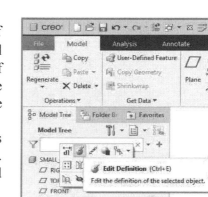

Figure 5-5 Glasses Icon

With the sketch highlighted, *pick* the **Revolve** tool 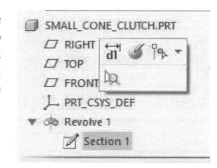 from the top of the screen. The Revolve tab will appear. Be sure solid is selected. You will use the internal centerline (InternalCL) as the axis of rotation. Instead of depth when extruding a feature, revolves have an angle of rotation for the section. A complete 360-degree rotation of the section is necessary to create the clutch. See Figure 5-4.

Select the **glasses** icon (Figure 5-5) to view the revolved section in its finished form. Use the mouse to view the part from different locations. When all appears correct, *select* the **green checkmark** to build the revolved feature.

If you need to edit the revolved section, *highlight* **Revolve 1** in the model tree, then press and hold down the RMB until a pop-up menu appears. *Pick* **Edit Definition** icon (Figure 5-6) from this menu to reactivate the Revolve tab (Figure 5-4).

Figure 5-6 Edit Definition icon

If you need to change the sketch or its dimensions, you can edit the sketch by *picking* the arrow sign in front of **Revolve 1** to get Sketch 1 to show up. With **Section 1** highlighted, press down and hold the RMB until a pop-up menu appears. *Select* the **Edit Definition** icon (Figure 5-7) from this pop-up menu to get back into sketcher. When finished, *pick* the **checkmark** to exit sketcher and keep the sketch.

The keyway is added using an extrude cut with the left end of the clutch selected as the sketching plane. The hole in the center of the clutch is added as a reference for sketching (Figure 5-8). The rectangle tool is used to create a square 0.125 inches by 0.125 inches. To show design intent, the width of

Figure 5-7 Edit Definition of Sketch

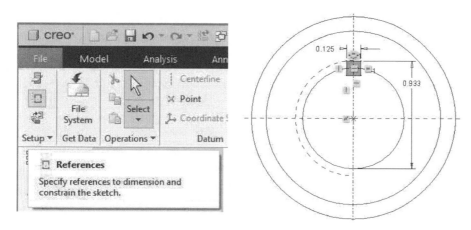

Figure 5-8 Sketch References and Keyway Dimensioned

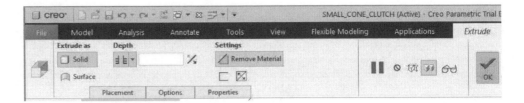

Figure 5-9   ExtrudeCutTab

the keyway and the distance across the hole to the top of the keyway are dimensioned. The machinist will use these two dimensions when cutting the keyway. See Figure 5-8.

When finished, exit sketcher by *picking* the **checkmark**. With **Sketch 2** highlighted, *select* the **extrude** tool. *Select* solid with a depth of through the part. *Select* the **Remove Material** option from the dashboard. See Figure 5-9. Make sure the yellow arrows point into the part, then *select* the **glasses** to view the cut. If all looks correct (Figure 5-10), *select* the **green checkmark** to accept the feature.

The tapered clutch is complete.

The second example is a 4-hole hub as shown in Figure 5-11. This basic hub without the four holes, fillets, and rounds can be created using four different extrudes or it can be made using the revolve tool once.

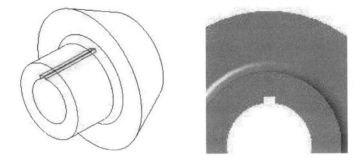

Figure 5-10   Keyway Added to Part

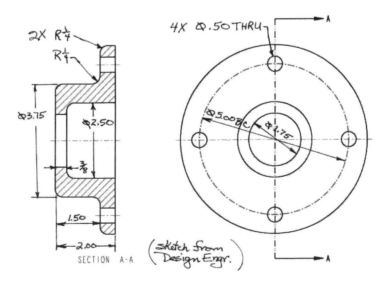

Figure 5-11   Engineering Sketch of Four-Hole Hub

Using the extrude tool the building procedure might look like this. First, draw the large diameter of the hub, and then extrude it 0.50 inches thick. See Figure 5-12. Note that this dimension doesn't represent design intent.

Second, draw the smaller hub portion using the large hub's surface as the sketching plane and extrude it 1.50 inches (Figure 5-13).

Third, using the RIGHT datum plane as the sketching plane, sketch the large hole on the top of the hub, and then do an extruded cut to a depth of 1.50 inches. See Figure 5-14.

Fourth, using the left surface of the smaller hub as the sketching plane, sketch the small hole on the left side of the hub, and then do an extruded cut through the part (Figure 5-15).

Now all that is left is to round the edges and add the four holes.

The procedure might look like this if we created it as a revolved feature. First, we would sketch a horizontal centerline and the upper cross-section of the hub (Figure 5-16), and then revolve it through 360 degrees as shown in Figure 5-17. We are done in one step with the sketch representing the engineer's design intent.

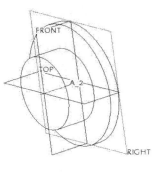

Figure 5-13 Extrude Hub

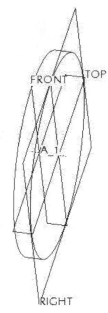

Figure 5-12 Extrude Large Diameter

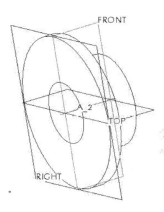

Figure 5-14   Add Large Hole

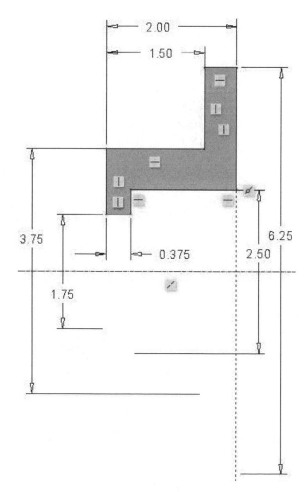

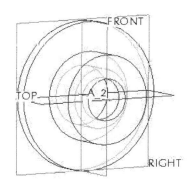

Figure 5-15 Add Small Hole

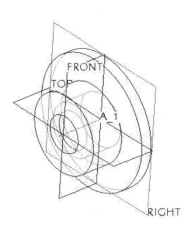

Figure 5-16   Sketched Cross-section   **Figure 5-17**   Revolved Hub

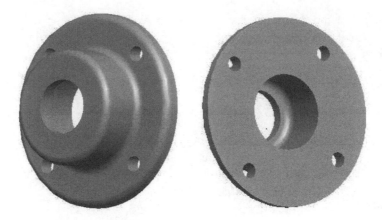

Figure 5-18   Finished Four-Hole Hub

Now all that is left is to round the edges and add the four holes. See Figure 5-18.

Besides using the Revolve command, Creo Parametric has one other command for handling cylindrical features, the sketched hole option.

Hole Options

Creo Parametric provides three hole types: simple (straight), standard threaded, and sketched holes. Simple holes have a straight profile throughout the length of the feature such as drilled, bored, or reamed holes. Standard holes contain either standard English or metric threads. Sketched holes are used to create unique profiles, such as counterbore, spot-faced, or countersink holes. See Figure 5-19.

The Sketched Hole option requires the user to sketch a vertical profile of the hole. Unlike the normal sketching environment, the hole sketcher does not provide

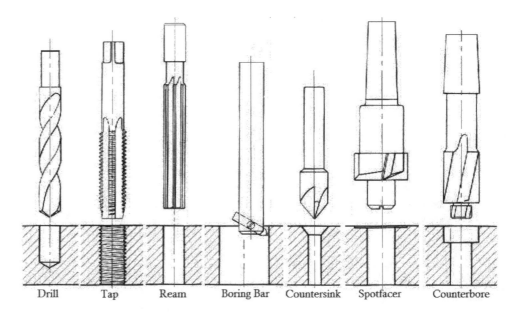

Drill   Tap   Ream   Boring Bar   Countersink   Spotfacer   Counterbore

Figure 5-19   Hole Making Tools

an option for specifying references. A sketched hole is created originally independent of any specific part features and later placed according to the hole placement. See Figure 5-20.

A vertical centerline is required when sketching a hole. Within sketcher, use the Centerline icon to create the vertical centerline. All sketched entities must be created on one side of this vertical centerline and the section must be closed. The centerline cannot serve as part of the hole's profile. Also, at least one sketched line must lie perpendicular to the centerline. This line will be aligned with the placement plane when placing the hole. For sketched holes with multiple perpendicular lines, the uppermost line within sketcher will serve as the placement reference.

Hole Placement

In addition to the three types of holes, Creo Parametric provides five placement options: Linear, Coaxial, Radial, Diameter, and On Point. The Linear option is used to locate a hole from two reference edges (Figure 5-21). The Coaxial option is used to locate a hole centerline coincident with an existing axis. The Radial option is used to locate a hole at a distance from an axis and at an angle from a reference plane (Figure 5-22). The hole's distance from the reference axis is defined by a radius. With the Diameter option, the hole's distance from the reference axis is defined by a diameter. The Radial and Diameter hole placement options are typically used with the Pattern command to create a radial pattern of holes. For radial patterned holes, the angular dimension is used to create the pattern (Figure 5-23).

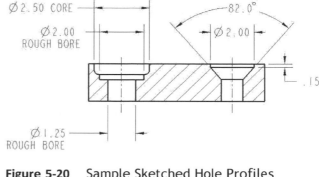

Figure 5-20 Sample Sketched Hole Profiles

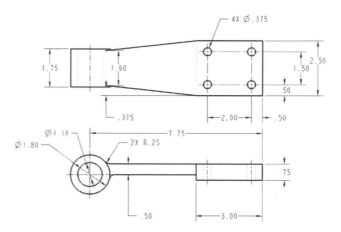

Figure 5-21 Linear Hole Placement for the Four Simple Holes

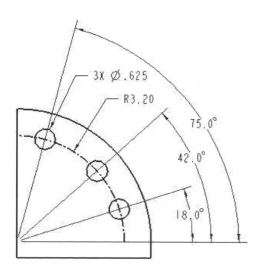

Figure 5-22 Radial Placement of Simple Holes

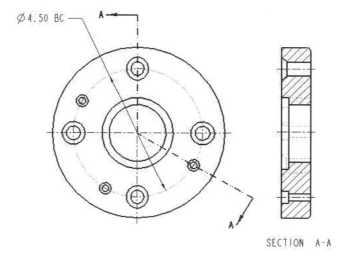

Figure 5-23 Diameter Placement of Sketched Holes around Edge and Axial Placement Large Sketched Center Hole

▶ **Revolves Practice**

New part creation follows a standard procedure in Creo Parametric. This procedure is used for revolved thin-walled and solid parts. This text will make several assumptions when you are creating a new part. We will use the default template file when creating a new part. This template contains the FRONT, TOP, and RIGHT side datum planes along with an XYZ coordinate system locating the origin (0, 0, 0). We will use the default system of units, IPMS, inch-pound mass-second. Sometimes we will use the IPS system or the mmNS system. When we are doing stress analysis, we will use the IPS, inch-pound-second, system of units because stresses make more sense in these units. New parts will not be assigned a material property, so if an analysis is done requiring a material property, a material must be assigned by the user.

There are several ways to revolve a feature. Let's begin by starting Creo Parametric and setting the working directory.

Step 1: Start Creo Parametric by *double-clicking* with the LMB on the ***Creo Parametric*** icon on the desktop, or from the Program list: Creo Parametric.

Step 2: Set your working directory by *selecting* the ***Select Working Directory*** icon in the Home ribbon. Locate your working directory, and then *pick* **OK.** See Figure 5-24.

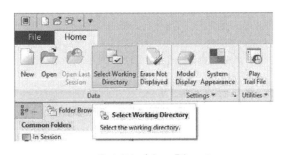
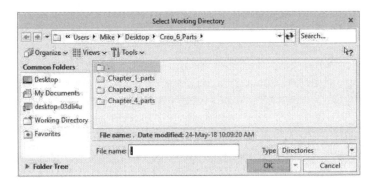

Figure 5-24   Set Working Directory

Figure 5-25   New Object

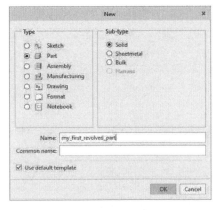

Figure 5-26   New Part

Step 3: Create a new object by *picking* the ***New*** icon at the top of the screen (Figure 5-25), or by *selecting* **File>New.**

Step 4: *Select* **Part** from the window and name it "my_first_revolved_part". The sub-type should be Solid. Make sure the default template is *checked*, and then *pick* **OK.** See Figure 5-26.

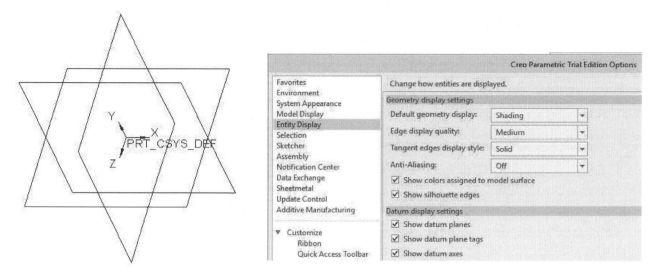

Figure 5-27 FRONT, TOP, and RIGHT Side Plane

An XYZ coordinate system will appear in the middle of the screen along with FRONT, TOP, and RIGHT side planes. See Figure 5-27. The words FRONT, TOP, and RIGHT will appear if this option is set. (**File>Options>Entity Display,** *check* **Show datum plane tags.**)

We will create a 3D revolved part by sketching a 2D shape, then rotating it about an axis. In this case, we are going to create a step shaft.

Design Intent

Design a 7.50-inch long step shaft with a 1.00-inch diameter 2.00 inches long, a 1.75-inch diameter 3.00 inches long, and the remaining length 1.25 inches in diameter. The ends of the step shaft have 0.06-inch x 45° chamfers. The places where the smaller diameters meet the 1.75-inch diameter need to be rounded using a 0.06-inch radius to reduce stress concentrations. An SH-125 retaining ring needs a groove (0.056-inch wide by 0.037-inch deep) placed on the 1.25-inch diameter shaft 2.00 inches from the 1.75-inch diameter. See Figure 5-28.

Figure 5-28 Step Shaft

Step 5: We need to set the units for our step shaft to the IPS (Inch-Pound-Second) system of units. Use **File>Prepare>Model Properties.** See Figure 5-29.

Step 6: The following Model Properties window appears as shown in Figure 5-30. *Pick* the word **change** on the right end of the Units row. Note that a material property is not assigned to the part at this time.

Step 7: The Units Manager window (Figure 5-31) appears. If the system of units, **IPS,** is not selected as shown below, *pick* it from the list, then *pick* the ⟶ Set... button.

Figure 5-29   Model Properties

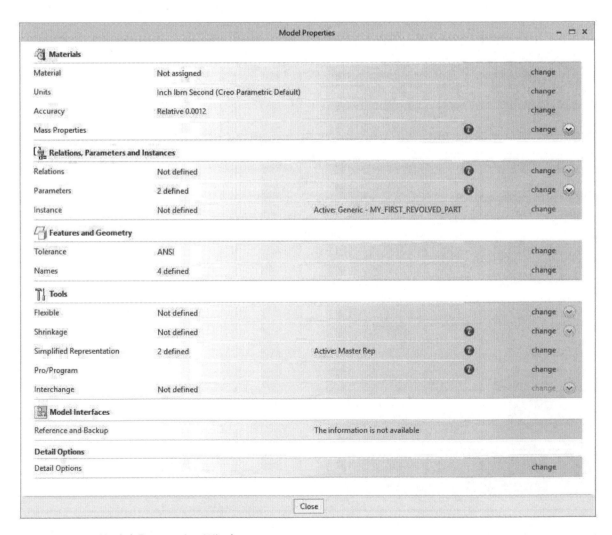

Figure 5-30   Model Properties Window

Step 8: After *picking* the [➜ Set...] button, the following window appears. *Select* Convert dimensions (for example 1″ becomes 25.4mm), then *pick* **OK.** See Figure 5-32.

Step 9: *Pick* **Close** on the Units Manager window. *Pick* **Close** on the Model Properties window.

Step 10: [◖◗ Revolve] *Select* the **Revolve** tool located in the |Model| tab ribbon.

Step 11: The upper left corner of the screen should look similar to Figure 5-33. The two icons above the word Placement are for creating a **solid** or a **surface**. Be sure the block looking icon (solid) is selected. Note that on your screen the word Placement is red. This means that this section needs additional information to continue.

Figure 5-31 Units Manager Window

Since we know that the revolved section needs a sketch to revolve, we can go directly into sketcher by *picking* the FRONT datum plane with the LMB. If you do this, jump to step 14. If not, proceed with steps 12 and 13.

Step 12: *Pick* the |Placement| tab. When you *pick* the word Placement, the following appears. Note that on your screen "• **Select 1 item**" has a red dot (Figure 5-34), thus it needs additional information before you can continue. *Pick* the **Define...** button.

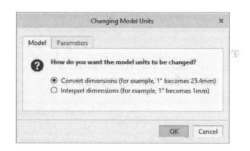

Figure 5-32 Changing Model Units

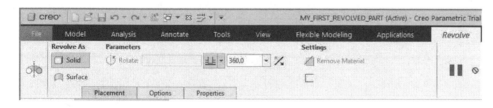

Figure 5-33 Placement Needs More Information

Figure 5-34 Pick Define Button

Step 13: Move the cursor onto the FRONT plane so that it turns green. *Select* it by clicking the left mouse button (LMB). The RIGHT plane turns blue and the area in the upper right corner of the screen shows the following sketch window (Figure 5-35). The FRONT plane will be the sketching plane. This is the plane we will use to create our 2D sketch. The sketch orientation indicates that the RIGHT plane will be toward the right side of the screen. This is what we want. *Pick* **Sketch.**

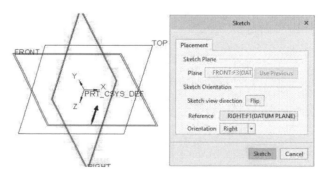

Figure 5-35 Sketch Plane and Orientation

Step 14: 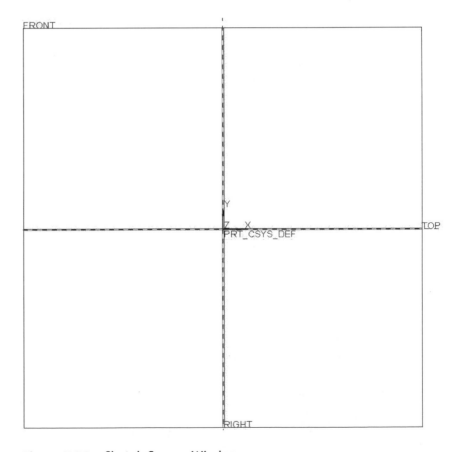 If needed, *pick* the **Sketch View** icon to orient the sketch plane so that it is parallel with the display screen.

Step 15: A new orientation (Figure 5-36) appears. In the middle of the screen, you can see the XYZ coordinate system. This is (0, 0, 0), our reference point in 3D space. We are sketching on the FRONT plane. The RIGHT plane is perpendicular to the screen and facing toward the right side of the screen. The TOP plane is perpendicular to the screen and facing upward.

Design Intent

The intersection of the FRONT and TOP datum planes will be the centerline for the step shaft. The origin (0, 0, 0) will be the reference point for the first step change in diameter.

The sketcher screen may be colored. For the purposes of this textbook, the sketcher screen will be shown as dark lines on a white background as shown below. Note that a new set of tools is available at the top of the screen under the Sketch tab. These are the sketcher tools described previously. We will *select* the centerline icon above the word Datum. A small X appears at the end of the mouse cursor.

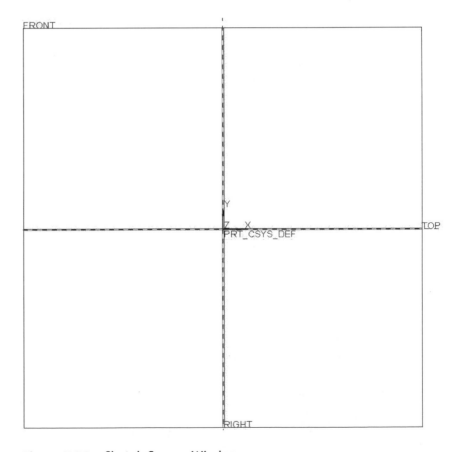

Figure 5-36   Sketch Screen Window

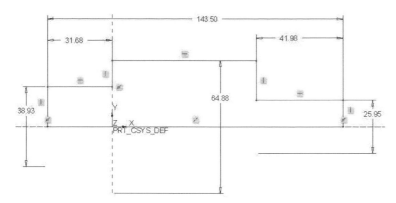

Figure 5-37 Initial Sketch

Step 16: [Centerline] Sketch a horizontal geometry centerline aligned with the TOP datum plane.

Step 17: [icon] Sketch the step shaft's upper section profile with the <u>first step change in diameter at the or</u>igin. Begin sketching the front view section as shown in Figure 5-37 starting to the left of the origin and drawing a vertical line first. To draw a line, use the LMB and *select* the centerline left of the origin. Move the mouse cursor vertically. When the line looks long enough, *select* the location on the screen using the LMB. A new line will begin at this point. Move the mouse cursor to the right. *Select* a location on the RIGHT datum plane, then continue drawing lines until the upper section of the step shaft is complete. The section must be closed so be sure to return to your initial starting point by adding a line on top of the centerline.

Step 18: When finished drawing the lines, click the MMB twice to exit drawing a line mode or move the mouse cursor to the **Select** icon at the top of the screen and *pick* it.

Sketcher places enough dimensions on your sketch to completely define it. Your sketch may look similar to the one pictured in Figure 5-37. Don't worry about the physical size of the step shaft at this time. The actual values of the dimensions are not important at this time. Note that the default dimensions do not match our design intent so we must use the dimension tool to add the appropriate dimensions. In my case, the 38.29-inch dimension does not match design intent. We need to add the overall length of the step shaft. (V = this line is vertical, H = this line is horizontal.)

The dimmed dimensions on your sketch are not the correct lengths and are referred to as weak dimensions. Here is where the power of parameter modeling shows up. The lines in the sketch will change size to match the dimensions you enter.

Step 19: [icon] To get all the dimmed dimensions drawn to approximately the correct value, highlight the entire sketch by drawing an imaginary box around it. *Pick* a point above and to the left of the sketch, hold the LMB (left mouse button) down, and then drag the mouse to a point below and to the right of the sketch. Release the LMB. All selected entities turn red. *Select* the **Modify** icon from the top of the screen.

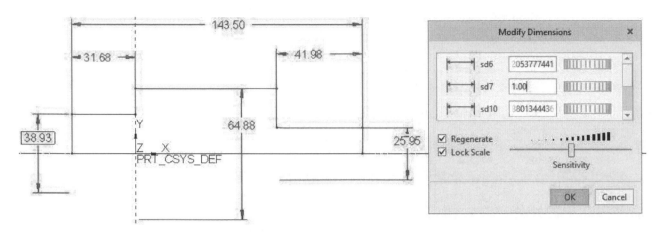

Figure 5-38   Lock Scale and Modify Dimensions

Step 20: This brings up the following window. Your dimensions will be different because no two people can sketch a view using exactly the same dimensions. Clicking on any of the dimensions in this window highlights the appropriate dimension on your sketch. *Select* the **Lock Scale** checkbox, and then locate the left diameter dimension of the step shaft. Type 1.00 <Enter> since this is the desired diameter for the left portion of the step shaft, then *pick* the **OK**. *Pick* the **Refit** icon. See Figure 5-38.

Step 21: ⟨↔⟩ *Select* the ***normal dimension*** tool and add the overall length of the step shaft. Note that a weak dimension will disappear when this dimension is added. To change an existing dimension, *double-click* on it with LMB, type the new value, then press <Enter>. Your sketch should appear as shown in Figure 5-39 when you are finished.

Step 22: ⟨⟩ If necessary, *select* the ***Shade Closed Loops*** icon at the top of the screen. For a revolved section, the sketch must be a closed area. This tool lets you see if you have a problem with your sketch.

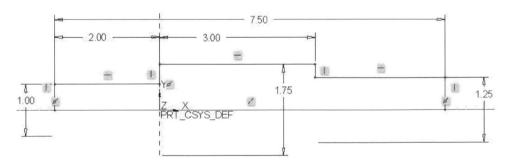

Figure 5-39   Step Shaft Sketch

Step 23: *Select* the ***Feature Requirements*** icon at the top of the window. This will verify that your sketch is complete and ready to be revolved. See Figure 5-40.

Step 24: *Pick* **Close** to close the Feature Requirements window.

Step 25: *Pick* the ***green checkmark*** at the top of the window to exit from sketcher and revolve the sketch around the created centerline. Figure 5-41 shows the Revolve Dashboard.

If the Revolve tab ribbon doesn't list "InternalCL" as the axis of rotation (Figure 5-41), then you did not create a geometry centerline that would be visible outside of sketcher. In this case, *pick* the word Placement, then *pick* the Edit... button to re-enter sketcher. Change the construction centerline to a geometry centerline, then exit sketcher again.

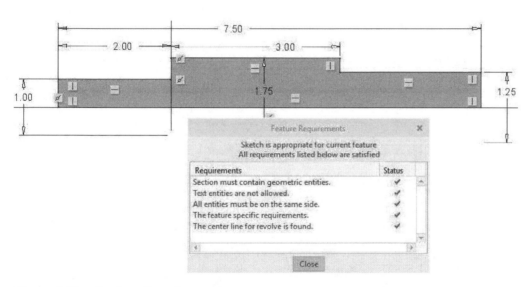

Figure 5-40   Feature Requirements

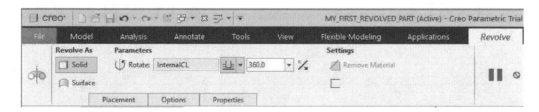

Figure 5-41   Revolve Tab

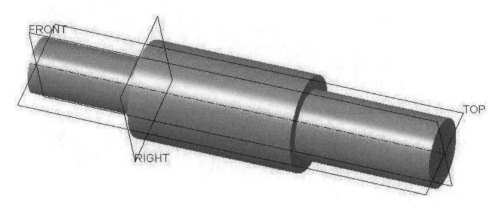

Figure 5-42   Revolved Step Shaft

Step 26:   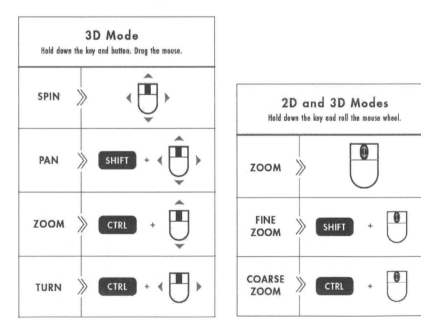 *Pick* the ***glasses*** icon to view the revolved section. See Figure 5-42.

Step 27:   Use the mouse to move the step shaft around. Practice using all of these options again. See Figure 5-43.

Step 28:   *Pick* the ***Refit*** icon in the graphics toolbar.

Step 29:   *Pick* the ***green checkmark*** to accept the feature build.

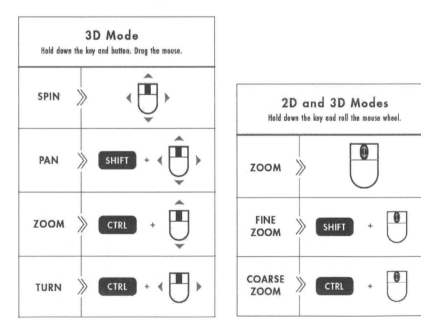

Figure 5-43   Mouse Button Functions

Note: The zoom when spinning the mouse wheel works only if the "Scroll inactive windows when I hover over them" feature is turned OFF in Windows 10.

Step 30: *Turn off **Plane Display*** to hide the FRONT, TOP, and RIGHT datum planes. See Figure 5-44.

Let's look at the different model display options again. Become familiar with each. See Figure 5-45.

Step 31: *Pick **No Hidden*** to remove all hidden lines and show only visible lines.

Step 32: *Pick **Hidden Line*** to show the hidden lines as faint lines.

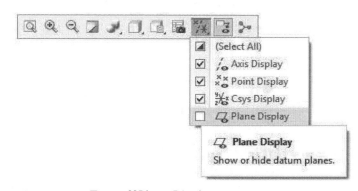

Figure 5-44 Turn off Plane Display

Step 33: *Pick **Wireframe*** to show all hidden lines as solid lines.

Step 34: *Pick **Shading with Edges*** to see the edges of the object drawn with black lines.

Step 35: *Pick **Shading*** to show all surfaces as shaded planes.

Step 36: *Pick **Shading with Reflections*** to add light sources and shadows. The number and location of the light sources can be changed. We will not do that now.

We will continue with our design.

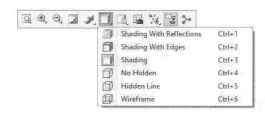

Figure 5-45 Part Viewing Mode

Step 37: Save your part. *Pick **OK*** in the Save Object window.

Step 38: Set the model back to ***Hidden line*** mode. Turn on ***Plane Display*** to show the FRONT, TOP, and RIGHT datum planes.

Step 39: *Select **Revolve 1*** in the model tree. When a pop-up menu appears, *pick* the **Edit Dimensions icon** from the list. All dimensions used to create the revolved section should reappear on the screen. See Figure 5-46.

Step 40: (Optional) *Select **File>Print>Print*** to make a hard copy of your step shaft with these dimensions showing. *Pick **OK.*** *Pick **OK.***

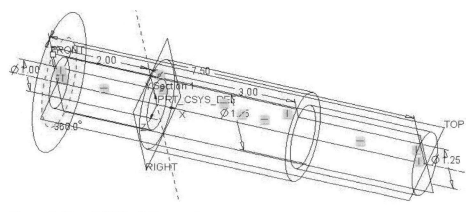

Figure 5-46 Edit Shaft Parameters

Design intent requires that the ends of the step shaft have 0.06-inch x 45° chamfers. The places where the smaller diameters meet the 1.75-inch diameter need to be rounded using a 0.06-inch radius to reduce stress concentrations. An SH-125 retaining ring needs a groove (0.060-inch wide by 0.040-inch deep and 1.17 inches in diameter) placed on the 1.25-inch diameter shaft 2.00 inches from the 1.75-inch diameter.

Step 41: *Select* the ***Chamfer*** tool in the Model tab. *Pick* the right end of the step shaft, hold down the <Ctrl> key, then *pick* the other three edges. Set the size of the chamfer to 0.06 inches. See Figure 5-47. *Pick* the ***green checkmark*** to accept the features.

Step 42: *Select* the ***Round*** tool in the Model tab. *Pick* the edge where the 1.25-inch diameter meets the 1.75-inch diameter's wall. Rotate the shaft so the 1.00-inch diameter is toward you. Hold down the <Ctrl> key and *pick* the edge where the 1.00-inch diameter meets the 1.75-inch diameter's wall. Set the value to 0.06 inches. See Figure 5-48. *Pick* the ***green checkmark*** to accept the features.

Step 43: *Pick* the ***Revolve*** tool in the Model tab ribbon.

Step 44: *Pick* the FRONT plane as the sketching plane. See Figure 5-49.

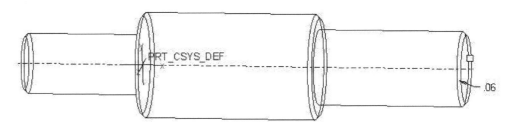

Figure 5-47 Adding Equal-sized Chamfers

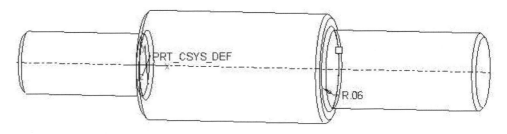

Figure 5-48 Add Equal-sized Rounds

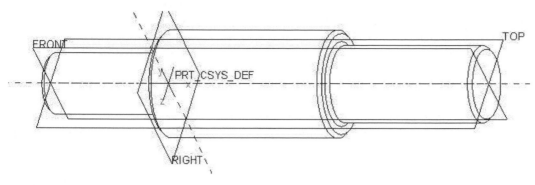

Figure 5-49 Sketch Plane and Orientation

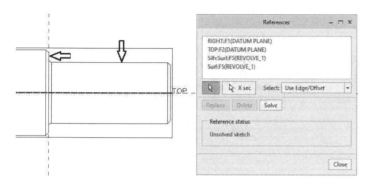

Figure 5-50 Add Sketcher References

Step 45: *Select* the ***References*** icon from the upper left Sketcher tab ribbon.

Step 46: If needed, *pick* the ***Sketch View*** icon to orient the sketch plane so that it is parallel with the display screen.

Step 47: *Pick* the vertical wall where the 1.75-inch diameter meets the 1.25-inch diameter, and then *pick* the 1.25-inch diameter's top horizontal line. See Figure 5-50.

Step 48: *Pick* the **Close** button. >**File>Options>Sketcher** set 3 decimal places. *Pick* **OK**. *Pick* **No** when asked if you want to save settings to a configuration file.

Step 49: *Select* the ***Line*** tool. Draw an open top box that represents the retaining ring's groove as shown in Figure 5-51. (You may use a closed rectangle instead.) Set the values per the design intent.

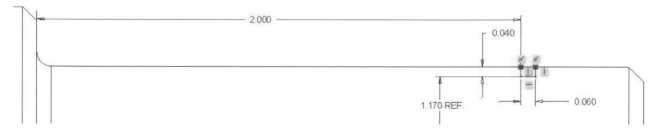

Figure 5-51 Sketch Retaining Ring Groove

Figure 5-52 Revolve as Solid and Remove Material

Step 50: *Pick* the **checkmark** to exit from sketcher and keep the sketch.

Step 51: *Select* the step shaft's axis as the axis of rotation using the LMB.

Step 52: *Pick* the **Revolve as Solid** icon. (Revolve as Surface is the default for open sections.)

Step 53: *Pick* the **Remove Material** icon (Figure 5-52) to make this revolved feature a revolved cut.

Step 54: *Pick* the **glasses** to view the cut feature. Be sure you did steps 51, 52, and 53.

Step 55: When all seems correct, *pick* the **green checkmark**. See Figure 5-53.

Step 56: Save the step shaft. See Figure 5-54.

Step 57: (Optional) *Select* **File>Print>Print** to make a hard copy of your step shaft with **Point, Csys**, and **Plane Displays** off. See Figure 5-54. *Pick* **OK**. *Pick* **OK**.

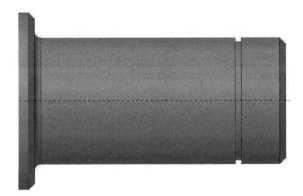

Figure 5-53 Retaining Ring Groove

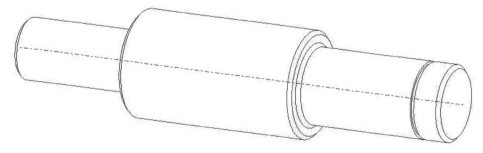

Figure 5-54 Revolved Step Shaft with Groove

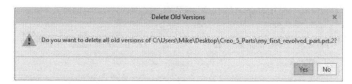

Figure 5-55   Pick Yes button

Step 58: *Select* **File >Manage File >Delete Old Versions.**

Step 59: *Pick* the **Yes** button. See Figure 5-55.

Design Intent

The shaft of a small tachometer is to be attached to the right end of the step shaft. The engineer decides to attach the tachometer using a 6-degree taper and friction. Add a tapered hole to the right end of the 1.25-inch diameter with a large diameter of 0.375 inches and a depth of 1.00 inches.

Step 60: ⬚ Hole *Select* the **Hole** tool in the Model tab ribbon.

Step 61: ⬚ *Select* the **sketched hole** option from the Hole tab ribbon. See Figure 5-56.

Step 62: ⬚ *Select* the **Activate Sketcher** icon in the Hole Tab ribbon (Figure 5-57).

Step63: ⬚ *Select* the geometry **centerline** tool above the Datum area of the ribbon, then create a vertical centerline anywhere on the screen.

Step64: ⬚ *Select* the **line** tool, then sketch the tapered hole's right side profile as a closed section. See Figure 5-58.

Figure 5-56 Sketched Hole Icon

Figure 5-57 Activate Sketcher

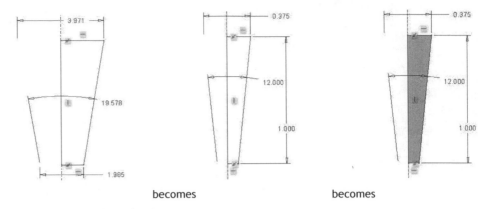

becomes   becomes

Figure 5-58 Define the Taper

Step 65: ⟷ *Select* the ***dimensioning*** tool, then add the overall length of the taper. If the angle dimension or the large diameter dimension disappears, re-add them. Change the angle to 12 degrees, the overall length to 1.00 inches, and the large diameter to 0.375 inches. See Figure 5-58.

Step 66: *Verify* that the section is closed.

Step 67: *Pick* the ***green checkmark*** to exit from sketcher.

Step 68: *Pick* the red word **Placement**.

Step 69: *Select* the step shaft's center axis. Hold down the <Ctrl> key and *pick* the right end of the step shaft. (The end with the cut groove.) The word Placement should turn black meaning it has enough information to continue. If the tapered hole is outside the step shaft, *pick* the **Flip** button to change its direction.

Step 70: *Select* Hidden Line mode.

Step 71: *Pick* the ***glasses*** icon to view the tapered hole in the right end of the step shaft. Verify the tapered hole is positioned correctly. See Figure 5-59.

Step 72: *Pick* the ***green checkmark*** to accept the feature.

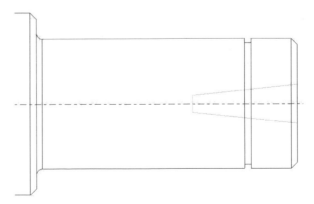

Figure 5-59 Show Taper

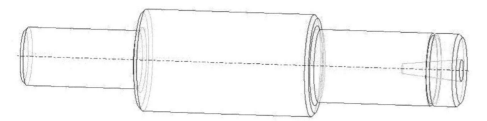

Figure 5-60   Finished Step Shaft

Step 73: Save the step shaft.

Step 74: *Select* **File >Manage File >Delete Old Versions.** *Pick Yes* button.

Step 75: With *Hidden Lines* still selected, *select* **File>Print>Print** to make a hard copy of your step shaft to prove you have completed this practice. *Pick* **OK.** *Pick* **OK.** See Figure 5-60.

▶ Revolves Exercise

A clutch is needed for a 15-horsepower one-cylinder, two-cycle engine that develops its maximum torque at 3400 revolutions per minute.

Design Intent

Design a cone clutch with a maximum outside diameter of 7 inches and a cone angle of 12°. (Typical cone angles are 10°, 12°, and 15°.) Assume the coefficient of friction between the cone and the cup material is 0.32 and the maximum allowable pressure that can be exerted on the braking material is 50 psi. Assume a 1.00-inch diameter shaft can handle the maximum engine torque. Assume a safety factor of 2. See Figure 5-61.

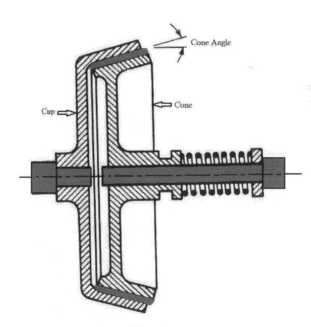

Figure 5-61   Cone Clutch

Assumptions:
   Shaft diameter = 1.00 inches
   Cone/cup angle = 12 degrees
   Brake material average coefficient of friction, $f = 0.32$
   Uniform wear on brake material
   Brake material thickness on cone = 0.125 inches
   Cone Large diameter, $Dia_{cone} = 6.00$ inches
   Width of brake material in contact = 1.00 inches
   Factor of safety for brake = 2.0

Calculations lead us to:

Cone's material large diameter in contact with cup,

$$D = Dia_{cone} + 2 \cdot (Brake\_material\_thickness) \cdot \cos(Cone\_Angle) = 6.244 \ inches$$

Cone small diameter at contact,

$$dia_{cone} = dia_{come} - 2 \cdot Width_{cone} \cdot \sin(Cone\_Angle) = 5.584 \ inches$$

Cone's material small diameter at contact with cup,

$$d = D - 2 \cdot Width_{cone} \cdot \sin(Cone\_Angle) = 5.829 \ inches$$

From 30 hp at 3400 r.p.m., Maximum Torque,

$$H = \frac{\Gamma \cdot n}{63025} = 556 \ in. \ lbs.$$

Maximum pressure on braking material must be less than 50 psi,

$$P_a = \frac{8 \cdot \Gamma \cdot \sin(Cone\_Angle)}{\pi \cdot f \cdot d(D^2 - d^2)} = 31.5 \ psi \le 50 \ psi$$

Maximum Force to be applied by spring,

$$F = \frac{\pi \cdot pa \cdot d(D - d)}{2} = 120 \ lbs.$$

Design OK.

More assumptions:

Diameter of cone hub = 2.00 inches
Length of cone hub = 2.24 inches
Cone hub splined to the shaft so it can slide on shaft
Width of shifting groove = 0.250 inches
Depth of shifting groove = 0.200 inches
Diameter of cup hub = 2.00 inches Length of
cup hub = 1.25 inches
Cup hub keyed to the shaft
Cup locked in place on the shaft with a setscrew opposite the keyway

Now that our design is complete, let's create the cone of our clutch.

Step 1:   Start Creo Parametric by *double-clicking* with the LMB on the ***Creo Parametric*** icon on the desktop, or from the Program list: Creo Parametric.

Step 2:   Set your working directory by *selecting* the ***Select Working Directory*** icon in the Home ribbon. Locate your working directory, and then *pick* **OK.**

Step 3:   Create a new part by *picking* the ***New*** icon at the top of the screen, or by *selecting* **File>New.**

Step 4:   *Select* **Part,** then name the part "Clutch_Cup_15hp". Be sure the "Use default template" box is *checked. Pick* **OK.**

Step 5:   *Select* the ***Revolve*** tool.

Step 6:   *Select* the FRONT datum plane as the sketch plane.

Step 7: 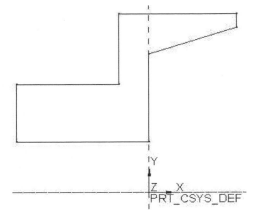 If needed, *pick* the **Sketch View** icon to orient the sketch plane so that it is parallel with the display screen.

Step 8: Create a horizontal geometry centerline (found above in the Datum section) through the origin since all revolved sections must revolve around a predetermined centerline.

Step 9: Sketch the general shape of the upper half of the clutch cup as shown in Figure 5-62.

Step 10: Use the *dimension* tool to add dimensions that represent the design intent. Define all of the diameters, the width of the hub and the overall width of the clutch. To create diameter dimensions, *pick* the point, *pick* the centerline, *pick* the first point again, then use the MMB to place the dimension.

Step 11: Box in the entire sketch and dimensions, then *select* the **modify dimension** tool. *Check* the Lock Scale box (Figure 5-63) before you type 6.25 <Enter> for the cup diameter. *Pick* **OK**. *Pick* the **Refit** icon.

Figure 5-62 Sketched Cross-section

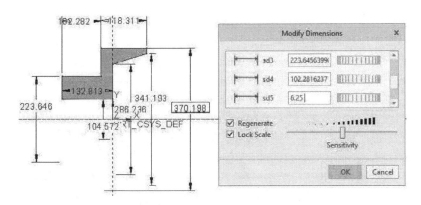

Figure 5-63 Lock Scale and Modify Dimensions

Step 12: Adjust the rest of the dimensions to match Figure 5-64. Add a reference dimension to verify that the slanted length of the cup is greater than the 1.00-inch width of the cone.

Step 13: Verify that the sketch is a closed area. Correct the sketch if necessary.

Step 14: Exit sketcher and keep the sketch.

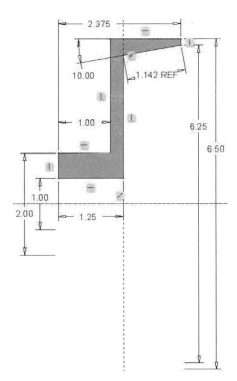

Figure 5-64 Design Intent of Cone Clutch Cup

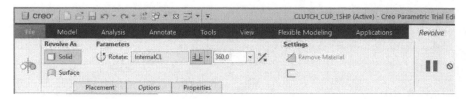

Figure 5-65 Revolve Tab Ribbon

Step 15:   The Revolve tab should appear as shown in Figure 5-65. Make changes if necessary.

If "Select 1 item" appears (Figure 5-66) after the rotational centerline indicator, then you didn't create a geometry centerline when you created Sketch 1. The construction centerline in sketcher does not appear outside of sketcher. If you used the construction centerline or forgot the centerline, *pick* the Placement tab, then the **Edit...** button.

This will place you back in sketcher. If you drew the construction centerline, it might use it. If not, then highlight the centerline, then *select* **Toggle Construction** (Figure 5-67) to change the centerline to a geometry centerline. Exit sketcher, then return to step 15.

Step 16:   *Select* the ***glasses*** icon, then move the cup around to see if the cup appears as shown in Figure 5-68. If so, *select* the ***Green Checkmark***. If you accidentally exit from revolve mode, *highlight* **Revolve 1** in the model tree, press and hold down RMB until a pop-up menu appears, then *select* **Edit Definition** icon to return to revolve mode. (Note that the MMB is the same as the <Enter> key or *picking* the ***green checkmark***.)

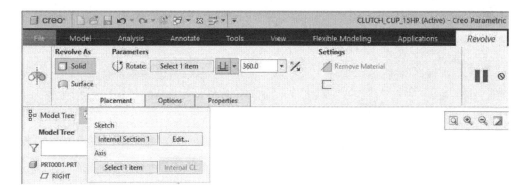

Figure 5-66 Axis of Rotation Not Selected

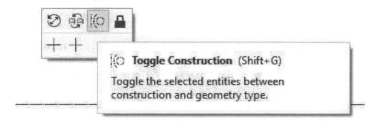

Figure 5-67 Change to Geometry Centerline

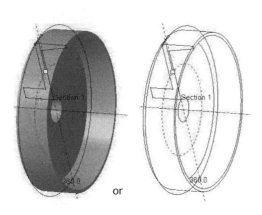

or

Figure 5-68 Revolved Feature

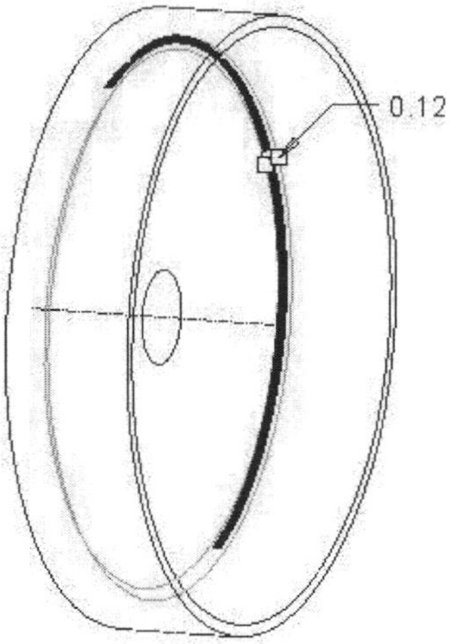

Figure 5-69 Add Round to Clutch Cup

Step 17: File>Save. *Pick* **OK.**

The corner where the slanted surface meets the vertical wall of the cup will have high stresses if we leave it as a sharp corner.

Design Intent

Lower the stresses at the critical diameter changes by rounding the corners. Inside the cup use a 0.12-inch fillet. Between the hub and the cup use a 0.25-inch fillet.

Step 18: ⬡ Round Use the **Round** tool to create a rounded corner where the slanted surface of the cup meets the vertical wall of the cup. See Figure 5-69. *Select* the edge, then set the radius to 0.12 inches. *Pick* the **green checkmark**.

Step 19: Use the **Round** tool again to round the edge where the hub meets the large diameter of the cup. See Figure 5-70. *Select* the edge, then set the radius to 0.25 inches. Press the MMB or *pick* the **green checkmark** to accept the modification.

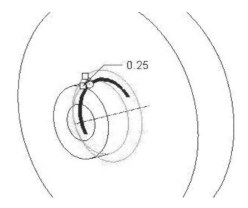

Figure 5-70   Add Round to Hub

Design Intent

Add a keyway to allow for the transmission of torque using a square key.

We will use a key to transfer torque between the cup and its shaft. According to Table 5-1, we should use a ¼-inch by ¼-inch square key for a 1.00-inch diameter shaft. The keyway should be ⅛ inch deep.

Table 5-1 Recommended Key Sizes

Shaft Diameter		Key Size		Keyway
Over (inch)	Including (inch)	w (inch)	h (inch)	Depth (inch)
5/16	7/16	3/32	3/32	3/64
7/16	9/16	1/8	3/32	3/64
		1/8	1/8	1/16
9/16	7/8	3/16	1/8	1/16
		3/16	3/16	3/32
7/8	1 1/4	1/4	3/16	3/32
		1/4	1/4	1/8
1 1/4	1 3/8	5/16	1/4	1/8
		5/16	5/16	5/32
1 3/8	1 3/4	3/8	1/4	1/8
		3/8	3/8	3/16
1 3/4	2 1/4	1/2	3/8	3/16
		1/2	1/2	1/4
2 1/4	2 3/4	5/8	7/16	7/32
		5/8	5/8	5/16
2 3/4	3 1/4	3/4	1/2	1/4
		3/4	3/4	3/8

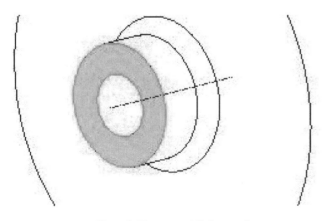

Step 20: Use the ***extrude*** tool to make a cutout in the hub for the square key. Note that the word **Placement** is red on your screen. This indicates that more information is needed.

Step 21: *Select* the leftmost surface of the hub. Highlighted in Figure 5-71.

Step 22: *Select* **References** in the [Sketch] tab ribbon.

Figure 5-71   Sketch Plane and Orientation

Step 23: *Select* the circle that represents the 1.00-inch diameter hole in the hub and the FRONT datum plane from the model tree. *Pick* **Close.**

Step 24: If needed, *pick* the **Sketch View** icon to orient the sketch plane so that it is parallel with the display screen if necessary. Orient the sketch so FRONT datum plane is Vertical as shown in Figure 5-72.

Step 25: Use the **Zoom In** tool so that the two hub circles fill the screen.

Step 26: Draw a vertical construction centerline through the center of the hole along the FRONT datum plane.

Step 27: Use the **Rectangle** tool to draw the outline of the square key in the hub, 0.25 inches per side.

Step 28: If the rectangle is not symmetric about the vertical centerline, use the **symmetric** tool to make the keyway symmetrical.

Step 29: Verify that you have a closed section. See Figure 5-72.

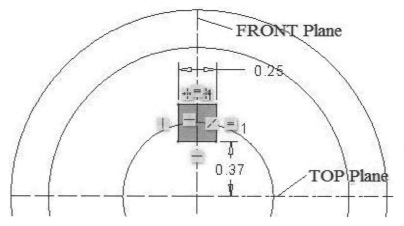

Figure 5-72 Keyway Sketch

Design Intent

Although the step above will create the appropriate keyway it does not represent the design intent. The keyway will not be dimensioned using the 0.125-inch value because this value cannot be measured by the machinist or quality control. Instead, the dimension from the opposite side of the hole's diameter to the top, flat surface of the keyway will be measured. Let's correct our design before continuing.

Calculations:

D = diameter of hole = 1.00 inches
W = width of key or keyway = 0.25 inches
H = distance from top of keyway to opposite side of the hole

$$H = \frac{D + W + \sqrt{D^2 - W^2}}{2} = \frac{1.00 + 0.25 + \sqrt{1.00^2 - 0.25^2}}{2} = 1.109 \text{ inches}$$

Step 30: **File>Options.** *Select* **Sketcher** in the left window. Set the number of decimal places to 3. *Pick* **OK.** When the Creo Parametric Options appears, *pick* **NO.**

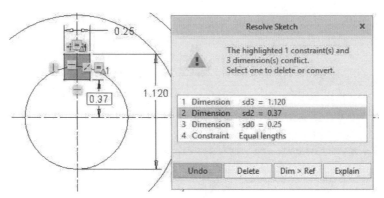

Figure 5-73   Delete Unwanted Dimension

Step 31: 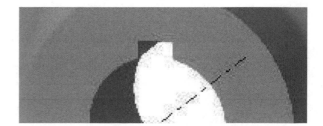Use the ***normal dimension*** tool to create a dimension from the top of the keyway to the bottom of the hole.

Step 32: If the Resolve sketch window appears, highlight the 0.37-inch dimension (yours may be different), then *select* the **Delete** button (Figure 5-73) to delete this dimension.

Step 33: Press the <Enter> key to accept the value of 1.109 inches. If you want to make a little bit of clearance for the key, change the value of 1.109 inches to 1.110 inches.

Step 34: Exit sketcher by *picking* the ***green checkmark***.

Step 35: *Select* ***Solid***, ***Intersect with all surfaces***, ***Remove Material***. If the section does not go through the part, *pick* the ***Change Depth Direction*** icon. See Figure 5-74.

Step 36: *Select* the ***glasses*** icon to see the keyway. If everything is OK, *pick* the ***green checkmark*** or press the MMB. See Figure 5-75.

Step 37: File>Save.

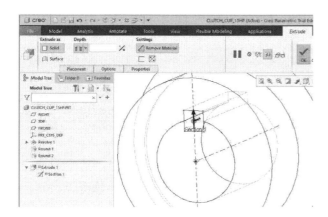

Figure 5-74   Remove Material

Figure 5-75   Keyway Cutout

Design Intent

To make the clutch cup easier to handle, round the three outside edges using a 0.03-inch radius.

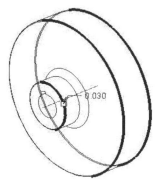

Step 38: [Round] Use the **Round** tool to round the three remaining outside sharp edges (Figure 5-76). Use a radius of 0.03 inches. *Pick* the first edge, then hold down the <Ctrl> key and *pick* the other two edges. This will make the three edges a common set. *Pick* the **green checkmark**.

Figure 5-76 Round Edges

Design Intent

Add a 1/16-inch by 45-degree chamfer at each end of the hole to make inserting the shaft easier at assembly time.

Step 39: [Chamfer] Use the **chamfer** tool to add a 0.06-inch 45 degree (D x D) chamfer to the outer edges of the hole. Hold down the <Ctrl> key before *picking* the second edge. See Figure 5-77. Do not chamfer the keyway. Press MMB to accept the modifications.

Step 40: **File>Save.** See Figure 5-78.

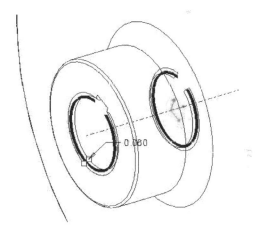

Figure 5-77 Add Chamfers

Figure 5-78 Clutch Cup

Design Intent

Add a threaded hole for the setscrew opposite the keyway which will hold the clutch cup firmly in position on the shaft. For a 1.00-inch diameter shaft, a ¼-20 UNC cup-point socket setscrew is recommended.

Step 41:   🔲 Hole   Use the *Hole* tool to create a threaded hole in the hub opposite the keyway. First, *select* the **standard threaded hole**. *Select* the outer rounded surface of the hub for the hole placement. *Pick* the word **Placement.** Use the LMB to *select* the area in the offset reference box. *Select* the leftmost surface of the hub (see Figure 5-79), hold down the <Ctrl> key, and *pick* the FRONT datum plane. With type "Radial" set, make the angle 0.0 or 90 degrees from the FRONT datum plane and the axial distance from the right edge surface 0.50 inches.

Step 42:   *Select* **UNC, ¼-20,** and **drill to selected point, curve, or surface.** *Select* the surface of the hole as the surface. See Figure 5-80.

Step 43:   *Select* the word **Shape.** Change the depth of the threads to 0.6 inches. See Figure 5-80.

Step 44:   *Select* the **glasses** to see the resulting threaded hole (Figure 5-81). If everything looks good, *pick* the **green checkmark**.

Step 45:   🔲   *Select* **Refit** to get the clutch cup centered properly for printing.

Step 46:   **File>Save.**

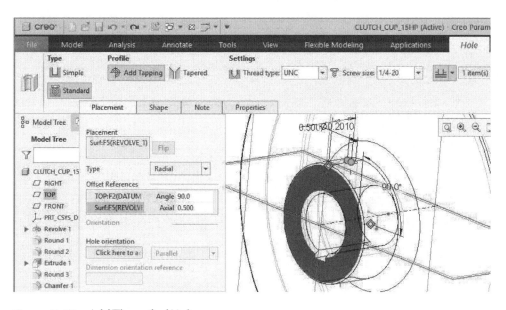

Figure 5-79   Add Threaded Hole

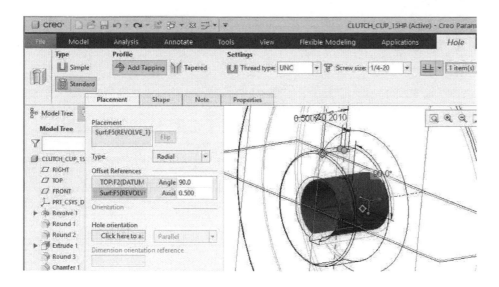

Figure 5-80 Threaded Hole Shape

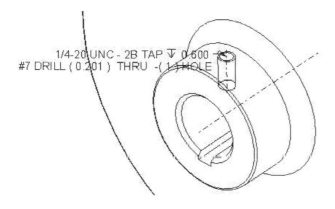

1/4-20 UNC - 2B TAP ↧ 0.600
#7 DRILL (0.201) THRU -(1) HOLE

Figure 5-81 Threaded Hole

Step 47: **File>Manage File>Delete Old Versions.** *Pick* **Yes**.

Step 48: *Select* **Hidden line** mode. Orient the clutch cup similar to that shown below.

Step 49: Turn off Plane Display and Csys Display.

Step 50: **File>Print>Print** to prove that you have completed this exercise. *Pick* **OK.** *Pick* **OK.** See Figure 5-82.

Step 51: Exit from Creo Parametric.

End of Revolves Exercise.

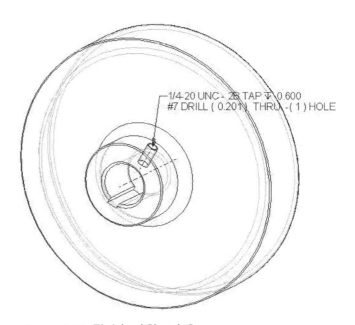

Figure 5-82 Finished Clutch Cup

▶ Review Questions

1. What feature is required of all revolved sections?

2. Can the sketched section be either open or closed when it is used to create a revolved protrusion?

3. Can the sketched section be either open or closed when it is used to create a revolved cut?

4. Can you use either a radial or a diametric dimension when creating a revolved protrusion?

5. How do you create a diametric dimension on a revolved sketch?

6. Describe the difference between a radial hole placement and a linear hole placement.

7. What feature is required to revolve a pattern?

8. How do you create a tapped or threaded hole?

9. What does the revolve icon look like?

10. Name the three hole options.

11. Name the three hole placements.

12. What can be done with the Render tab?

13. Besides selecting the green checkmark with the LMB, describe another way to accept an action.

14. What datum plane is used to sketch a hole?

15. Can a sketched hole be created around a construction centerline and/or a geometric centerline?

Revolves Problems

5.1 Design the "**step_pulley**" (Figure 5-83). The origin needs to be on the centerline at the left end of the pulley. **Design change:** increase the 4.00-inch inside diameter to 4.02 inches. All dimensions are in inches.

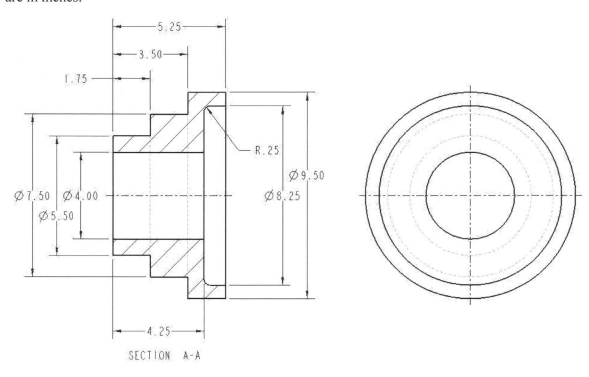

Figure 5-83 Problem 1—Step Pulley

5.2 Design the "**tapered_axle**" (Figure 5-84). The origin needs to be on the centerline at the center of the axle. The 1/2-inch diameter hole goes through the origin. Don't forget to create the two slanted keyways. **Design change:** decrease 4.00-inch diameter to 3.50 inches. All dimensions are in inches.

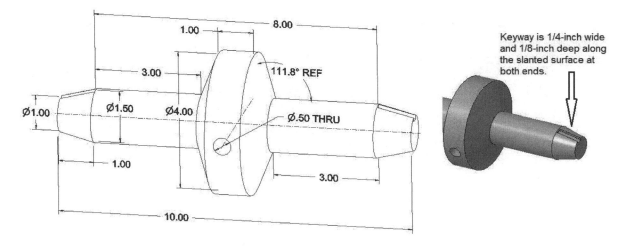

Figure 5-84 Problem 2—Tapered Axle

5.3 Design the "**Axle**" (Figure 5-85). The origin needs to be on the centerline at the location where the diameter changes from 1.1178 inches to 0.7871 inches. When designing this axle we will use its MMC (Maximum Material Condition). Use 3-decimal places in sketcher. All dimensions are in inches.

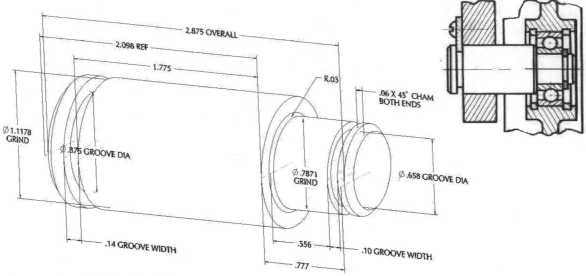

Figure 5-85   Problem 3—Axle

5.4 Design "**Rod_Head**" (Figure 5-86). The origin is on the centerline at the left end of the part. Use 3-decimal places in sketcher. **Design change:** increase the diameter of the 1/2-inch hole to 0.505 inches. The diameter on the right end is also 1.375 inches. All dimensions are in inches.

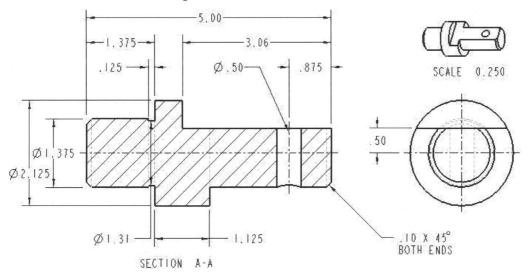

Figure 5-86   Problem 4—Rod_Head

5.5   Design the "**V-belt_Pulley**" (Figure 5-87). The origin is on the centerline at the front of the hub. Add a .03 x 45° chamfer to both ends of the center hole. **Design change:** increase the width of the keyway to 0.252 inches. All dimensions are in inches.

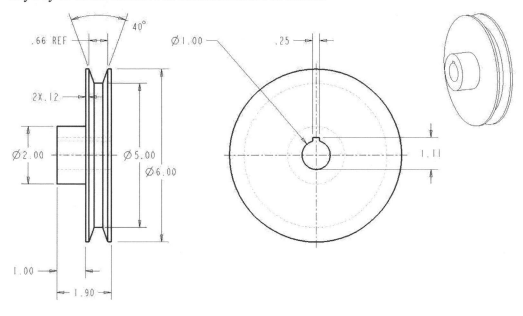

Figure 5-87   Problem 5—V-belt Pulley

5.6   Design the "**Tapered_wedge**" (Figure 5-88). A roll pin goes into the 0.15-inch diameter hole. The large portion of the taper starts at 1.25 inches and tapers at a 16° angle. The top slot is for a 0.25 x 0.25-inch square key. **Design change:** increase the width of the keyway to 0.252 inches. All dimensions are in inches.

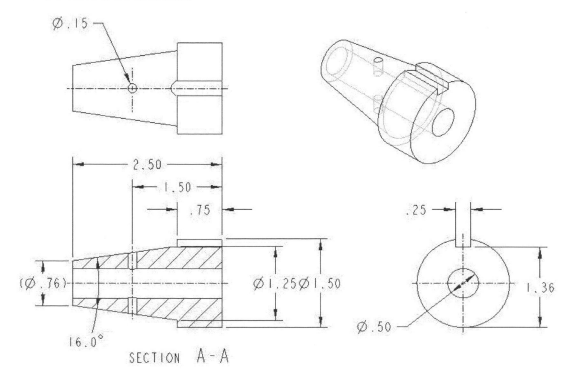

Figure 5-88   Problem 6—Tapered Wedge

5.7   Design the "**clutch_cone**" portion of a cone clutch (Figure 5-89) as described in the Revolves Exercise. You can leave the spline shape out of the cone's shaft hole at this time. We can add it in the next section which covers patterns. Add .06 x 45° chamfer to both ends of the center hole. Break all sharp edges using a .03-inch radius. ***Design change:*** change the 0.25-inch wide slot to 0.26 inches. All dimensions are in inches.

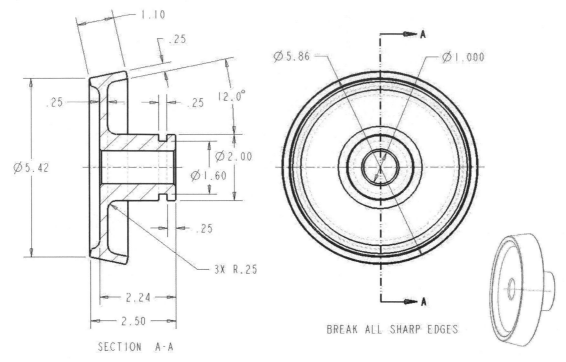

Figure 5-89   Problem 7—Cone Portion

5.8   Design the "**step_pulley_mm**" (Figure 5-90). The origin needs to be on the centerline at the left end of the pulley. **Design change:** decrease the 100 mm diameter to 98 mm to create a force fit. All dimensions are in millimeters.

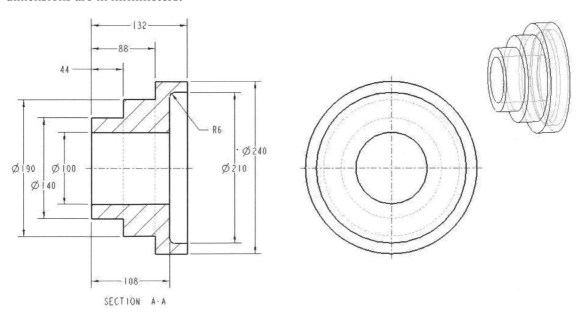

Figure 5-90   Problem 8—Step Pulley (metric)

5.9 Design the "**Axle_mm**" (Figure 5-91). The origin needs to be on the centerline at the location where the diameter changes from 29.94 mm to 20.01 mm. When designing/creating this axle we will use its MMC (Maximum Material Condition). All dimensions are in millimeters.

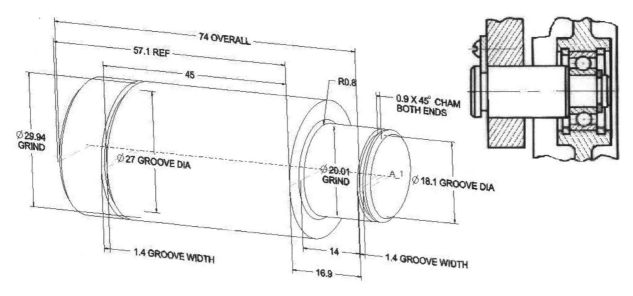

Figure 5-91 Problem 9—Axle (metric)

5.10 Design "**Rod_Head_mm**" (Figure 5-92). The origin is on the centerline at the left end of the part. **Design change:** decrease the 13-mm diameter hole to 10 mm. All Dimensions are in millimeters.

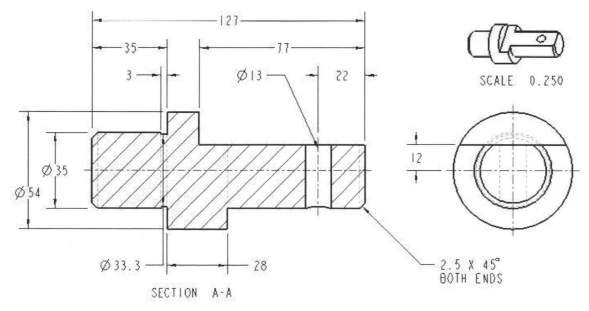

Figure 5-92 Problem 10—Rod_Head (metric)

5.11 Challenge problem - Design the **"idler_pulley"** (Figure 5-93). A view with hidden lines and a section view are shown below. In the section view rounds and fillets are not shown. Note that the 4.50 diameter portion of the idler pulley is rounded using a 10-inch radius arc center 0.84 inches left of the right end of the pulley. Create the pulley using the revolve feature, then add the fillets and the rounds. The outer lip of the idler pulley needs a 1/16-inch round. The corner between the outer lip and the rounded portion of the pulley needs a 1/16-inch fillet. The interior sharp corners need a 1/4-inch fillet to reduce the stresses as noted in the section view. The far right corner should be rounded using a 1/8-inch fillet. All dimensions are in inches.

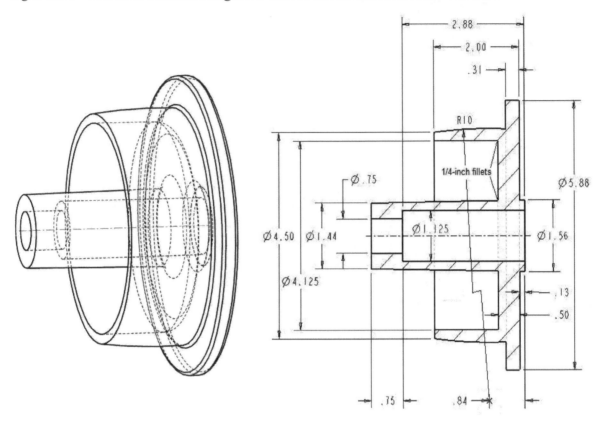

Figure 5-93   Problem 11—Idler Pulley

5.12 Design the **"tapered_pin"** (Figure 5-94). The origin is on the centerline at the big end of the pin. The pin is 3.00 inches long with a 12-degree taper. The big end of the taper pin has a 1.00-inch diameter. The top edge of the groove is located 2.50 inches from the big end. The groove is 0.06 inches wide and is machined to a diameter of 0.350 inches. A 0.03-inch radius round is added to both ends of the tapered pin. All dimensions are in inches.

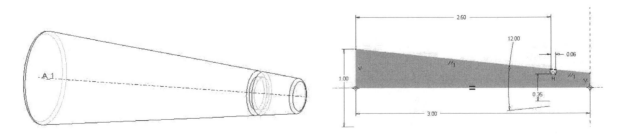

Figure 5-94   Problem 12—Tapered Pin

PATTERNS

Objectives

▶ Creating patterns explored (1D and 2D)

▶ Changing the pattern size with each instance

▶ Practice creating 1D and 2D linear patterns in parts

▶ Practice creating axial patterns in parts

▶ Pattern creation exercise according to design intent

▶ Pattern style design problems to reinforce concepts

Patterns Explored

Parts often contain repeated features of the same size and shape. In Creo Parametric these repeated features are called patterns. The first instance of the pattern is called the pattern leader. Not only can patterns duplicate the exact size and shape of a feature, but they can change the size and shape at each instance along the way. Patterns can be one-dimensional or two-dimensional. There are eight kinds of patterns available. They are dimension, direction, axis, fill, table, reference, curve, and point. We will look at several of these in this section.

If a number of features need to be patterned, then the set of features will need to be made into a group first. For example, a protrusion with a cut and a round are to be patterned. First, the three features are made into a group, and then the group is patterned.

Let's look at a one-dimensional dimension pattern. In order to use the dimension pattern option to create duplicate features, the pattern leader must be placed with a location dimension. This location dimension is selected and then incremented to create the other features in the pattern.

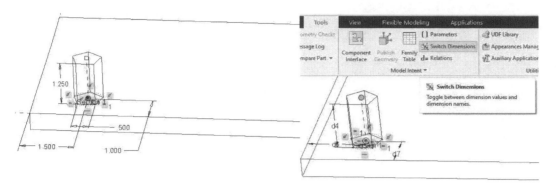

Figure 6-1   Location Dimensions

Above (Figure 6-1) we have a base plate with a pentagon located 1.500 inches (Locate_X) from the left edge and 1.000 inch (Locate_Y) from the front edge. The pentagon is .500 inches (flat_size) across a flat and 1.00 inch (Height) high. In order to make the dimensions more meaningful, we can name them as shown on the right above. Note that "X," "Y," and "Z" are reserved names so they are not available as dimension names. To name a dimension, *pick* it with the LMB. A Dimension tab appears at the top of the screen. In the box to the right of Name such as "d4," enter the new name for the dimension. In order to see the dimension name on the screen, *select* the **Switch Dimensions** icon at the top of the screen. *Pick* this option again to change the dimensions back to numerical values.

To create a pattern of four hexagons along the front of the base plate, we need to highlight the pentagon extrusion in the model tree (**Extrude 2**), and then *pick* the ***Pattern*** tool at the top of the screen. The first box is the pattern type, which is Dimension. The 1ˢᵗ Direction box is shaded indicating that it is active. *Select* the Locate_X (1.50 inch) dimension on the part using the LMB. When a text box appears near the dimension, type "2.00" <Enter> to set the spacing between patterns. If you *pick* the Dimension tab you can see the increment between patterns as 2.00 inches. The Number of members box includes the original item. Set this value to 4 <Enter>. You should see four black dots that represent the locations of the pattern. See Figure 6-2. *Pick* the ***green checkmark*** to accept the pattern.

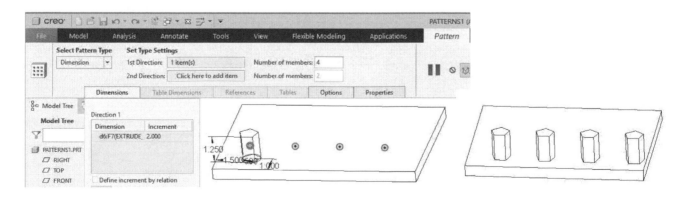

Figure 6-2   Pattern Placement

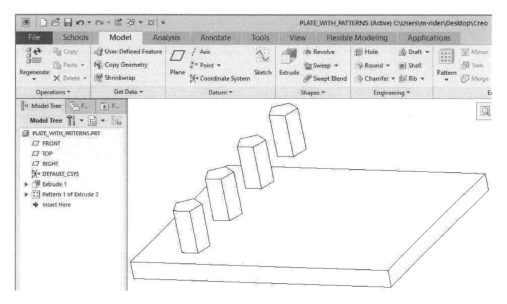

Figure 6-3 Pattern of the Part

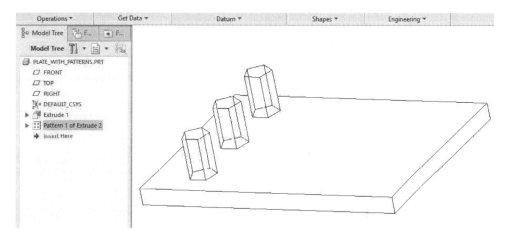

Figure 6-4 Pattern of Three

Select **Pattern 1 of Extrude 2** in the model tree. *Pick* the **Edit Definition** icon from this menu. *Select* the "1 item(s)" area, then press down and hold the RMB. *Pick* the option **Remove.** *Select* the 1.00 (Locate_Y) dimension. Leave the number of patterns at 4 and the spacing at 2.00 inches. *Pick* the **green checkmark**. See Figure 6-3. The fourth pentagon has been located in free space and does not touch the base plate. We can fix the problem by making the spacing closer or reducing the number of patterns to 3. See Figure 6-4.

We can change the size of the pattern leader by adding a size dimension to the pattern control. For example, we want each pentagon in the row to grow by ½ inch. *Select* the **Edit Definition** icon on the pattern. Note that 1 item is selected for pattern control. Hold down the <Ctrl> key and *pick* the Height dimension of the pentagon. The pattern control now lists 2 item(s). Set the Height increment to 0.5 inches. Now each pentagon in the pattern grows in height by ½ inch. *Pick* the ***green checkmark***. See Figure 6-5.

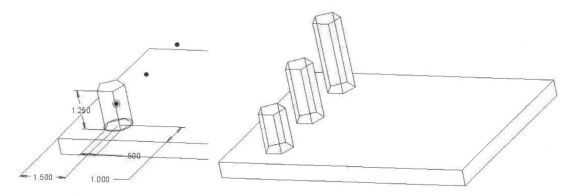

Figure 6-5   Pattern with Two Variables

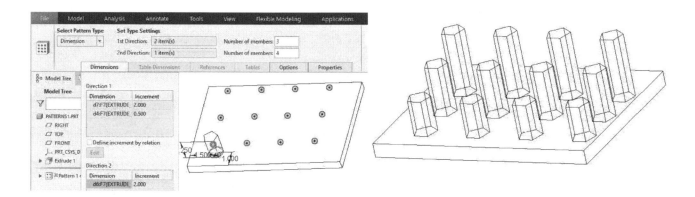

Figure 6-6   Two-dimensional Pattern

Now let's look at a 2-dimensional pattern. Select the **Edit Definition** icon on the pattern again. Using the LMB *pick* the 2nd Direction box (Click here to add item). *Pick* the Locate_X (1.50 inch) dimension and set its increment to 2.00 inches. Set the number of patterns to 4. Change the Locate_Y (1.00 inch) increment to 1.50 inches. You should see 12 black dots nicely spaced on the base plate. See Figure 6-6. *Picking* the **green checkmark** builds the pattern as shown below.

Now let's look at the second type of pattern, the **direction** pattern. The part shown below has a 1.250-inch diameter cylinder 2.000 inches high located tan-

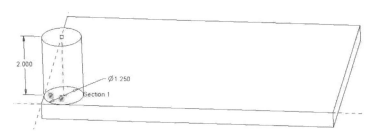

Figure 6-7   Cylinder with Edge Constraints

gent to the two edges. There are no location dimensions present since the cylinder is located using the tangent constraints. In this case, we want to create a pattern of 4 cylinders evenly spaced along the front of the 9-inch long base plate. See Figure 6-7.

Select **Pattern** from the Model tab. In the pull-down menu above the word Dimensions, *select* the **Direction** option, then pick the front edge of the base plate. Set

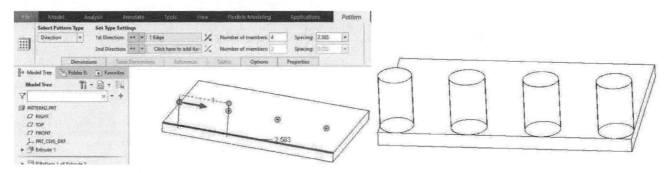

Figure 6-8 Pattern Driven by Direction

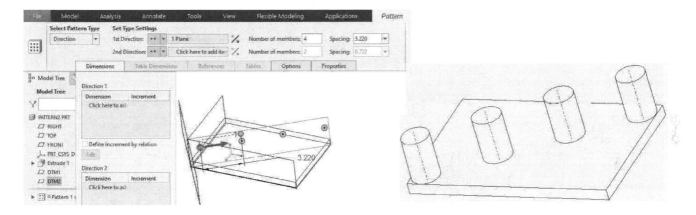

Figure 6-9 Normal to Datum Pattern

the number of patterns to 4. Set the increment between patterns to (9-1.25)/3 or 2.58 inches. See Figure 6-8. Note that you can enter an equation in the spacing between patterns area. Creo Parametric will do the math for you. *Pick* the **green checkmark** to build the pattern.

Let's create four cylinders evenly spaced along the diagonal of the 9-inch long by 6-inch wide base plate. We will use normal to a datum plane as the pattern direction. See Figure 6-9.

Note: The diagonal would be $\sqrt{9^2 + 6^2} = 10.82 inches$. The center of the cylinder is located $0.625 * \sqrt{2} = 0.88$ *inches* from the corner of the base plate. The increment between patterns would be:

$$\frac{\sqrt{9^2 + 6^2} - 2*\left(0.625*\sqrt{2}\right)}{3} = \frac{9.05}{3} = 3.02$$

We will create a datum plane through the corner of the base plate at a 58.5-degree angle, $\tan^{-1}\left(\frac{9-1.25}{6-1.25}\right)$, and then create a second datum plane normal to the first datum plane.

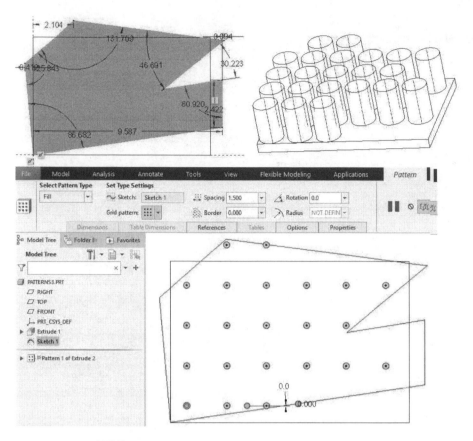

Figure 6-10   Fill Pattern

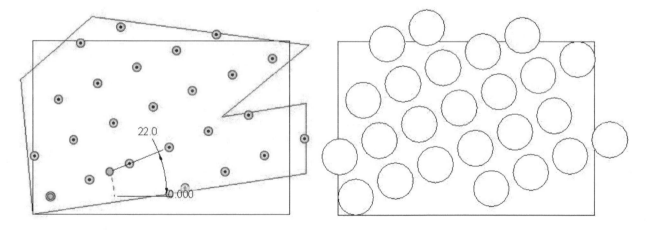

Figure 6-11   Angled Fill Pattern

The **fill** pattern option allows you to fill a sketched area with the pattern leader. This sketched area can be of any shape. In the TOP view, the intersection of the horizontal and vertical dashed lines represents the location of the pattern leader. Creo Parametric will fill the sketched area with the pattern leader according to the incremental spacing. See Figure 6-10. Each black dot represents a pattern feature.

The pattern can also be created at an angle as shown in Figure 6-11. In this case, the same sketched area and pattern spacing were used, but the pattern was filled in at an 22-degree angle from the horizontal.

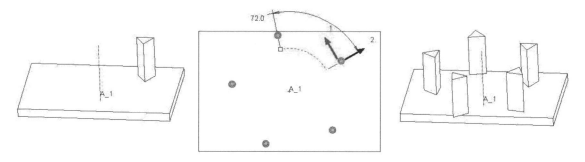

Figure 6-12   Axis Pattern

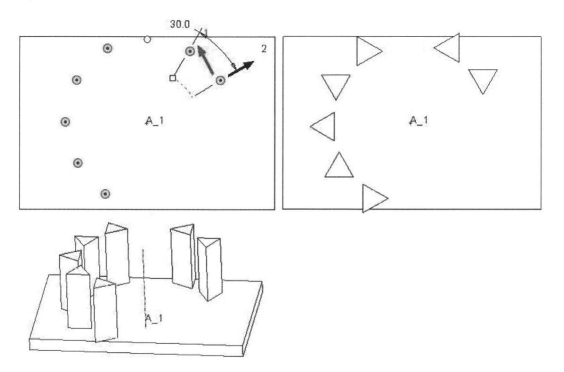

Figure 6-13   Axis Pattern through 240 Degrees (One turned OFF)

Let's look at the **Axis** pattern option next. This option is typically used with parts containing revolved sections, but it is not limited to this. Here are a triangle and an axis located on the base plate. See Figure 6-12. We highlight the extruded triangle in the model tree, then *pick* the **Pattern** tool. We set the pattern type to **Axis,** then *select* the axis in the middle of the base plate. We set the number of patterns to 5 and the angle to 72 degrees.

Next, we edit the pattern to create eight instances every 30 degrees around the axis. See Figure 6-13. Note that the pattern does not have to be a complete circle, 8*30° = 240 degrees. In the TOP view to the right, notice that the pattern leader is rotated by an additional 30 degrees each time it is placed.

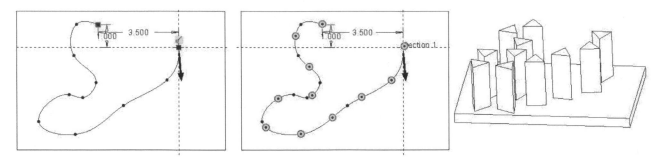

Figure 6-14   Curve Pattern

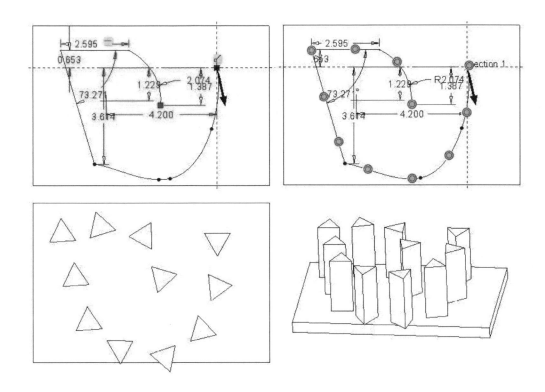

Figure 6-15   Curve Pattern Made Up of Spline, Arc, and Lines

The **curve** pattern option (Figure 6-14) uses a sketched curve to place the pattern leader at regular intervolves. Below we sketched a spline curve as seen from the TOP view. Always start sketching at the location of the pattern leader. When you finish sketching, exit sketcher, then build the pattern along the sketched curve. Any type of curve can be used.

This time the curve is made up of a spline, two straight lines, and a circular arc. The spacing was left at 1.50 inches. See Figure 6-15.

There are many other features available when using patterns. This should give you a basic idea of what can be done using patterns in Creo Parametric.

Patterns Practice (Linear)

The picture in Figure 6-16 is an antique erector set plate with many holes. If you were to create this part in Creo Parametric then individually create each hole in the plate, it would take a long time. We can use the linear pattern feature of Creo Parametric and quickly create all the holes.

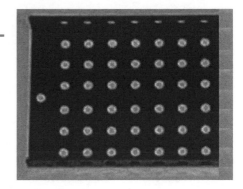

Figure 6-16   Erector Set Plate

Design Intent

Create the erector set plate to scale. The plate is 4.00 inches long and 3.00 inches wide, made from 0.035-inch thick steel. The holes are 0.166 inches in diameter and spaced ½ inch apart in both directions. The single hole on the left end is ½ inch away from the last row and centered on the plate. The two edges are bent over creating a ½-inch lip with the same hole spacing. Assume the origin is in the middle of the plate containing all the holes.

The plate is best created using the extrude function. Let's begin by starting Creo Parametric and setting the working directory.

Step 1:   Start Creo Parametric by *double-clicking* with the LMB on the **Creo Parametric** icon on the desktop, or from the Program list: Creo Parametric.

Step 2:   Set your working directory, by *selecting* the **Select Working Directory** icon in the Home ribbon. Locate your working directory, and then *pick* **OK.**

Step 3:   Create a new object by *picking* the **New** icon at the top of the screen, or by *selecting* **File>New.**

Step 4:   *Select* **Part** from the window and name it "erectorset_plate_3x4". The sub-type should be Solid. Make sure the default template is *checked*, and then *pick* **OK.**

An XYZ coordinate system will appear in the middle of the screen along with FRONT, TOP, and RIGHT side planes. The words FRONT, TOP, and RIGHT will appear if the appropriate option is set. (File>Option>Entity Display, *check* Show datum plane tags.) We will create the erector plate by sketching its 2D cross-section, then extruding it to a length of 4.00 inches.

Figure 6-17   Model Properties

Step 5: We need to set the units for our erector plate to the IPS (Inch-Pound-Second) system of units. Use: **File>Prepare>Model Properties.** See Figure 6-17.

Step 6: The following Model Properties window appears as shown in Figure 6-18. *Pick* the word **change** on the right end of the Units row. Note that a material property is not assigned to the part at this time.

Step 7: The Units Manager window (Figure 6-19) appears. If the system of units, **IPS,** is not selected as shown below, *pick* it from the list, then *pick* the ![Set] button.

Step 8: After *picking* the ![Set] button, the following window appears. *Select* Convert dimensions (for example 1″ becomes 25.4mm), then *pick* **OK.** See Figure 6-20. *Pick* **Close** on the Units Manager window. *Pick* **Close** on the Model Properties window.

Step 9: **File> Options.** *Pick* Sketcher. *Set* the number of decimal places to 3. *Pick* **OK.** When asked if you want to save settings to a configuration file, *pick* **No.**

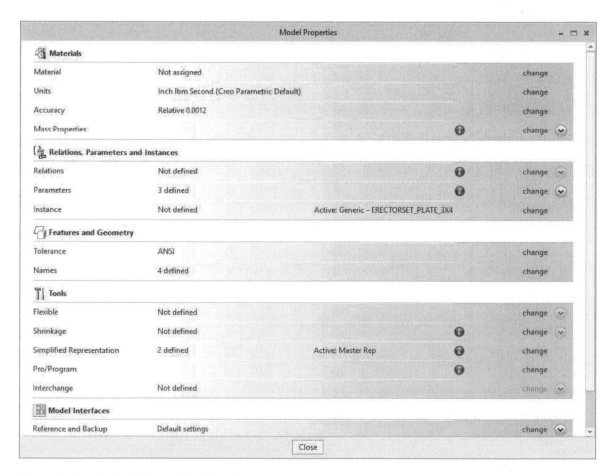

Figure 6-18   Model Properties Window

We could create the sketch first, then extrude it using the extrude tool. Let's *pick* the **Extrude** tool, then make the sketch as we did earlier in this textbook.

Step 10: *Pick* the **extrude** tool from the top of the screen.

Step 11: Move the cursor onto the RIGHT plane so that it turns green. *Select* it by clicking the left mouse button (LMB). *Pick* **sketcher setup**, then make sure the TOP plane is oriented toward the TOP of the screen. *Pick* **Sketch**. See Figure 6-21.

Step 12: If needed, *pick* the **Sketch View** icon to orient the sketch plane so that it is parallel with the display screen.

Step 13: Sketch a vertical construction centerline through the origin.

Step 14: Sketch the cross-section of the erector plate in Figure 6-22.

Step 15: *Pick* the **green checkmark** to exit from sketcher and keep the sketch. Use constraints to reduce the number of dimensions needed to completely define the sketch.

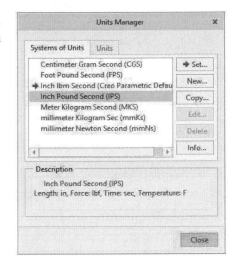

Figure 6-19   Units Manager Window

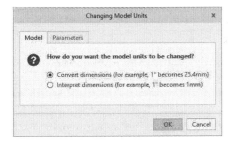

Figure 6-20   Changing Model Units

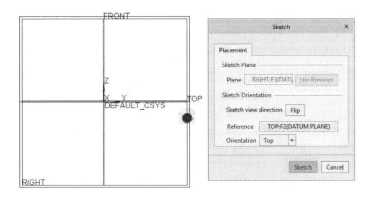

Figure 6-21   Sketcher Plane and Orientation

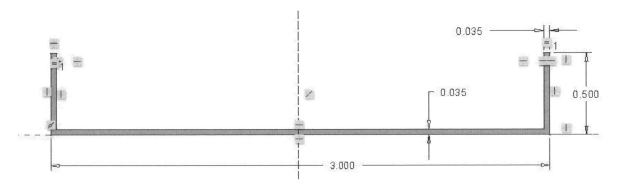

Figure 6-22   Cross-section of Erector Plate

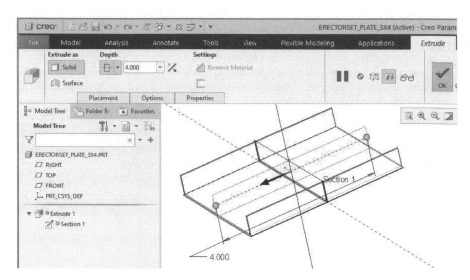

Figure 6-23   Extrude Erector Plate

Step 16: 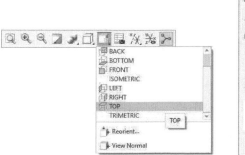 Set extrude to ½ the length in each direction, and then set the length of the extrusion to 4.00 inches. See Figure 6-23.

Step 17: *Pick* the ***green checkmark*** or press the MMB to accept the extrusion.

Step 18: *Save* the part. *Pick* **OK** in the Save Object window.

Step 19: *Pick* TOP from the ***Named View List*** in the graphics window (Figure 6-24) to orient the plate so it is ready for the next step.

Step 20: *Select **No Hidden*** from the graphics toolbar at the top of the window if not already in No Hidden Line mode.

Step 21: *Select* the ***hole*** tool from the Model ribbon, and then place a hole 0.166 inches in diameter ¼ inch from the bottom and right side in the top view. Move an offset reference collector (broken diamond) to the bottom edge of the view and the other offset reference collector to the right edge. Set the hole's depth to through all. See Figure 6-25.

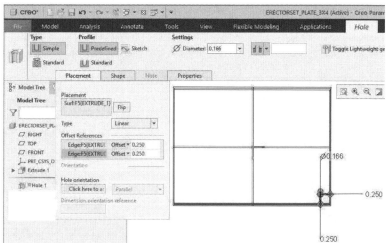

Figure 6-24   Named View List   **Figure 6-25**   Add Hole

Step 22: *Pick* the **glasses** to preview the hole before accepting it and exiting the hole tool feature. If everything looks OK, *pick* the **green checkmark** to accept the feature.

Now we are ready to pattern the hole 42 times. This will not include the single hole on the left side in the top view or the two sets of seven holes in the side lips.

Step 23: With Hole 1 highlighted in the model tree, *pick* the **pattern** tool in the Editing area.

Step 24: *Pick* the vertical dimension 0.250 of the hole we just created. When a box shows up near the dimension, type "0.50" <Enter> for the spacing. (You should see two black dots on the screen.)

Step 25: Using the LMB *select* the blank box to the right of 2nd Direction. *Pick* the horizontal dimension 0.250 of the hole we just created. When a box shows up near the dimension, type "0.500" <Enter> for the spacing. (You should see four black dots on the screen.)

Step 26: At the top of the screen, change the number of instances for the first direction from 2 to 6 <Enter>. Change the number of instances for the second d from 2 to 7<Enter>. See Figure 6-26. (There should be 42 black dots visible on the plate. These are the 42 holes we are about to create in the plate.)

Step 27: *Pick* the **green checkmark** to build the 42-hole pattern.

Step 28: *Save* the part.

Step 29: Select the **hole** tool again. Locate this hole on the left side of the plate near its middle vertically. Drag one green diamond to the right edge of the plate and the other broken diamond to the FRONT datum plane. *Select* the word **Placement** to bring up the placement menu. Set

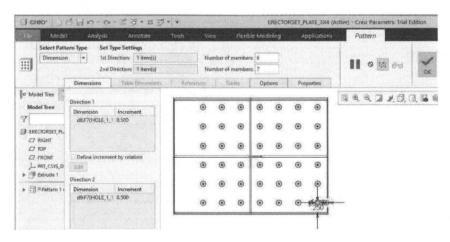

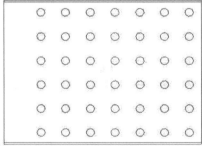

Figure 6-26 Hole Pattern Created

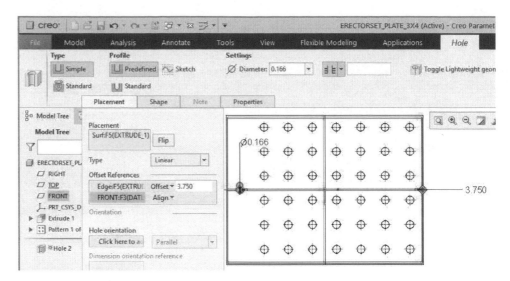

Figure 6-27   Add Another Hole

the offset to 3.750 inches for the right edge. Align the placement for the FRONT datum plane. Set the diameter to 0.166 inches through. See Figure 6-27.

Step 30: *Pick* the ***green checkmark*** to build the hole.

Step 31: *Save* the part.

This completes the top hole pattern. Now we need to create the side lips' hole pattern. We will do this the same way. First, create one hole, and then pattern it in one direction every ½ inch for a total of seven holes. If we create the first hole through the part, then we can create the hole pattern in both lips at the same time.

Step 32: From the Named View List, *pick* the FRONT view.

Step 33: Using the ***hole*** tool, create a hole 0.166 inches in diameter located ¼ inch from the right edge and the bottom edge. Tilt the part to verify that the hole goes through both side lips (Figure 6-28). *Pick* the ***green checkmark*** to build the hole.

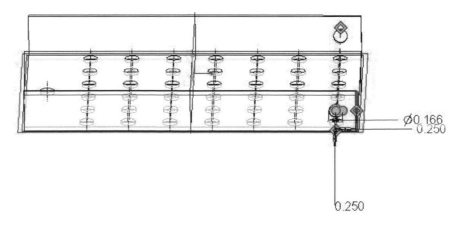

Figure 6-28   Add Side Hole through Part

Step 34: 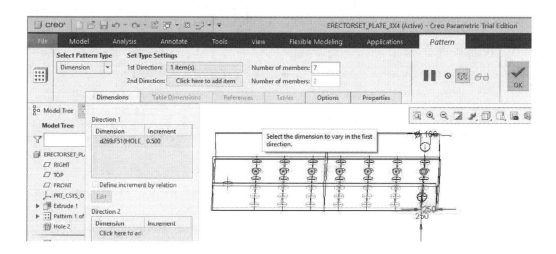 With Hole 3 highlighted in the model tree, *pick* the ***pattern*** tool again.

Step 35: *Select* the horizontal 0.250-inch dimension for the newly created hole. When the rectangular box appears, type "0.500" <Enter> to set the spacing.

Step 36: Set the quantity to 7 <Enter>. You should see seven black dots on the screen. See Figure 6-29.

Step 37: *Pick* the ***green checkmark*** to build the hole pattern.

Step 38: From ***Named View List***, *pick* the **Standard Orientation.** See Figure 6-30.

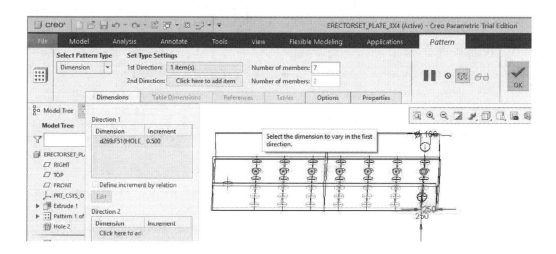

Figure 6-29 Pattern for Side Holes

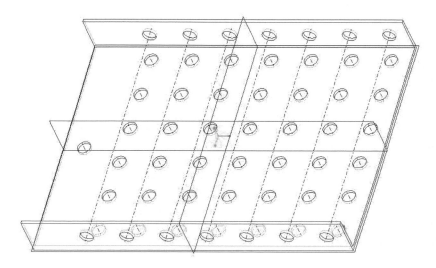

Figure 6-30 Complete Erector Set Plate

Step 39: *Save* the part.

Step 40: File >Manage File >Delete Old Versions. *Pick* **Yes.**

Step 41: Print the part to prove that you have completed this exercise.

This completes the antique 3″ × 4″ erector set plate.

▶ Patterns Practice (Axial)

In this practice session we will design a hollow ground, 7.25-inch diameter circular saw blade. Cutting the teeth in the saw blade will be done with the axial pattern tool.

Design Intent

Design a 7.25-inch diameter hollow ground, 72-tooth saw blade with a 0.625-inch diameter hole for mounting. The hollow ground saw blade is used for satin-smooth crosscuts, rips, and miters in wood. The blade is 0.056 inches thick. Its hub is 2.50 inches in diameter and 0.100 inches thick. In the picture, a small ¼-inch hole was added to the saw blade to balance it for high-speed operation. We will not create this hole. See Figure 6-31.

The hollow ground saw blade blank is best created as a revolved feature. Let's begin by starting Creo Parametric and setting the working directory. (If you have difficulty following the instructions, refer back to a previous practice session.)

Step 1: Start Creo Parametric by *double-clicking* with the LMB on the ***Creo Parametric*** icon on the desktop, or from the Program list: Creo Parametric.

Figure 6-31   72-Tooth Saw Blade

Step 2: Set your working directory, by *selecting* the ***Select Working Directory*** icon in the Home ribbon. Locate your working directory, and then *pick* **OK.**

Step 3: Create a new object by *picking* the ***New*** icon at the top of the screen, or by *selecting* **File>New.**

Step 4: *Select* **Part** from the window and name it "circular_saw_blade." The sub-type should be Solid. Make sure the default template is *checked*, and then *pick* **OK.**

Step 5: We need to set the units for our step shaft to the IPS (Inch-Pound-Second) system of units. Use: **File>Prepare>Model Properties.**

Step 6: When the Model Properties window appears, *pick* the word **change** on the right end of the Units row. Note that a material property is not assigned to the part at this time.

Step 7: When the Units Manager window appears, if the system of units, *IPS,* is not selected as shown below, *pick* it from the list, then *pick* the ⟦ Set... ⟧ button.

Step 8: After *picking* the ⟦ Set... ⟧ button, a new window will appear. *Select* Convert dimensions (for example 1″ becomes 25.4mm), then *pick* **OK.**

Step 9: *Pick* **Close** on the Units Manager window. *Pick* **Close** on the Model Properties window.

Let's create the sketch after selecting the revolve tool. (We could sketch the cross-section, then pick the *revolve* tool.)

Step 10: ⟦ ⊙ Revolve ⟧ *Pick* the *revolve* tool from the top of the screen. *Pick* the ⟦Placement⟧ tab, then *pick* the **Define...** button.

Step 11: Move the cursor onto the RIGHT plane and *select it.* IMPORTANT: Be sure the Sketch window in the upper right corner of the screen shows the TOP plane oriented toward the top of the screen. *Pick* the **Sketch** button.

Step 12: 🔲 *Pick* **sketcher setup**, then make sure the TOP plane is oriented toward the top. *Pick* **Sketch**.

Step 13: 🔲 If needed, *pick* the **Sketch View** icon to orient the sketch plane so that it is parallel with the display screen.

Step 14: ⟦Centerline⟧ Sketch a horizontal geometry centerline through the origin.

Step 15: ⟦ ⋮ ⟧ Sketch a vertical construction centerline through the origin.

Step 16: **File> Options.** *Pick* **Sketcher**. *Set* the number of decimal places to 3. *Pick* **OK**. When asked if you want to save settings to a configuration file, *pick* **No.**

Step 17: Sketch the following cross-section of the saw blade, and then pick the **checkmark** to exit from sketcher and revolve the section. Remember to create the sketch with the center of the saw blade at the origin. See Figure 6-32. Verify that solid is selected and the depth is set at 360 degrees. The revolved feature should be using the **InternalCL** as its axis of rotation. See Figure 6-33.

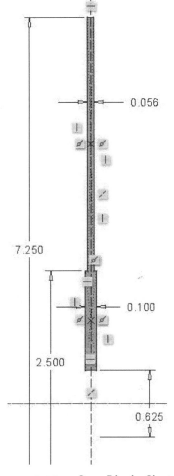

Figure 6-32   Saw Blade Sketch

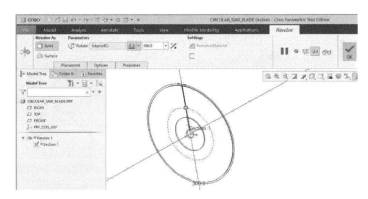

Figure 6-33   Revolve Tab and Part

Step 18: *Pick* the **green checkmark** to accept the revolved feature.

Step 19: *Save* the part. *Pick* **OK** in the Save Object window.

Now, we need to cut the 72 teeth into the outer edge of the saw blade. We will do this by cutting one tooth, then patterning it every 5 degrees around the center axis of the saw blade.

Step 20: *Select* the **sketch** tool from the top of the screen.

Step 21: *Select* the large diameter as the sketching plane with the RIGHT datum plane oriented toward the right. *Pick* the **Sketch** button. See Figure 6-34.

Step 22: If needed, *pick* the **Sketch View** icon to orient the sketch plane so that it is parallel with the display screen.

Step 23: *Select* the **construction centerline** tool, then draw a centerline through the origin and at a 20-degree angle and a second construction centerline through the origin 5 degrees counterclockwise from the first one.

Step 24: *Select* the **References** tool from the Sketch ribbon. *Pick* the outside diameter of the saw blade as a reference. See Figure 6-35. *Pick* the **close** button to close the References window.

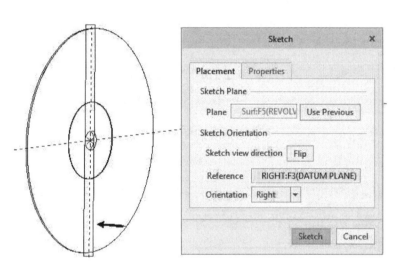

Figure 6-34   Sketch Plane and Orientation

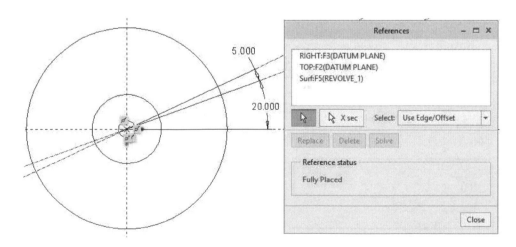

Figure 6-35   Add Sketcher Reference

Step 25: Draw two solid lines that will represent the cutout for a sawtooth as shown in Figure 6-36. One solid line lies along the upper radial construction line and the other solid line is 55 degrees from the second construction line.

Step 26: Zoom in on the sawtooth.

Step 27: *Select* the **Project** icon from the top of the screen. *Pick* the outside diameter of the saw blade blank with the LMB. *Pick* the Close button.

Step 28: *Select* the **Delete Segment** tool from the top of the screen. Press and hold the LMB while dragging the cursor through the arcs not between the 20 and 25 degree lines. Verify that the section is closed. See Figure 6-36.

Step 29: Pick the *green checkmark* to exit from sketcher.

Step 30: With the Sketch highlighted in the model tree, *pick* the *extrude* tool. *Select* extrude as a *solid* with a depth as *intersect with all surfaces* (or through), and *remove material*. See Figure 6-37. If the yellow arrow points away from the saw blade, *pick* the *arrow* icon near the depth box to change its direction.

Step 31: *Pick* the *green checkmark* to accept the cut extrusion.

Step 32: *Save* the part.

Step 33: With **Extrude 1** highlighted in the model tree, *select* the *Pattern* tool.

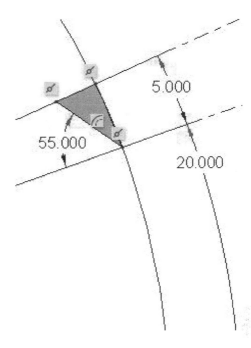

Figure 6-36   Closed Cross-section

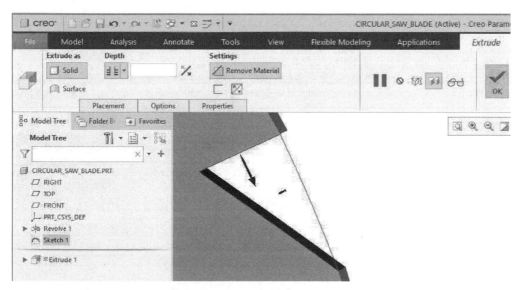

Figure 6-37 Extruded Cut

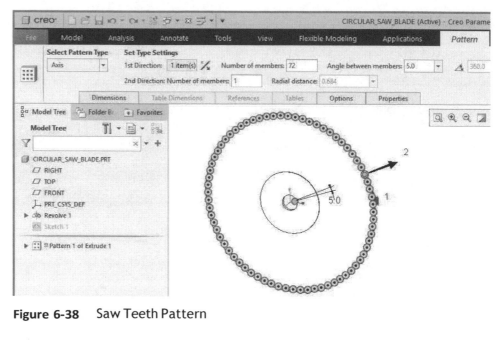

Figure 6-38   Saw Teeth Pattern

Step 34: *Select* the **Axis** option above the word Dimensions.

Step 35: Reorient the saw blade so you can see and *pick* the Z-axis or the A_1 axis from Revolve 1. Set the number of patterns to 72. Set the angle between pattern members to 5 degrees <Enter>. See Figure 6-38.

Step 36: *Pick* the **green checkmark** to build the pattern (Figure 6-39).

Step 37: *Save* the part.

Now we need to add the second deeper cut, thus removing every sixth tooth.

Step 38: *Select* the **sketch** tool. *Select* the **Use Previous** button in the Sketch window. See Figure 6-40.

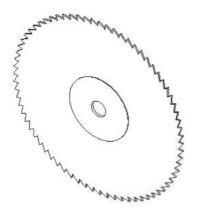

Figure 6-39   Saw Blade with Teeth

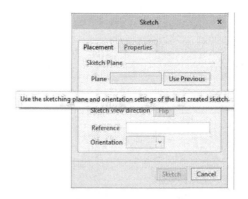

Figure 6-40   Use Previous Sketch Placement

Step 39: 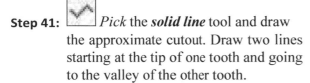 If needed, *pick* the **Sketch View** icon to orient the sketch plane so that it is parallel with the display screen.

Step 40: References *Select* the line that makes up the back of a sawtooth cut and the four points of the two nearby saw teeth as shown in Figure 6-41. It doesn't matter which tooth you *pick* as the tooth to cut out. *Pick* the **Close** button to close the References window.

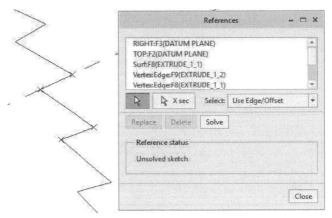

Figure 6-41 Add Sketcher References

Step 41: *Pick* the **solid line** tool and draw the approximate cutout. Draw two lines starting at the tip of one tooth and going to the valley of the other tooth.

Step 42: *Pick* the Circular Trim **Fillet** tool and create a round at the bottom of the cutout.

Step 43: Add dimensions to match the design intent of 25 degrees and a radius of 0.12 inches. See Figure 6-42.

Step 44: *Select* the **3-point arc** from the pop-up menu (as before in step 27) and create an arc that will totally enclose the tooth. This can be done by *picking* the tip of one tooth and the valley of the next tooth, then *picking* the tip of the sawtooth.

Step 45: Verify that the section is closed. See Figure 6-43.

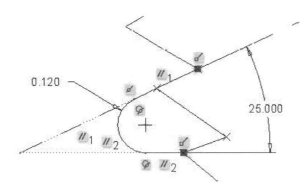

Figure 6-42 Design Intent Dimensions

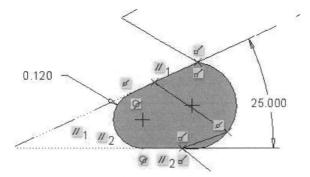

Figure 6-43 Closed Sketcher Section

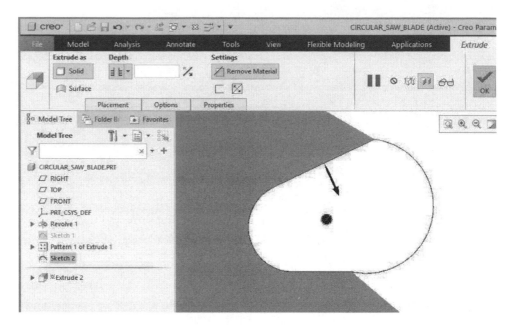

Figure 6-44   Remove Material Extrude Ribbon

Step 46: *Pick* the **green checkmark** to exit from sketcher and keep the sketch.

Step 47: With the Sketch highlighted in the model tree, *select* the **extrude** tool. *Select* **solid** extrude, **through the entire part,** and **remove material.** *Pick* the **arrow** icon if the cut is not inside the part. See Figure 6-44.

Step 48: Verify the cutout.

Step 49: *Pick* the **green checkmark** to accept the build. See Figure 6-45.

Step 50: **Save** the part.

Step 51: With Extrude 2 highlighted in the model tree, *select* the **Pattern** tool.

Step 52: *Select* the **Axis** option above the word Dimensions.

Step 53: Orient the saw blade so you can *pick* the Z-axis or A_1 axis. Set the number of patterns to 12. Set the angle between pattern members to 30 degrees <Enter>. See Figure 6-46.

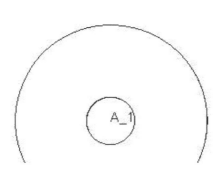

Figure 6-45   Saw Blade Cutout

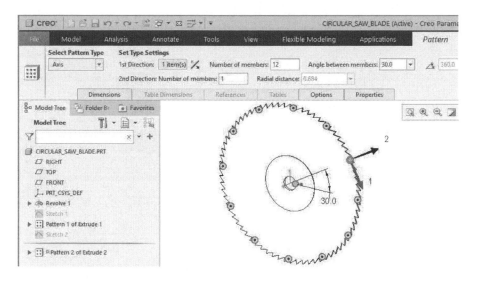

Figure 6-46 Saw Blade Cutout Pattern

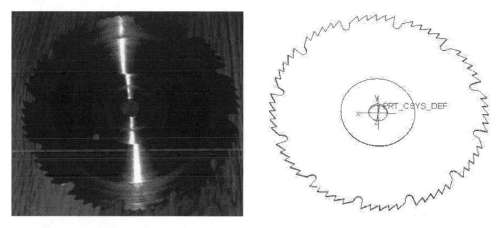

Figure 6-47 Finished Saw Blade

Step 54: *Pick* the **green checkmark** to accept the pattern. See Figure 6-47.

The small hole is not shown in our saw blade because its size and location must be determined during the balancing process. If the saw blade is made from perfectly homogeneous steel, then the balance hole would not be necessary.

Step 55: *Save* the part.

Step 56: **File >Manage File >Delete Old Versions.** *Pick* the **Yes** button.
(Note that we saved the part frequently during its construction. This is a good practice and you should do this when working on any part creation.)

Step 57: Print the part to prove that you completed this practice.

This completes the circular saw blade design session.

▶ **Patterns Exercise**

The dual-toothed sprocket shown in Figure 6-48 is to be designed. We will use the revolve tool to create the sprocket blank, and the axial pattern tool to cut the sprocket teeth. We will use the extrude tool to cut the keyway.

Design Intent

Design the dual-toothed, 17-tooth #40 and 21-tooth #50 sprocket shown above. The smaller sprocket is 3.00 inches in diameter and ¼-inch thick. The larger sprocket is 4.50 inches in diameter and 5/16 of an inch thick. The bore is 1.25 inches in diameter with the outside hub diameter equal to 1.75 inches. The hub protrudes past the larger sprocket by ¼ inch so that the larger sprocket can be welded to the hub. The smaller sprocket is welded to the hub on the inner side. The keyway is 5/16 inches wide with a distance of 1.36 inches from the top of the keyway to the opposite side of the hole. Because the hub is only 1.75 inches in diameter, the keyway is not going to be cut to its standard depth. Place the origin at the outer surface of the smaller sprocket and the center of the hole.

The sprocket blank is best created as a revolved feature. Let's begin by starting Creo Parametric and setting the working directory. (If you have difficulty following the instructions below, refer back to a previous practice session.)

Step 1:   Start Creo Parametric by *double-clicking* with the LMB on the **Creo Parametric** icon on the desktop, or from the Program list: Creo Parametric.

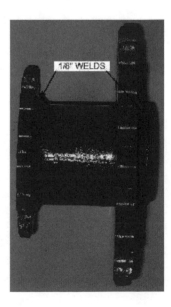

Figure 6-48   Dual-toothed Sprocket

Step 2: Set your working directory by *selecting* the **Select Working Directory** icon in the Home ribbon. Locate your working directory, and then *pick* **OK.**

Step 3: Create a new object by *picking* the **New** icon at the top of the screen, or by *selecting* **File>New.**

Step 4: *Select* **Part** from the window and name it "dual-toothed_sprocket". The sub-type should be Solid. Make sure the default template is *checked*, and then *pick* **OK.**

Step 5: We need to set the units for our step shaft to the IPS (Inch-Pound-Second) system of units. Use: **File>Prepare>Model Properties.**

Step 6: The Model Properties window appears. *Pick* the word **change** on the right end of the Units row.

Step 7: The Units Manager window appears. If the system of units, **IPS**, is not selected as shown below, *pick* it from the list, then *pick* the [Set...] button.

Step 8: After picking the [Set...] button, the changing model units window appears. *Select* Convert dimensions (for example 1″ becomes 25.4mm), then *pick* **OK.**

Step 9: *Pick* **Close** on the Units Manager window.

Step 10: With the Model Properties window still open, *pick* the word **change** on the right end of the Material row.

Step 11: When the Materials window (Figure 6-49) appears, *double-click* on "Standard-Materials_Granta-Design", then *double-click* on "Ferrous_metals", then *double-click* on "stccl_low_carbon" to copy the material properties of this steel to the sprocket.

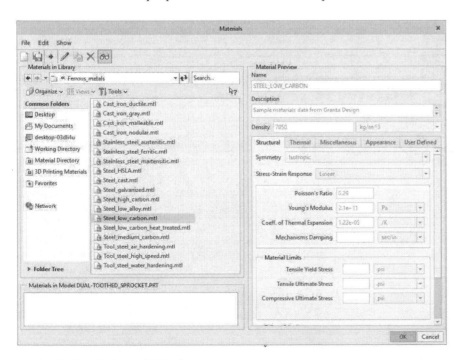

Figure 6-49 Material Window

Step 12: The material selected will appear in the lower left corner of the Materials window. *Pick* **OK** to close the Materials window.

Step 13: *Pick* Close on the Model Properties window.

Step 14: *Pick* the **Revolve** tool from the top of the screen.

Step 15: Move the cursor onto the FRONT plane so that it turns blue. *Select* it by clicking the LMB.

Step 16: If needed, *pick* the ***Sketch View*** icon to orient the sketch plane so that it is parallel with the display screen.

Step 17: Sketch a horizontal geometry centerline through the origin. (Do not use the construction centerline since we want this centerline to appear outside of sketcher.)

Step 18: Sketch the cross-section of the dual-toothed sprocket blank in the first quadrant (any size).

Step 19: Verify that you have created a closed cross-section.

Step 20: Use the ***Normal dimensioning*** tool to show design intent. See Figure 6-50.

Step 21: Use the ***Modify dimension*** tool with Lock Scale *checked* to resize the sketch.

Step 22: Adjust the dimensions to reflect the values shown in Figure 6-50.

Step 23: When everything is correct, *pick* the ***green checkmark*** to exit from sketcher.

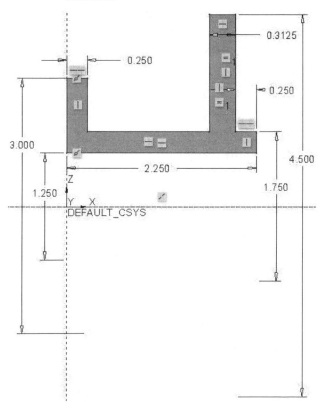

Figure 6-50 Design Intent Sketched and Scaled

Step 24: Be sure InternalCL is selected as the axis of rotation, and the value is set at 360 degrees.

Step 25: When everything looks OK, *pick* the **green checkmark** to accept the feature. See Figure 6-51.

Step 26: **File>Save.** *Pick* **OK** in the Save Object window.

Step 27: We need to calculate the working diameter of the smaller sprocket that will be fitted with #40 roller chain. No. 40 roller chain has a pitch of 0.500 inches (40/80), thus the working diameter is:

$$d = \frac{N * Pitch}{\pi} = \frac{17 * 0.500}{\pi} = 2.706 \text{ inches}$$

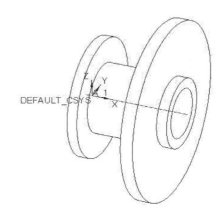

Figure 6-51 Revolved Sprocket Blank

Step 28: *Pick* the **extrude** tool.

Step 29: Move the cursor onto the RIGHT plane so that it turns blue. *Select* it by clicking the LMB. *Pick* the **Sketch View** icon.

Step 30: If the smaller diameter is not visible, *pick* Sketch Setup icon, *pick* the **Flip** button, then *pick* the **Sketch** button to return to sketcher.

Step 31: 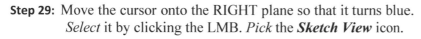 *Pick* the **Sketcher References** icon, then pick the outer surface of the smaller sprocket in the 1st quadrant. *Pick* **Close** to close the References window.

Step 32: Draw a construction centerline at 30 degrees from the horizontal.

Step 33: Draw a circle centered at the origin and set its diameter equal to 2.706 inches.

Step 34: *Select* the circle just drawn. In the pop-up menu *select* the Construction icon as shown in Figure 6-52. The circle will be redrawn using dots.

Step 35: Draw a small circle at the intersection of the 30-degree construction centerline and the construction circle. Set its radius to 0.14 inches (or the diameter to 0.28).

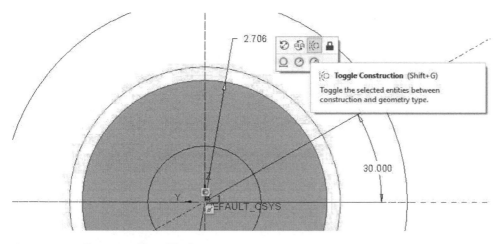

Figure 6-52 Construction Circle

Step 36: Draw a horizontal solid line from the bottom of the just drawn circle to the outside edge of the small sprocket. Draw a 60-degree line from the small circle to the outside edge of the small sprocket.

Step 37: Erase the portions of the small circle inside the two solid lines using the ***Delete Segment*** tool. See Figure 6-53.

Step 38: ☐ *Select* the **Project** icon from the top of the screen. *Pick* the outside diameter circle of the gear blank. Use the **Delete Segment** tool to delete the arcs outside this area.

Step 39: Verify that you have created a closed section. See Figure 6-54.

Step 40: *Pick* the ***green checkmark*** to exit from sketcher.

Step 41: Use the ***arrow*** icon to change the direction of the extrusion if necessary.

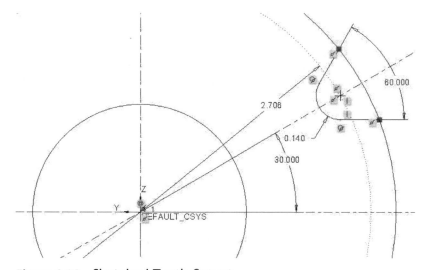

Figure 6-53   Sketched Tooth Cutout

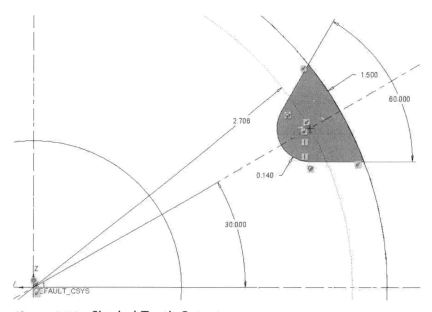

Figure 6-54   Shaded Tooth Cutout

Step 42: ⊥⊥ *Select **extrude to selected point, curve, plane, or surface,** then* pick *the back surface of the small sprocket. See Figure 6-55.*

Step 43: Verify that you have cut out the appropriate section.

Step 44: *Pick the **green checkmark** to accept the cut feature.*

Step 45: With **Extrude 1** highlighted in the model tree, *pick* the **Pattern** tool.

Step 46: Above the word Dimensions, *select* **Axis** from the pull-down menu. *Pick* the x-axis (or the A_1 axis) as the axis of revolution.

Step 47: Set the number of patterns to 17. *Pick* the **Angle** icon and verify that the total angle is 360 degrees. See Figure 6-56. You should see 17 equally spaced dots around the small sprocket blank.

Figure 6-55 Extrude to Selected Surface

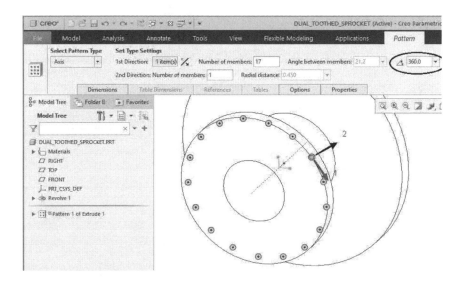

Figure 6-56 Teeth Cutout Pattern

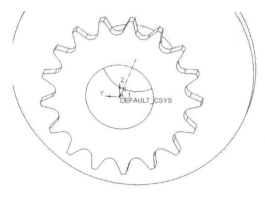

Figure 6-57   Small Sprocket with Teeth

Step 48: *Pick* the **green checkmark** to accept the pattern. See Figure 6-57.

Step 49: *Pick* the **Save** file icon.

Step 50: Calculate the working diameter of the large sprocket. The pitch for #50 roller chain is 0.625 inches.

$$d = \frac{N * Pitch}{\pi} = \frac{21 * 0.625}{\pi} = 4.178 \text{ inches}$$

Step 51: [□] *Select* the **datum plane** tool from the Model ribbon. *Pick* the back side of the large sprocket blank. *Pick* the ⎢Properties⎥ tab and name the datum plane "BACK" <Enter>. *Pick* the **OK** button. See Figure 6-58. (We will use this datum plane as the sketching plane when creating the large sprocket's tooth profile.)

Step 52: [Extrude] *Pick* the **extrude** tool.

Step 53: Move the cursor to the BACK datum plane. *Select* it by clicking the LMB.

Step 54: [⧉] If needed, *pick* the **Sketch View** icon to orient the sketch plane so that it is parallel with the display screen.

Step 55: [▣] Use the **References** tool to add the outer circle of the sprocket blank as a reference by picking a point in the 1ˢᵗ quadrant. *Pick* **Close**.

Step 56: Draw a construction centerline at 30 degrees from the horizontal.

Step 57: Draw a circle centered at the origin and set its diameter equal to 4.178 inches.

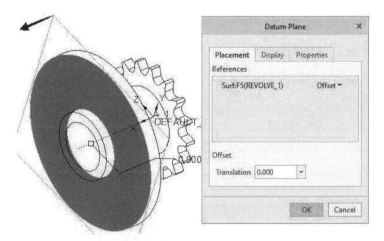

Figure 6-58   Back Datum Plane

Step 58: *Select* the circle just drawn. When a pop-up menu appears, *select* the **Construction** icon. See Figure 6-59.

Step 59: Draw a circle at the intersection of the 30-degree construction centerline and the construction circle. Set its radius to 0.19 inches (0.38 in diameter).

Step 60: Draw a horizontal solid line from the bottom of the just drawn circle to the outside edge of the large sprocket blank. Draw a 60-degree line from the small circle to the outside edge of the large sprocket blank.

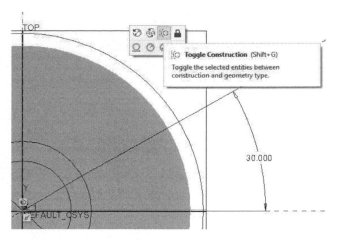

Figure 6-59   Change to Construction Circle

Step 61: Erase the portions of the small circle inside the two solid lines using the *Delete Segments* tool. See Figure 6-60.

Step 62: ⬜ *Pick* the *Project* icon. *Pick* the outer circle. *Pick* **Close**. Use the *Delete Segment* tool to remove unwanted portions of the larger circle.

Step 63: Verify that you have created a closed section. See Figure 6-61.

Step 64: *Pick* the *green checkmark* to exit from sketcher.

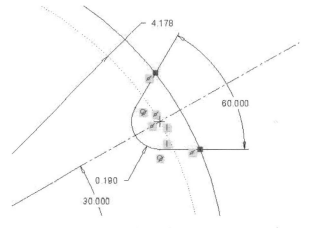

Figure 6-60   Large Sprocket Cutout

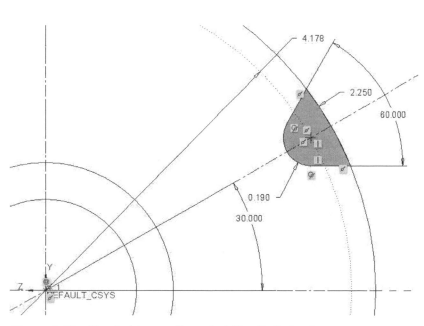

Figure 6-61 Shaded Large Sprocket Tooth Cutout

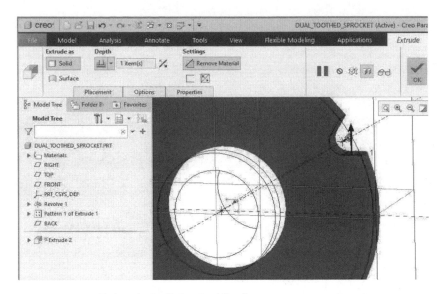

Figure 6-62   Extrude to Selected Surface

Step 65: Use the *arrow* icon to change the direction of the extrusion if necessary.

Step 66: ⊥⊥ *Select **extrude to selected point, curve, plane, or surface,** then pick* the opposite surface of the large sprocket. See Figure 6-62.

Step 67: Verify that you have cut out the appropriate section.

Step 68: *Pick* the **green checkmark** to accept the cut feature.

Step 69: With **Extrude 2** highlighted in the model tree, *pick* the **pattern** tool.

Step 70: Above the word Dimensions, *select* **Axis** from the pull-down menu. *Pick* the x-axis (or the A_1 axis) as the axis of revolution.

Step 71: Set the number of patterns to 21. *Pick* the **Angle** icon and verify that the total angle is 360 degrees. See Figure 6-63. You should see 21 equally spaced black dots around the large sprocket.

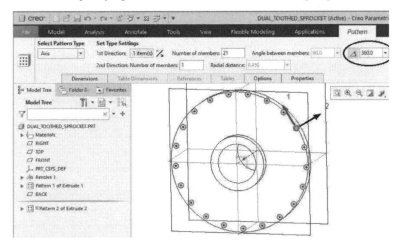

Figure 6-63   Large Sprocket Tooth Pattern

Step 72: *Pick* the ***green checkmark*** to accept the pattern. See Figure 6-64.

Step 73: *Pick* the ***Save*** file icon.

Now let's add the keyway.

Step 74: *Pick* the ***extrude*** tool.

Step 75: Move the cursor onto the FRONT surface so that it turns blue. *Select* it by clicking the LMB.

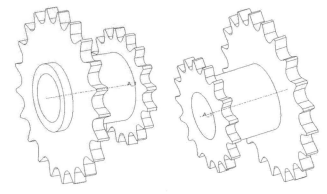

Figure 6-64 Dual Sprocket with Teeth

Step 76: If needed, *pick* the ***Sketch View*** icon to orient the sketch plane so that it is parallel with the display screen.

Step 77: Add the 1.25-inch diameter hole in the sprocket as a sketcher ***Reference***. *Pick* **Close** to close the References window.

Step 78: Draw a vertical construction centerline through the origin.

Step 79: Draw a circle on top of the sprocket hole (1.25 inches in diameter).

Step 80: Draw three solid lines above the circle to form the keyway.

Step 81: Use the ***symmetric constraint*** tool to force the keyway to be symmetric about the vertical construction centerline.

Step 82: Delete the segments of the circle outside of the keyway.

Step 83: Use the ***normal dimensioning*** tool to dimension the keyway per the design intent. The keyway is 0.312 inches wide. The distance from the top of the keyway to the opposite side of the hole is 1.36 inches.

Step 84: Verify that the section is closed. See Figure 6-65.

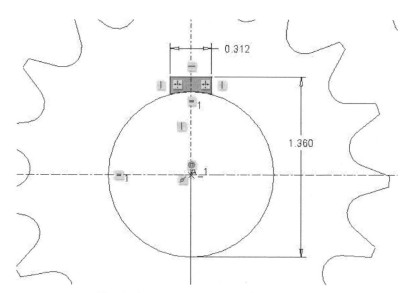

Figure 6-65 Shaded Keyway Cutout

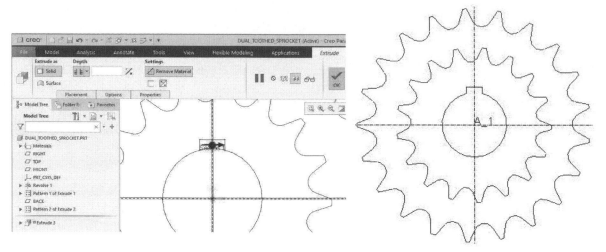

Figure 6-66   Keyway Cutout

Figure 6-67   Dual Sprocket with Keyway

Step 85: *Pick* the **green checkmark** to exit from sketcher and keep the sketch.

Step 86: *Select **through all surfaces, remove material,** and the **arrow** icon if the cut does not intersect the sprocket. See Figure 6-66.*

Step 87: *Pick* the **green checkmark** to accept the extrusion cut of the keyway. See Figure 6-67.

Step 88: *Pick* the **Save** file icon.

Step 89: `Chamfer` *Pick* the **Chamfer** tool to simulate the welds between the two sprockets and the hub. *Pick* the outside edge where the large sprocket touches the hub, then hold down <Ctrl> and *pick* the inside edge where the small sprocket touches the hub. Set the value to 0.12 inches. *Pick* the **green checkmark**. See Figure 6-68.

Step 90: Rename the set of chamfers as "WELDS" in the model tree.

We are almost done; however, when we look at the original dual-toothed sprocket and ours we notice that the teeth are chamfered on the original sprocket. If we do this now, we will have to pick the edge of every tooth on the sprocket. If we chamfered the edge before we cut the teeth, we would need only to pick four edges. Let's back our model up to the point before we cut the first sprocket tooth and add a chamfer of 0.03 by 45 degrees. See Figure 6-69.

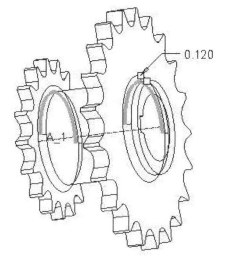

Figure 6-68   Add Welds to Dual Sprocket

Figure 6-69   Sprocket Tooth with Chamfer

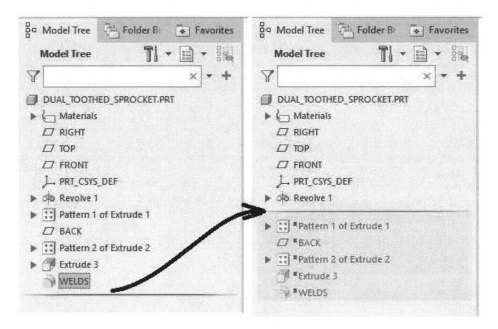

Figure 6-70   Drag Insert Here in Model Tree

Step 91: *Pick* the *Save* file icon.

Step 92: *Pick* the **green line** in the model tree with the LMB. *Press and hold down* the LMB, then drag the line to the area just below Revolve 1 as shown in Figure 6-70. Release the LMB to back the model up to just the Revolve 1 blank.

Step 93: Use the *Chamfer* tool to add a 0.03-inch by 45-degree chamfer to the outside diameter edges of the sprocket blanks. *Pick* the first edge, then press down and hold the <Ctrl> key while *picking* the other three edges. See Figure 6-71.

Step 94: *Pick* the **green checkmark** to accept the chamfer features.

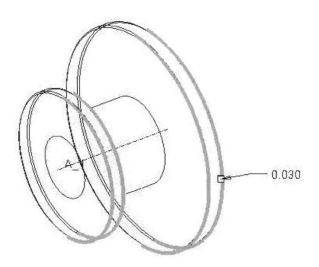

Figure 6-71   Add Chamfers to Sprocket Blank

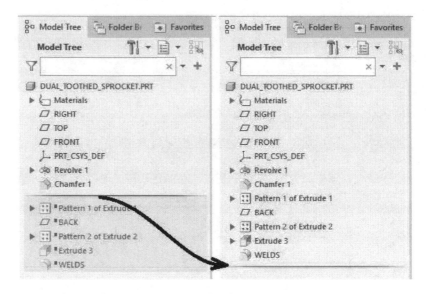

Figure 6-72   Drag Insert Here Back Down

Step 95:   *Pick* the **green line** in the model tree with the LMB. *Press and hold down* the LMB, then drag the line to the bottom of the list as shown in Figure 6-72. Release the LMB to update the model.

The dual-toothed sprocket should appear as shown in Figure 6-73.

Step 96:   *Pick* the ***Save*** file icon.

Step 97:   **File> Manage File> Delete Old Versions.** *Pick* the ***Yes*** button.

Step 98:   [Q] Fit to screen.

Step 99:   *Pick **Hidden line*** or ***No Hidden**,* turn off the spin center, *select* **File> Print> Print** to make a hard copy of the dual-toothed sprocket. *Pick* **OK** twice.

Step 100:   **File> Close.**

Step 101:   Exit from Creo Parametric.

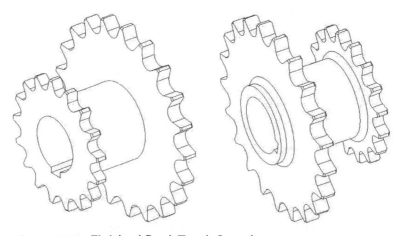

Figure 6-73   Finished Dual-Tooth Sprocket

Review Questions

1. What is the first feature of a pattern called?

2. What dimensions are available for patterning a feature?

3. What is the difference between a dependent and an independent copy? Which is more common?

4. How do you create a pattern of several related features?

5. What happens if you mirror one instance of a patterned feature?

6. Can you make a pattern of patterns?

7. What are the eight types of patterns available in Creo Parametric?

8. What feature is required for a one-dimensional pattern?

9. How do you make one instance of a pattern disappear (not shown)?

10. If you want to change the number of instances in a pattern, do you use Edit or Edit Definition?

11. What does the fill pattern option do?

12. Can you pattern an open section?

13. Name at least four different parts that can be created using the pattern feature of Creo Parametric. What type of pattern would be used to create these parts?

Patterns Problems

6.1   Design "**Packing_Ring**" (Figure 6-74). The origin is on the centerline at the back of the packing ring. The three holes are evenly spaced on a 2.00-inch diameter bolt circle and go through just the tabs. Note that there is a 2.50-inch diameter cutout that is 5/16-inch thick. All dimensions are in inches.

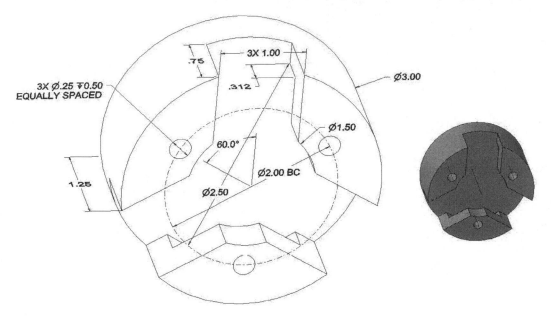

Figure 74   Problem 1—Packing Ring

6.2   Design "**Bearing_232**" (Figure 6-75). The origin is on the centerline at the left end of the bearing. The four holes are evenly spaced on a 7.26-inch diameter bolt circle. **Design change:** increase the 7.25-inch bolt circle to 7.50 inches. All dimensions are in inches. Note: one round is R.19 inches; all the rest of them are R.12 inches.

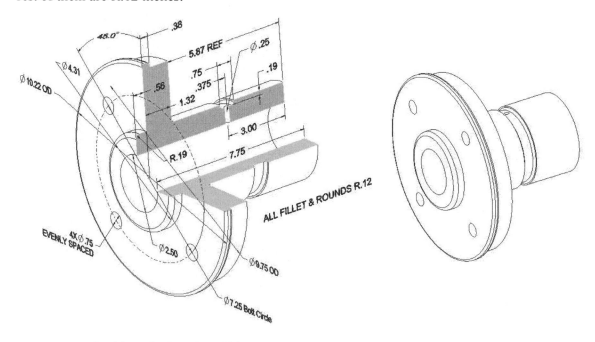

Figure 6-75   Problem 2—Bearing_232

6.3   Design the "**Truck_Wheel**" (Figure 6-76). The origin is on the centerline at the front of the wheel. Make the center hole 0.937 inches in diameter. Add .06-inch fillets and rounds after revolving the cross-section. The three holes are evenly spaced on a 3.375-inch diameter bolt circle. Use maximum material condition (MMC) for the center hole. All dimensions are in inches.

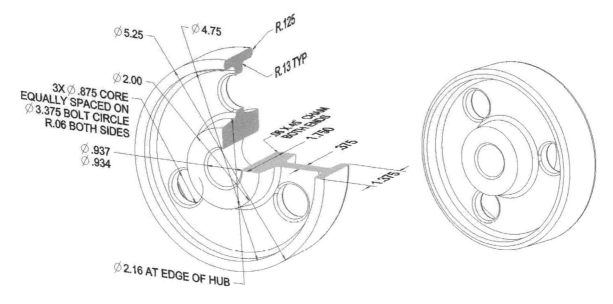

Figure 6-76 Problem 3—Truck Wheel

6.4   Design the "**Fixed_Bearing_Cup**" (Figure 6-77). The origin is on the centerline at the back of the part. Add fillets and rounds last. Don't round required sharp edges. The five holes are evenly spaced on a 1.25—inch diameter bolt circle. The outside diameter of the back side of the hub is 3.00 inches. All dimensions are in inches.

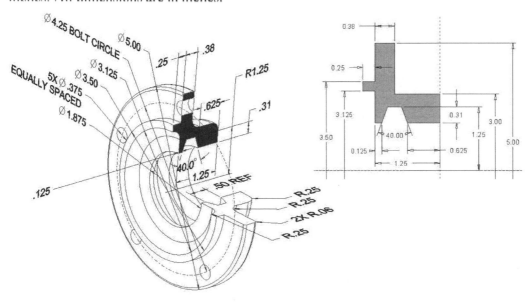

Figure 77 Problem 4—Fixed Bearing Cup

6.5 Design a "**heat_sink**" (Figure 6-78) similar to the one shown here, but as one solid aluminum extrusion instead of a group of pieces bolted together. The base is 1.81 inches by 3.00 inches with a thickness of 0.375 inches. Each fin is 0.06 inches thick, 3.00 inches long, and protrudes above the base plate by 1.25 inches. There is a 0.19-inch gap between each of the eight fins. After creating one fin, create the others using the linear pattern tool. All dimensions are in inches.

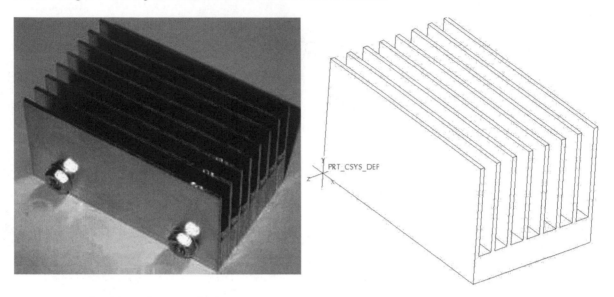

Figure 6-78   Problem 5—Heat Sink

6.6 Design a new "**shelf_support**" similar to the one shown in Figure 6-79. Make it 0.65 inches wide by 0.20 inches tall and made out of 0.056-inch steel. Make it 20 inches long with a 0.156-inch diameter hole every 4.50 inches. The first hole is 1.00 inch from the end. Place a 0.1-inch by 0.28-inch round end slot every ½ inch with the first slot ¾ of an inch from the first hole. Place two 0.10-inch diameter holes 0.2 inches apart and ½ inch from the first hole. Place this 2-hole pattern every 6 inches. Skip the 2-hole pattern if it lines up with one of the 0.156-inch diameter holes. (Hint: use Direction Pattern.) All dimensions are in inches.

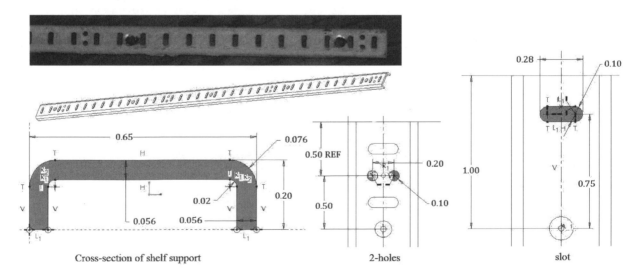

Figure 6-79   Problem 6—Shelf Support

6.7 Design the 4 wide by 10 high "**drawer_cabinet**" that is 6.00 inches deep (Figure 6-80). The cutout for each drawer is 2.00 inches high by 3.50 inches wide with ¼-inch spacing between drawers in both the vertical and horizontal directions. Create the cabinet by creating one of the drawer spaces, then use one pattern tool in two directions to get all 40 drawers. What is the overall height and width of the cabinet? All dimensions are in inches.

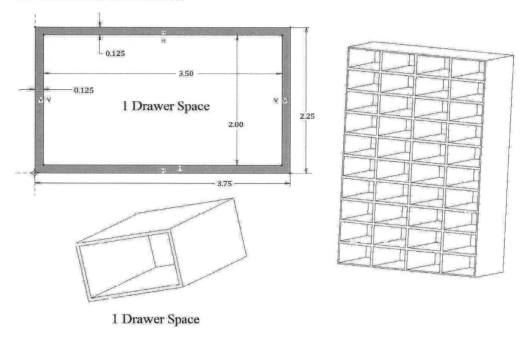

Figure 6-80 Problem 7—Drawer Cabinet

6.8 Design the straight-sided spline drive fitting for the "**clutch_cone**" (Figure 6-81) designed in problem 5.7 (page 152) and add it to the clutch cone. The fitting has 10 splines evenly spaced as shown below in Figure 6-81. For a 10-toothed spline that will slide when not under load and with an outside diameter of "D," the inside diameter is 0.860*D, and the width of the cutout is 0.156*D. The allowable torque per inch length of spline is $326*D^2$ in.lb/in. A 15 horsepower engine at 3400 r.p.m. generates 278 in.lbs. of torque. The cone's spline (Figure 5-89) will be 2.24 inches long and 1.00 inch in diameter; therefore, this spline can handle 730 in.lb. of torque. This provides a safety factor of 2.6. Add a 0.06-inch x 45° chamfer to both ends of the splined hole. All dimensions are in inches.

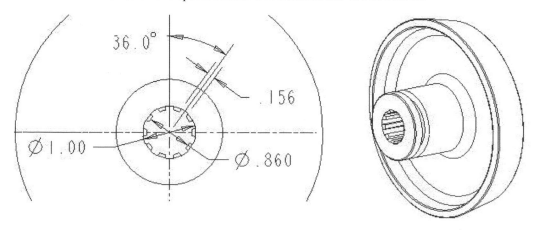

Figure 6-81 Problem 8—Spline Drive Fitting in clutch cone

6.9   Design "**Bearing_231_mm**" (Figure 6-82). The origin is on the centerline at the front end of the bearing. The four holes are evenly spaced on a 108-millimeter diameter bolt circle. Add a 3-millimeter by 45-degree chamfer at each end of the 38-millimeter diameter hole. Add fillets and rounds last. **Design change:** decrease the 108 mm bolt circle to 106 mm. All dimensions are in millimeters.

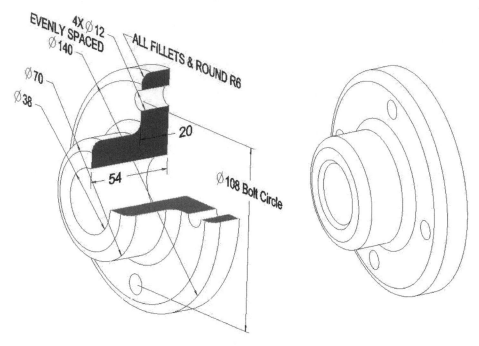

Figure 6-82 Problem 9—Bearing_231 (metric)

6.10   Design the "**Truck_Wheel_mm**" (Figure 6-83). The origin is on the centerline at the front of the wheel. Make the center hole 24 millimeters in diameter. Add the fillets and rounds before revolving the cross-section. The three 22-millimeter holes are evenly spaced on an 85-millimeter diameter bolt circle. **Design change:** Five 22-mm diameter holes evenly spaced are required instead of three. All dimensions are in millimeters.

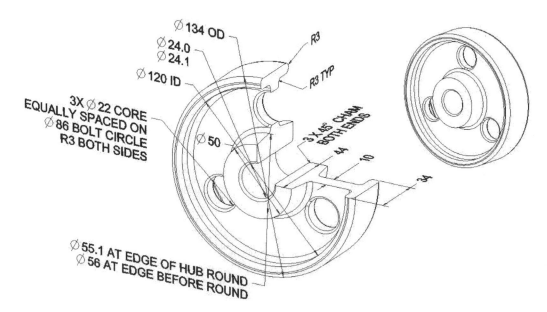

Figure 6-83 Problem 10—Truck Wheel (metric)

DIMENSIONING

Objectives

▶ Introduction to dimensioning

▶ Do's and Don'ts of dimensioning

▶ Create a drawing format in Creo Parametric

▶ Create a 3-view drawing template in Creo Parametric

▶ Create a 2-view drawing template in Creo Parametric

▶ Create a 1-view drawing template in Creo Parametric

▶ Modifying drawing options

Do's and Don'ts of Dimensioning

Background

Before the 1800s an inch was defined as the width of a man's thumb, and a foot was simply the length of a man's foot. (In 1793 France adopted the metric system based on the meter.) In 1824 the English Parliament defined the length of a yard. A foot was one-third of a yard, and an inch was one-twelfth of a foot. From these specifications, graduated rulers were developed. Until the twentieth century, common fractions were adequate for dimensions, but as machines became more complicated, more accurate specifications were required. It became necessary to express dimensions as decimal numbers.

Part Size and Shape

In addition to a complete shape description of a part, an engineering drawing must give a complete size description. In the early 1900s, the design and production functions were located in one factory. In some cases, the complete part creation was carried out by the same worker. Design drawings were nothing more than assembly drawings without dimensions. The worker used a scale directly on the drawing to obtain the required dimensions. It was up to the worker to make sure the parts fit together properly. If a question came up, the worker would consult the designer who was always nearby. Under these manufacturing conditions, it was not necessary for drawings to carry detailed dimensions or notes.

The modern methods of size description came into existence with the need for interchangeable parts. Detailed drawings must be dimensioned so that a worker can produce mating parts that can be assembled in a separate factory or when used as replacement parts by the customer. The need for precision manufacturing and the controlling of sizes for interchangeability shifted from the machinist to the designing engineer. The worker no longer used his judgment in engineering matters, but rather he executed the instructions given on the detailed drawings. Thus, it was necessary for design engineers to become familiar with materials and the processes of the shop.

The drawings must show the part in its completed form and contain all information necessary to make it. Therefore, when dimensioning a detailed drawing, the engineer must keep in mind the finished part, the shop processes required, and the design intent of the part. Dimensions need to be given that are required by the worker when making the part. There is no reason to provide dimensions to points or surfaces that are not accessible to the worker.

Dimensions must not be duplicated on a drawing. Only dimensions needed to produce and inspect the part, as intended by the design engineer, should be shown. The beginner often makes the mistake of giving the dimensions he/she used to make the computer-generated part model or the drawing. These are not necessarily the dimensions required by the shop.

Drawings must be made to scale with the scale listed in the title block.

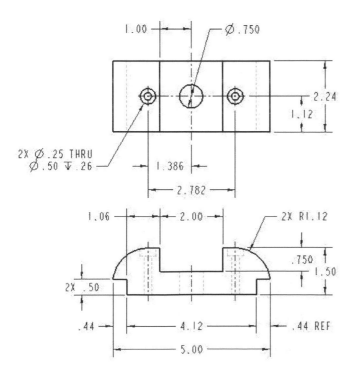

Figure 7-1   Properly Dimensioned Part

Learn to Dimension Properly

Dimensions are given in the form of linear distances, angles, and notes. The engineer must learn the skills of dimensioning; that is the type of lines to use, the spacing between dimensions, proper note creation, etc. A sample detailed part drawing is shown in Figure 7-1 without the border and title block.

The beginner must learn the rules for placing dimensions on a detailed drawing. The suggestions to follow provide a logical arrangement for maximum legibility. Part function is considered first; the shop processes second. The proper procedure is to dimension for design intent, then review the dimensions to see if any improvements in clarity can be made without affecting design intent.

A dimension line is a dark, solid line terminated by arrowheads, which indicates the direction and extent of a dimension. On detailed part drawings the dimension line is broken near the middle to provide an open space for the dimensional value.

The dimension lines closest to the part should be spaced at least ½ inch away, as shown in Figure 7-2. All other parallel dimension lines should be at least 3/8 inch apart. These are the defaults for drawing templates to be created. The spacing of dimension lines should be uniform throughout.

An extension line is a dark, solid line that "extends" from a point near the feature to the corresponding dimension. The dimension line meets the two extension lines at a right angle. A gap of about 1/16 of an inch should be left between the extension line and the part feature. The extension line should extend about 1/16 of an inch beyond the outermost dimension line. These are the defaults for Creo Parametric.

A centerline is a dark line composed of alternate long and short dashes and is used to represent the axes of symmetrical parts or to denote centers. As shown in Figure 7-3, do not create a gap when the centerline crosses the part outline. A centerline must end with a long dash. These are the defaults for Creo Parametric.

An example of the placement of dimension lines and extension lines is shown below. The shorter dimensions are closest to the part outline. Dimension lines must not cross extension lines. This would happen if the shorter dimensions were placed outside the longer dimensions. Do not do this. Note that it is acceptable for extension lines to cross. Extension lines should never be shortened. A dimension line must never coincide with or form a continuation of any part outline.

Do not allow dimension lines to cross. Dimensions should be lined up and grouped together as much as possible. In general, never dimension to hidden lines. In some cases, extension lines and centerlines may cross visible lines of the part. When this occurs, a gap must not be created.

A leader is a thin, solid line leading from a note or dimension to the specified feature. It is terminated by an arrowhead touching the part feature to which the note applies. An arrowhead must stop on a line, such as the edge of a hole. A leader should be an inclined straight line, except for its short horizontal shoulder extending from the mid-height of the beginning or end of the note.

A leader to a circle must be radial so that if extended it would pass through the center of the circle. A drawing presents a more pleasing appearance if leaders near each other are drawn parallel. Leaders should cross as few lines as possible, and should never cross each other. Leader lines should not be drawn parallel to nearby lines of the part drawing, allowed to pass through the corner of a part, be drawn excessively long, or drawn nearly horizontal or vertical. A leader line should be drawn at an angle between 20° and 70°. See Figure 7-4.

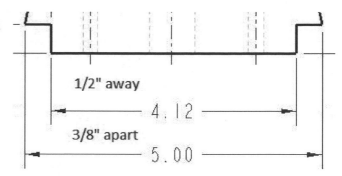

Figure 7-2   Dimension Line Spacing

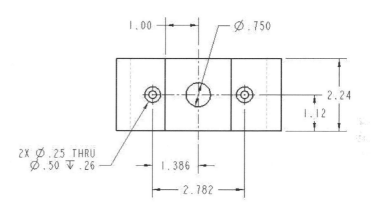

Figure 7-3   Centerlines

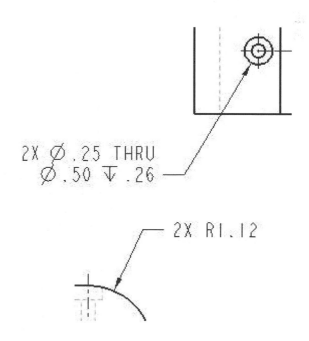

Figure 7-4   Leader Lines

Fractional and Decimal Dimensions

In the early 1900s, the worker scaled the design drawing without dimensions to obtain any necessary values, and he was responsible for making the parts fit together properly. When blueprinting became common, workers in separate factories would use the same drawings to make similar parts, thus it was necessary to dimension the drawings in common fractions and inches. The smallest dimension a worker could measure directly was 1/64 of an inch. This became the smallest division on the machinist's scale. When close fits were required, the drawing would carry a note, such as "running fit," or "force fit," and the worker would make a small adjustment to the size. Machinists were very skilled. Hand-built machines were often beautiful examples of precision workmanship.

Today there are many types of products where common fractional inches are used because extreme accuracy is not necessary. A boxcar drawing, in which the structure is very large, does not require great accuracy. Also, there are many parts where the ordinary machinist's scale is accurate enough.

As the manufacturing industry progressed, there became a greater demand for more accurate specifications of the important functional dimensions. Since it was hard to use smaller fractions, such as 1/128 of an inch, it became common practice to give dimensions as decimal values, such as 3.62, 1.375, or 2.1625. However, many relatively unimportant machine dimensions, standard nominal sizes of materials, drilled holes, standard threads, keyways, etc., were still expressed as whole numbers with common fractions.

Thus, a given detailed drawing could be dimensioned with whole numbers and common fractions, or with decimals, or with a combination of the two. The latter is common, especially with machined castings and forgings, where the rough dimensions need not be closer than ±1/64 of an inch. However, because of computer-aided design, the standard practice has moved toward the adoption of the decimal system only. In addition, the decimal system is compatible with numerically-controlled machines.

The requirements for accuracy have made the decimal dimensioning system the norm. The use of both common fractions and decimals on the same drawings has caused some confusion, thus there is a trend toward the use of the decimal system for all dimensions.

The decimal system, based upon the inch, has many of the advantages of the universal metric system and is compatible with most measuring devices and machine tools. American manufacturers have found that the decimal inch system, rather than the metric system, works without the need to scrap their inch type measuring devices.

In 1932, the Ford Motor Company adopted a complete decimal system. The shop scale chosen was divided on one edge into inches and tenths of an inch, and on the other edge into inches, tenths, and fiftieths of an inch. Thus, the smallest division was one-fiftieth, or .02 inches; two divisions were .04 inches, etc. When it was necessary to halve a value, the result will still be a two-place decimal. The automotive and aircraft industry used this system for many years.

Two-place decimals are used when tolerance of ±.01 inch or greater are allowed. Three or more decimal places are used for tolerance limits less than ±.01 inches. In the standard two-place decimal system, the second place should be an even digit (for example, .02, .04, and .06 are preferred over .01, .03, or .05). When a dimension is divided by two, such as in determining the radius from a diameter, the result is still a two-place decimal. Odd two-place decimals are allowed when required for design purposes.

On any drawing, decimal dimensions should be used wherever the degree of accuracy required is closer than 1/64 of an inch. For the decimal equivalents of common fractions, see Table 7-1.

Table 7-1 Fractions and Decimal Equivalence			
Fraction	**2-decimals**	**3-decimals**	**4-decimals**
1/64	0.02	0.016	0.0156
1/32	0.03	0.031	0.0313
3/64	0.05	0.047	0.0469
1/16	0.06	0.063	0.0625
5/64	0.08	0.078	0.0781
3/32	0.09	0.094	0.0938
7/64	0.11	0.109	0.1094
1/8	0.13	0.125	0.1250
9/64	0.14	0.141	0.1406
5/32	0.16	0.156	0.1563
11/64	0.17	0.172	0.1719
3/16	0.19	0.188	0.1875
13/64	0.20	0.203	0.2031
7/32	0.22	0.219	0.2188
15/64	0.23	0.234	0.2344
1/4	0.25	0.250	0.2500
17/64	0.27	0.266	0.2656
9/32	0.28	0.281	0.2813
19/64	0.30	0.297	0.2969
5/16	0.31	0.313	0.3125
21/64	0.33	0.328	0.3281
11/32	0.34	0.344	0.3438
23/64	0.36	0.359	0.3594
3/8	0.38	0.375	0.3750
25/64	0.39	0.391	0.3906
13/32	0.41	0.406	0.4063
27/64	0.42	0.422	0.4219
7/16	0.44	0.438	0.4375
29/64	0.45	0.453	0.4531
15/32	0.47	0.469	0.4688

Table 7-1 Fractions and Decimal Equivalence (Continued)

Fraction	2-decimals	3-decimals	4-decimals
31/64	0.48	0.484	0.4844
1/2	0.50	0.500	0.5000
33/64	0.52	0.516	0.5156
17/32	0.53	0.531	0.5313
35/64	0.55	0.547	0.5469
9/16	0.56	0.563	0.5625
37/64	0.58	0.578	0.5781
19/32	0.59	0.594	0.5938
39/64	0.61	0.609	0.6094
5/8	0.63	0.625	0.6250
41/64	0.64	0.641	0.6406
21/32	0.66	0.656	0.6563
43/64	0.67	0.672	0.6719
11/16	0.69	0.688	0.6875
45/64	0.70	0.703	0.7031
23/32	0.72	0.719	0.7188
47/64	0.73	0.734	0.7344
3/4	0.75	0.750	0.7500
49/64	0.77	0.766	0.7656
25/32	0.78	0.781	0.7813
51/64	0.80	0.797	0.7969
13/16	0.81	0.813	0.8125
53/64	0.83	0.828	0.8281
27/32	0.84	0.844	0.8438
55/64	0.86	0.859	0.8594
7/8	0.88	0.875	0.8750
57/64	0.89	0.891	0.8906
29/32	0.91	0.906	0.9063
59/64	0.92	0.922	0.9219
15/16	0.94	0.938	0.9375
61/64	0.95	0.953	0.9531
31/32	0.97	0.969	0.9688
63/64	0.98	0.984	0.9844
1	1.00	1.000	1.0000

In this system, common fractions may be used to indicate nominal sizes of materials, drilled holes, punched holes, threads, keyways, and other standard features. For example, ¼-20 UNC-2A; ¾ DRILL; or STOCK 1½ × 1½. If desired, decimals may be used for everything, including standard nominal sizes, such as .25-20 UNC-2B, or .75 DRILL.

When a decimal value is to be rounded to a lesser number of decimal places, the method to be used follows:

1. When the digit beyond the last kept digit is less than 5, the last digit kept does not change. Example: 1.1624 is rounded to three decimal places, or 1.162.

2. When the digit beyond the last kept digit is more than 5, the last kept digit is increased by 1. Example: 2.2768 is rounded to three decimal places, or 2.277.

3. When the digit beyond the last kept digit is exactly 5 with only zeros following, the kept digit if even, is unchanged; if odd is increased by 1. Example: 3.565 becomes 3.56 when rounded to two decimal places, and 3.375 becomes 3.38.

The reading direction of dimensions for the two systems has been approved by the American Standards Association or ASA. In the unidirectional system, all dimensions and notes are printed horizontally and are read from the bottom of the drawing. This is the preferred system. In the aligned system, all dimensions are aligned with the dimension lines so that they may be read from the bottom or from the right side of the drawing. Notes are always positioned horizontally.

Standard Sizes Preferred

Dimensions should be given, wherever possible, to make use of available materials, tools, parts, and gages. The dimensions for many commonly used machine elements, such as bolts, nails, keys, tapers, wire, pipes, sheet metal, chains, belts, and pins have been standardized. The design engineer must obtain these standard sizes from published handbooks, from the American Standards Association, or from the manufacturers' catalogs.

Detailed drawings of these standard parts are not made unless the parts are to be modified. They are drawn conventionally on the assembly drawing and are listed in the bill of materials. Common fractions are generally used to indicate the nominal sizes of standard parts. If the decimal system is used exclusively, all sizes are expressed in decimals; for example, .25 DRILL instead of ¼ DRILL.

Dimensioning Angles

Angles are dimensioned by means of the coordinate dimensions of the two legs of the right triangle, or by means of a linear dimension and an angle in degrees. Coordinate dimensions are suitable for work requiring a high degree of accuracy. Variations of the angle are hard to control because the amount of variation increases with the distance from the vertex. Methods of indicating various angles are shown in Figure 7-5.

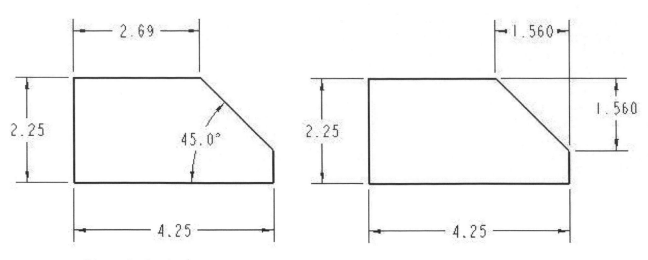

Figure 7-5   Dimensioning Angles

When degrees alone are indicated, the symbol ° is used. When minutes alone are given, the value should be preceded by 0°, for example, 0° 36□ However, it is preferred that an angle be given in degrees with one decimal place, such as 38.6°. Note that there is no gap where the extension lines cross in the right view and the smaller dimensions are closer to the part than the larger dimensions. See Figure 7-

Dimensioning Arcs

A circular arc is dimensioned in the view where its true shape is shown, as seen in Figure 7-6, by giving the numerical value of its radius, preceded by the letter **R**. The number of arcs with the same radius is indicated in front of the R-value. The center of the arc may be indicated by a small cross. When there is enough room, both the numerical value and the arrowhead may be placed inside the arc. For smaller arcs, the arrowhead may be left inside and the numerical value may be placed outside. For small arcs, both the arrowhead and the numerical value are placed outside the arc.

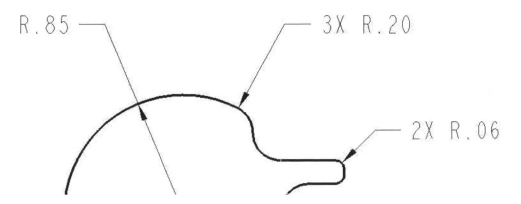

Figure 7-6   Dimensioning Arcs

Fillets and Rounds

Individual fillets and rounds are dimensioned as arcs. If there are only a few arcs and they are the same size, specifying one typical radius is sufficient. However, fillets and rounds are often quite numerous on a drawing and most of them are likely to be some standard size, such as R ¼. In these cases, it is customary to provide a general note on the drawing to cover all uniformly-sized fillets and rounds, thus: "FILLETS R ¼ AND ROUNDS R ⅛ UNLESS OTHERWISE SPECIFIED" or simply "ALL FILLETS AND ROUNDS R ¼." See Figure 7-7.

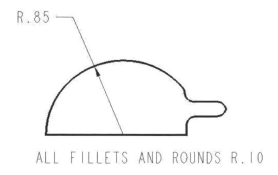

ALL FILLETS AND ROUNDS R.10

Figure 7-7 Fillets and Rounds

Finish Marks

A finish mark is used to indicate that a surface is to be machined or finished, such as on a rough casting or forging. To the die maker, a finish mark means that he must allow extra material for the rough workpiece. The number by the symbol is the surface roughness in micro-inches. A triangular symbol means that machining is mandatory. A surface finish of 8 micro-inches could be done by grinding or honing the surface. The finishing mark with the circle in the V indicates machining prohibited with a maximum surface roughness of 125 micro-inches. See Figure 7-8.

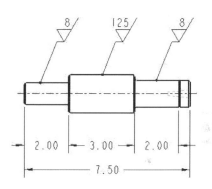

Figure 7-8 Finish Marks

Table 7-2 that follows lists the typical shop operations and their expected surface roughness when machining metal in both average micrometers and average micro-inches. To convert micro-inches to micrometers, divide micro-inches by 40.

A new system of roughness grades has come into being recent. It should be used on drawings for international suppliers and for new designs. It grades the surface roughness with a value from 1 to 12 with 12 being the roughest surface. N12 is equivalent to 2000 μ-inches or 50 μ meters; N11 is equivalent to 1000 μ-inches or 25 μ meters; N10 is equivalent to 500 μ-inches or 12.5 μ meters, etc., down to N1 being equivalent to 1 μ-inch or .025 μ meters.

Dimensions and Part Views

Dimensions should not be placed on a view unless clearness is promoted. The ideal dimensioning scheme is shown in Figure 7-9 where all dimensions are placed outside the part view. Compare this to the poor practice shown to its right. This is not to say that a dimension should never be placed on a part. Also, a leader line must not go through a corner of the part as shown on the right.

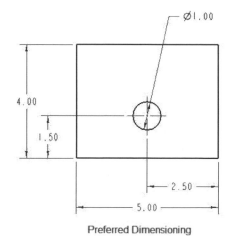

Preferred Dimensioning

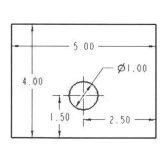

Poor Practice Dimensioning

Figure 7-9 Preferred and Poor Practice

Table 7-2 Operation Roughness Grades

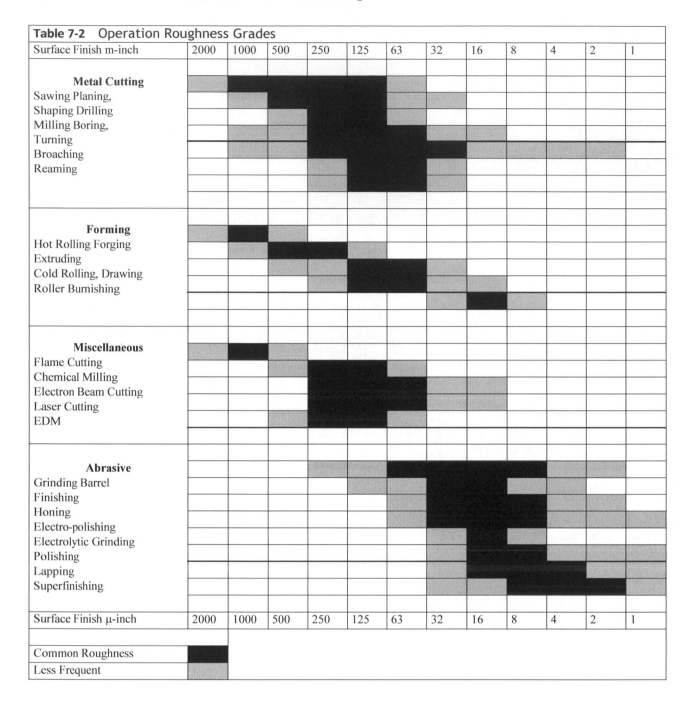

Surface Finish m-inch	2000	1000	500	250	125	63	32	16	8	4	2	1
Metal Cutting Sawing Planing, Shaping Drilling Milling Boring, Turning Broaching Reaming												
Forming Hot Rolling Forging Extruding Cold Rolling, Drawing Roller Burnishing												
Miscellaneous Flame Cutting Chemical Milling Electron Beam Cutting Laser Cutting EDM												
Abrasive Grinding Barrel Finishing Honing Electro-polishing Electrolytic Grinding Polishing Lapping Superfinishing												
Surface Finish μ-inch	2000	1000	500	250	125	63	32	16	8	4	2	1

Common Roughness ▬

Less Frequent ▤

When a dimension must be placed in a sectioned area, as shown in Figure 7-10, provide a blank space in the sectioned area for the dimensional value.

Contour Dimensioning

Views are drawn to describe the shapes of the various features of a part, and dimensions are given to define exact sizes and locations of those features. It follows

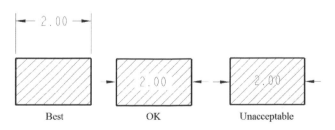

Figure 7-10   Dimensioning on a Sectioned View

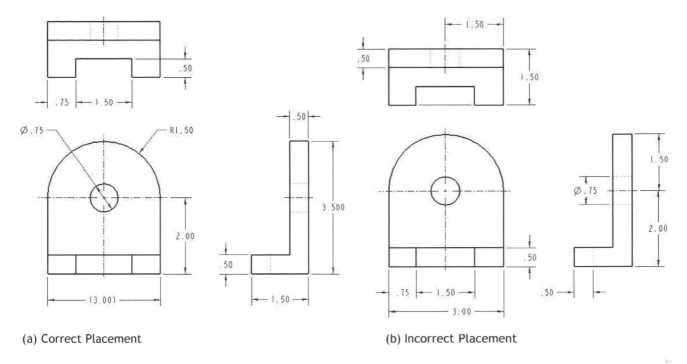

(a) Correct Placement   (b) Incorrect Placement

Figure 7-11 Contour Dimensioning

that a <u>dimension should be given where the shape is shown</u> true, that is, in the view where the contour is obvious (Figure 7-11a on the left). Incorrect placement of the dimensions is shown in Figure 7-11b on the right.

Individual dimensions should be attached directly to the contours that show their shape. This will prevent the attachment of dimensions to hidden lines. Also, this will prevent the attachment of a dimension to a visible line where the meaning is not clear.

Although the placement of notes for holes follows the contour principle, the <u>diameter for a solid cylindrical shape is given in the rectangular</u> view (Figure 7-12) so it can be located near the dimension for the cylinder's length.

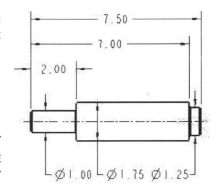

Figure 7-12 Dimensioning Solid Cylinders

Geometric Shapes

Mechanical parts are usually composed of simple geometric shapes, such as prisms, cylinders, cones, or spheres. They may be exterior (positive) or interior (negative) shapes. For example, a step shaft is made up of positive cylinders, while a counterbored hole is made up of negative cylinders.

These shapes result directly from the need to keep shapes as simple as possible, and from the requirements of the basic shop operations. Shapes with a plane surface can be produced by planing, shaping, or milling, while shapes having a cylindrical, conical, or spherical surface can be produced by turning, drilling, reaming, boring, or countersinking.

The dimensioning of engineering parts begins by giving the dimensions showing the sizes of the simple geometric shapes, then giving the dimensions locating the features with respect to each other. The former is called "size dimensions" and the latter is called "location dimensions." This method of

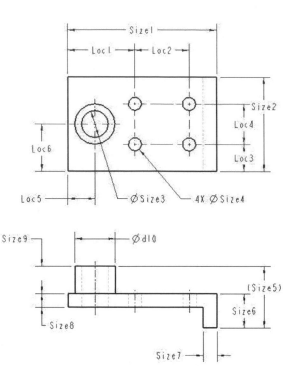

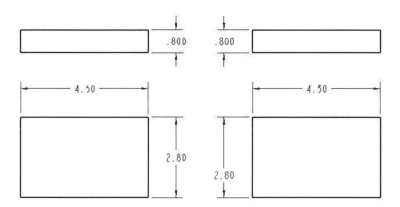

Figure 7-13   Size and Location Dimensions

geometric analysis is very helpful when dimensioning any part, but must be altered when there is a conflict with the function of the part.

A 2-view drawing of a part is shown in Figure 7-13. The geometric shape is dimensioned with size dimensions, and then the features are located with respect to each other using location dimensions. Note that a location dimension locates a three-dimensional feature and not a surface; otherwise, all dimensions would be classified as location dimensions.

Dimensioning Prisms

The right rectangular prism is the most common geometric shape. Front and top views are dimensioned as shown in Figure 7-14. The height and width are given in the front view, and the depth in the top view. A dimension between the views applies to both views and should not be placed elsewhere without good reason. The dimensions may be placed on the right side or the left side of the views.

Front and right side views are dimensioned, as shown in Figure 7-15. The height and width are given in the front

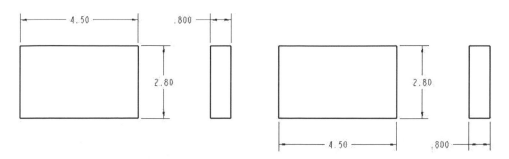

Figure 7-14   Dimensioning Prisms

Figure 7-15   Dimensioning Prisms

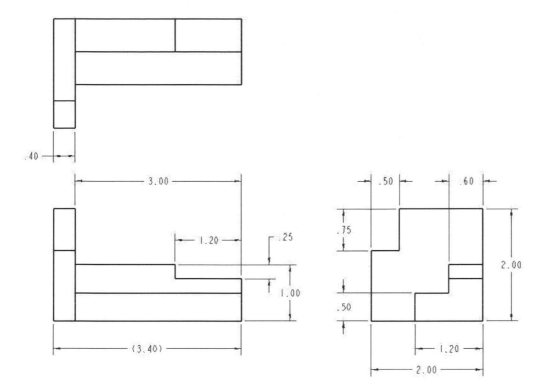

Figure 7-15 Contour Principle

view, and the depth in the right side view. A dimension between the views applies to both views and should not be placed elsewhere without good reason. The dimensions may be placed above or below the views.

The application of size dimensions to a machine part composed entirely of rectangular prisms is shown in Figure 7-16. The 3.40-inch dimension is listed as a reference dimension since it is not necessary.

Dimensioning Cylinders

The right circular cylinder is the next most common geometric shape and is commonly seen as a shaft or a hole. The general method of <u>dimensioning an external cylinder is to give</u> <u>both its diameter and its length in the rectangular view</u>. If the cylinder is drawn in a vertical position, the length or height of the cylinder may be given to its right or left, as seen in Figure 7-17.

If a cylinder is drawn in its horizontal position, its length may be given above the rectangular view or below it. An application showing the dimensioning of cylindrical shapes is shown in Figure 7-18.

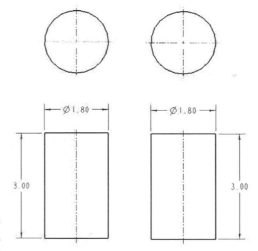

Figure 7-17 Dimensioning Vertical Cylinders

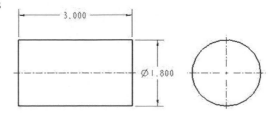

Figure 7-18 Dimensioning Horizontal Cylinders

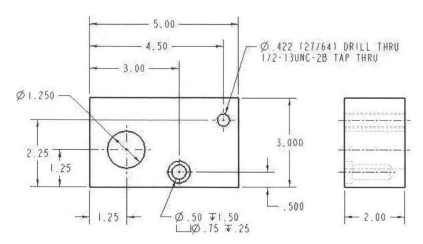

Figure 7-19   Dimensioning Holes

The radius of a cylinder must never be given since measuring tools are designed to check diameters. Small cylindrical holes, such as drilled, reamed, or bored holes, are typically dimensioned using notes specifying the diameter and the depth.

Dimensioning Holes

Holes that are to be drilled, bored, reamed, or punched are specified using standard notes. The order of items in a note should correspond to the order of the procedure in the shop when producing the hole. The note may include the shop processes. Two or more holes should be dimensioned using a single note that includes the number of holes preceding the first line of the note with the leader pointing to one of the holes.

As shown in Figure 7-19, the leader of the note should point to the circular view of the hole. It can point to the rectangular view only when clearness is promoted. When the circular view of the hole has two or more concentric circles, as with a counter-bored, countersunk, or tapped hole, the arrowhead should touch the outer circle. Note that the ½-inch drill note was placed on the view in order to keep it near the feature and to avoid having the leader line cross several extension lines.

The use of decimal fractions instead of common fractions to designate drill sizes has gained wide acceptance. For numbered or letter-size drills, it is preferred that the decimal size be given as well. For example, a number drill would be shown as **#20 (ø.161) DRILL,** or a letter size drill as **"K" (ø.281) DRILL.**

On part drawings where the parts are to be produced in large quantity, dimensions and notes may be given without specification to the shop process. Even though the shop operations are omitted, the tolerances dictate the shop processes needed.

DimensioningRound-EndShapes

The method for dimensioning round-end shapes (Figure 7-20) depends upon the degree of accuracy required. When precision is not necessary, the dimensioning method used is the one which is convenient for the shop.

The link to the right is to be cut from sheet metal, thus it should be dimensioned as it would be laid out in the shop, that is, by giving the hole's center-to-center distance and the radii at the ends. Only one radius dimension is necessary. If an overall dimension is shown, it must be shown as a reference dimension.

The pad on a casting with a milled slot in Figure 7-21 is dimensioned from center to center for the convenience of both the patternmaker and the machinist. An additional reason for the center-to-center distance is that it gives the total travel of the mill cutter, which can be easily controlled. The width dimension indicates the diameter of the mill cutter; hence, it would be inappropriate to give the radius of the machined slot.

When accuracy is required, the dimensioning method shown in Figure 7-22 is recommended. Overall lengths of rounded end shapes are given in each case, and radii are indicated, but without a specific value or listed as REF.

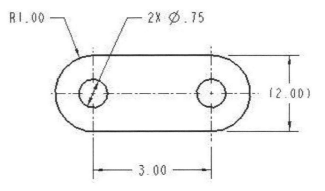

Figure 7-20   Dimensioning Rounded Ends

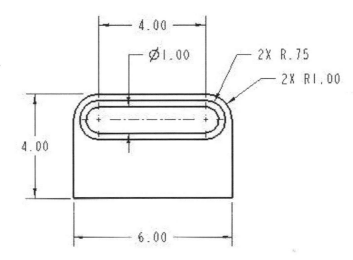

Figure 7-21   Dimensioning a Milled Slot

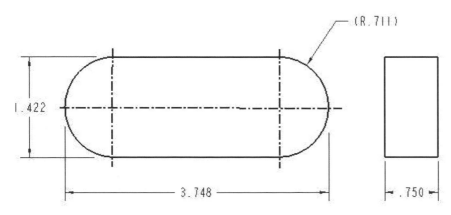

Figure 7-22 Dimensioning Rounded Ends

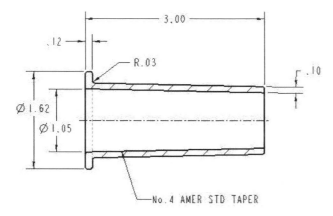

Figure 7-23   Dimensioning a Taper

Figure 7-24   Dimensioning a Standard Taper

Dimensioning Tapers

A taper is a conical surface on a shaft or a hole. The method of <u>dimensioning a taper is to provide the amount of taper per unit length in a note</u>: "TAPER .025 INCH per INCH," and then give the diameter at one end, along with the taper length. See Figure 7-23.

Standard machine tapers are used on machine spindles, shanks of tools, and pins. They are dimensioned on a drawing by dimensioning the larger diameter and the taper size in a note, such as "NO. 4 AMERICAN STANDARD TAPER." See Figure 7-24.

Dimensioning Threads

Local notes are used to specify thread dimensions. For a <u>tapped hole, the note should be attached to the circular view of the hole. The note is placed in the longitudinal view for external threads</u> where the threads are easily recognized. Table 7-3 is the ANSI Standard V-thread tap size table for Unified National Coarse and Fine Threads.

Table 7-3	Tap Drill Sizes				
Diameter		**Fine Threads**	**Tap Drill Size**	**Coarse Threads**	**Tap Drill Size**
#0		80	3/64	–	–
#1		72	No. 53	64	No. 53
#2		64	No. 50	56	No. 50
#3		56	No. 45	48	No. 47
#4		48	No. 42	40	No. 43
#5		44	No. 37	40	No. 38
#6		40	No. 33	32	No. 36
#8		36	No. 29	32	No. 29
#10		32	No. 21	24	No. 25
#12		28	No. 14	24	No. 16
1/4		28	No. 3	20	No. 7
5/16		24	I	18	F
3/8		24	Q	16	5/16
7/16		20	25/64	14	U
1/2		20	29/64	13	27/64
9/16		18	33/64	12	31/64
5/8		18	37/64	11	17/32
3/4		16	11/16	10	21/32
7/8		14	13/16	9	49/64
1		14	59/64	8	7/8

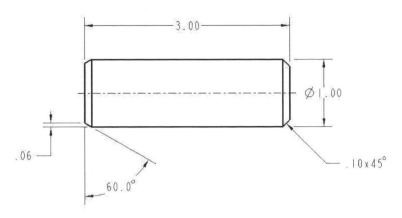

Figure 7-25   Dimensioning Chamfers

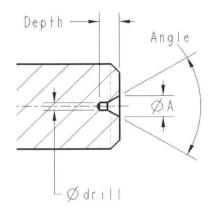

Figure 7-26   Shaft Center

Dimensioning Chamfers

A chamfer is a beveled or sloping edge. When the angle is not 45°, it is dimensioned by giving the length of one leg and the angle, as shown in Figure 7-25. A 45° chamfer is usually dimensioned using a note without the word CHAMFER or CHAM.

Shaft Centers

Shaft centers are required on shafts, spindles, and other conical parts for turning, grinding, or other rotational operations. Such a center may be dimensioned, as shown in Figure 7-26. Normally the shaft centers are produced by a combined drill and countersink. The angle is 60°. Table 7-4 shows the recommended shaft center sizes for given shaft diameters.

Table 7-4   Shaft Center Sizes

Shaft Diameter (inch)	Up to (inch)	A (inch)	Drill (inch)	Depth (inch)
0.1875	0.2499	0.08	0.047	0.090
0.2500	0.3749	0.09	0.047	0.103
0.3750	0.5124	0.13	0.167	0.042
0.5125	0.8124	0.19	0.078	0.188
0.8125	1.1249	0.25	0.094	0.229
1.1250	1.4999	0.31	0.156	0.292
1.5000	1.9999	0.38	0.094	0.400
2.0000	2.9999	0.41	0.219	0.355
3.0000	3.9999	0.50	0.219	0.462
4.0000	4.9999	0.56	0.219	0.516

Dimensioning Keyways

The method for dimensioning keyways for stock keys is shown in Figure 7-27. A dimension is used to center the keyway on the shaft or collar. The preferred method of dimensioning the depth of a keyway is to give the <u>dimension from the top or bottom of the keyway to the opposite side of the shaft or hole</u>. The method of computing such a dimension for stock keys is shown below.

The method for dimensioning keyways for Woodruff keys is shown in Figure 7-28. Values for the dimensions for Woodruff keys in a shaft can be found in the machinist's handbook. Reference dimensions are shown below because they are listed in the handbook, and are not normally specified on the drawing.

Location Dimensions

After the geometric shapes composing a structure have been dimensioned for size, location dimensions are specified to show the relative positions of these geometric shapes. Rectangular shapes are located with reference to their edges.

$$H = \frac{D - A + \sqrt{D^2 - A^2}}{2} = \frac{1.25 - 0.125 + \sqrt{1.25^2 - 0.125^2}}{2} = 1.185 \; inches$$

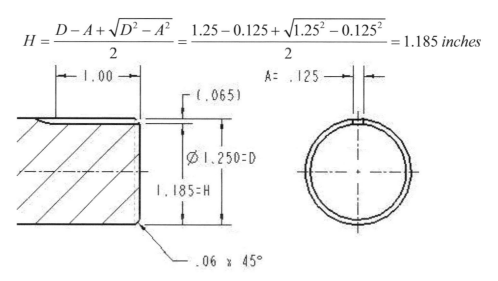

Figure 7-27 Dimensioning a Shaft Keyway

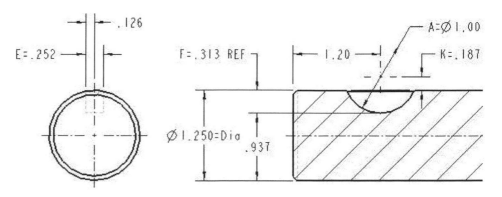

Figure 7-28 Dimensioning for a Woodruff Key

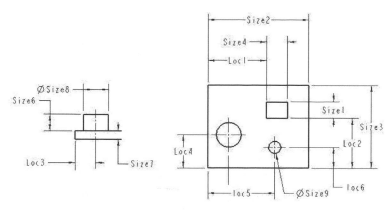

Figure 7-29 Locating Features

conical shapes are located by their centerlines. Location dimensions for a hole are given in the circular view of the hole, as seen in Figure 7-29.

Location dimensions should come from finished surfaces wherever possible because rough castings and forgings vary in size, and unfinished surfaces cannot be relied upon for accurate measurements. The starting dimension, used in locating the first machined surface on a rough casting, will come from a rough surface or from the center or the centerline of the rough part.

Location dimensions should reference a finished surface as a datum plane, or a centerline as a datum axis. When several cylindrical surfaces have the same centerline, do not locate them with respect to each other.

Holes equally spaced around a common center may be dimensioned by giving the bolt circle diameter and specifying "equally spaced" in the note (Figure 7-30). Holes unequally spaced are located by means of the bolt circle diameter plus the angular measurement reference to one of the datum planes. See Figure 7-31.

Where greater accuracy is required, coordinate dimensions should be given as shown in Figure 7-32. In this case, the diameter of the bolt circle is marked REF to indicate that it is to

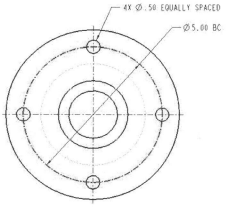

Figure 7-30 Equal Hole Spacing

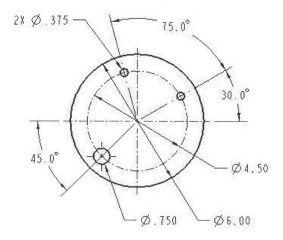

Figure 7-31 Unequal Hole Spacing

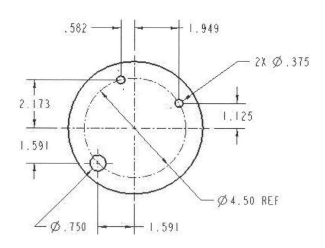

Figure 7-32 Accurately Locating Holes

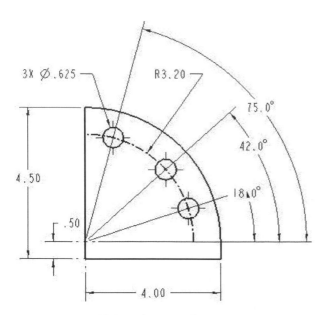

Figure 7-33   Holes along an Arc

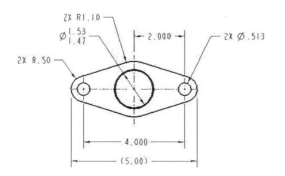

Figure 7-34   Design Intent

be used only as a reference dimension. Reference dimensions are given for information purposes only. They are not intended to be measured and do not govern the shop processes. They represent calculated dimensions and are often useful in showing design intent.

When several non-precision holes are located on a common arc, they are dimensioned by giving the radius of the arc and the angular measurements from a datum plane (Figure 7-33).

The three holes are on a common horizontal centerline. One dimension locates the right hole from the center; the other dimension gives the distance between the two small holes. See Figure 7-34. Note the omission of the dimension from the center to the left hole occurs because the distance between the two small holes is the design intent. If the relationship between the center hole and each of the small holes was the design intent, then show it and mark the overall dimension between the two small holes as REF.

Another example of locating holes by means of linear dimensions is shown in Figure 7-35. In this case, one measurement is made at an angle to the coordinate dimensions because of the design intent between the holes.

The holes shown in Figure 7-35 are located from two baselines or datums (left side and bottom). When all holes are located from a common datum, the sequence of measuring and machining operations is controlled, overall tolerance accumulations are avoided, and the design intent of the finished part is assured. The datum surfaces selected must be more accurate than any measurement made from them, and must be accessible during creation and setup to facilitate tool and fixture design. Thus, it may be necessary to specify the accuracy of the datum surfaces in terms of straightness, roundness, and flatness, thus Geometric Dimensioning and Tolerancing.

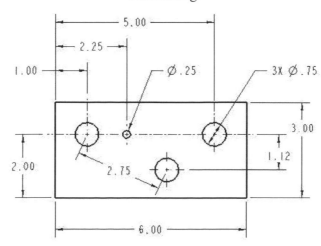

Figure 7-35   Locating Holes

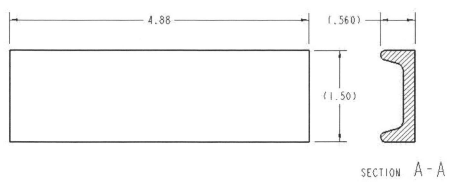

SECTION A-A

Figure 7-36 Dimensioning Purchased Parts

Modified Purchased Parts

In many machines, there are parts that are purchased and used as is. Sometimes the purchased parts need to be modified. For example, an assembly needs a C-channel of a given length. In this case, the detailed drawing shows only the dimensions needed to modify the purchased part. If other dimensions are shown in Figure 7-36, they are listed as reference dimensions.

Mating Dimensions

When dimensioning a single part, its relationship to its mating parts must be taken into account. For example, a Block fits into the slot of its Base. The dimensions common to both parts are mating dimensions. See Figure 7-37.

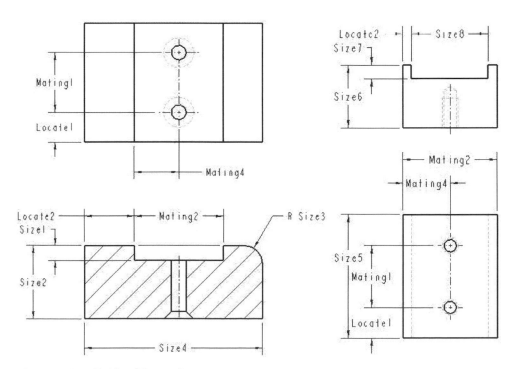

Figure 7-37 Mating Dimensions

These mating dimensions should be given on the detailed drawings in their corresponding locations. The other dimensions are not mating dimensions since they do not control the mating of the two parts. The actual values of the two corresponding mating dimensions may be different. For example, the width of the slot may be dimensioned several thousandths of an inch larger than the width of the block, but these are the mating dimensions based on a basic width. The mating dimensions need to be identified so that they can be specified in the corresponding locations on the two parts, and so that they can be given with the degree of accuracy necessary for the proper fitting of the two parts.

Notes

It is necessary to supplement the direct dimensions with notes. The notes must be brief and carefully worded so as to have only one interpretation. Notes should not be created in crowded areas on the drawing sheet. Avoid placing notes between views. Notes should not be placed closer to a different view as to suggest the note applies to the wrong view. Note leaders should be short and cross as few lines as possible. They should never run through the corner of a part or through any specific points or intersections on the part view.

Notes are classified as general notes when they apply to an entire drawing. They are referred to as local notes when they apply to specific features.

General notes should be placed in the lower right-hand corner of the drawing, above or to the left of the title block, or in a central position below the view to which they apply. Some general notes are "FINISH ALL OVER" or "BREAK ALL SHARP EDGES TO R 0.06" or "DRAFT ANGLES @ 3° UNLESS OTHERWISE SPECIFIED." On detailed drawings, the title block should show some general notes on material and implied tolerances.

Local notes (Figure 7-38) apply to specific features, and are connected by a leader line to the point where the operation is performed. Some local notes are "4X 0.25 DRILL" or "¼ X 45° CHAM." The leader must be attached to the front of the first word or just after the last word.

In general, leaders and local notes should not be located on the drawing until the dimensions are finalized. If notes are placed first, they typically have to be moved when locating the dimensions.

Certain commonly used abbreviations may be used in notes, as THD or MAX. All abbreviations should conform to the American Standard Association.

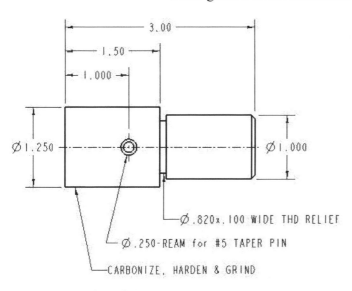

Figure 7-38   Local Notes

Dimensioning for Numerical Control

In general, the basic dimensioning practices listed here are compatible with the data requirements for numerically-controlled machines. However, to make the best use of this type of production, the designer must consult the manufacturing machine manuals before making the drawings. Certain considerations should be noted:

1. Coordinate dimensioning is required from three mutually perpendicular references or datum planes. These must be clearly identified on the drawing.

2. The reference planes may be located on or off the part, but preferably located so that all dimensions are positive.

3. All dimensions must be in decimals.

4. Angles should be specified by coordinate dimensions rather than degrees.

5. Standard tools, such as drills, reamers, or taps should be specified if known.

6. All tolerances should be determined by the design requirements of the part, and not by the capability of the manufacturing machine.

Checklist for Dimensioning

The following lists many of the situations where a new draftsman is likely to make a mistake when dimensioning. The beginner should check his/her detailed drawing against this list before submitting it to his/her supervisor.

1. Dimensions should be attached to the view where the shape is best shown (**contour principle**).

2. Each dimension should be given clearly so that it can be interpreted in only one way.

3. Dimensions should not be duplicated or the same information be given in two different ways, and no dimensions should be given except those needed to produce or inspect the part.

4. Dimensions should be given between points or surfaces which have a functional relation to each other or which control the location of mating parts.

5. Dimensions should be given so that the machinist does not have to calculate, scale, or assume any dimension.

6. Do not expect the worker to assume a feature is centered (such as a hole in a plate), but rather give a location dimension from one side. However, if the hole is to be centered on a symmetrical rough casting, reference the centerline in a note and omit the locating dimension to the centerline.

7. Avoid dimensioning to hidden lines. Create a sectioned view if necessary.

8. Dimensions applying to two adjacent views should be placed between the views unless clearness is promoted by placing the dimension outside one view.

9. A dimension should be attached to only one view.

10. Dimension lines should be spaced uniformly throughout the drawing. They should be at least ½ inch from the part outline and ⅜ inch apart.

11. Dimension lines should not cross, if avoidable.

12. The longer dimensions should be placed outside shorter dimensions so that dimension lines do not cross extension lines.

13. Dimension lines and extension lines should not cross; however, extension lines may cross each other.

14. A dimension line should never be drawn through a dimensional value. A dimensional value should never be on top of any line of the drawing.

15. No line of the drawing should be used as a dimension line or coincide with a dimension line.

16. A dimension line should never be joined end-to-end (chain fashion) with any parallel line of the drawing.

17. Dimensional values should be approximately centered between the arrowheads, except in a "stack" of dimensions where the dimensional values are to be "staggered."

18. Dimensional values should never be crowded or in any way made difficult to read.

19. Dimensional values should not be lettered over sectioned areas unless necessary, in which case a clear space should be left for the dimensional value.

20. In engineering drawings, omit all inch marks, except where necessary for clearness, such as 1″ VALVE.

21. An overall dimension should be present in the three major directions to give the overall size of the part. The overall dimension can be marked as REF if the other dimensions are needed to meet design intent.

22. Avoid a complete chain of detail dimensions. Instead, omit one dimension or add **REF** to one detail dimension or to the overall dimension.

23. A centerline may be extended and used as an extension line, in which case it is still drawn like a centerline.

24. Centerlines should not extend from view to view.

25. Leaders for holes should be straight, not curved, and pointing to the circular views of holes wherever possible.

26. Leaders should slope at an angle between 20 and 70 degrees from the horizontal. Vertical and horizontal leader lines are not acceptable.

27. Leaders should extend from the beginning or end of a note; the horizontal ¼-inch long "shoulder" should extend from the mid-height of the text.

28. Dimensional values for angles should be listed horizontally.

29. Notes should always be lettered horizontally.

30. Notes should be brief and clear, and the wording should use standard symbols, nomenclature, and abbreviations.

31. Finish marks should be placed on the edge views of all finished surfaces, including hidden edges and the contour and circular views of cylindrical surfaces.

32. Finish marks should be omitted on holes or other features where a note specifies a machining operation.

33. Finish marks should be omitted on parts made from rolled stock.

34. If a part is finished all over, omit all finish marks, and use the general note: FINISH ALL OVER, or FAO.

35. A solid cylinder is dimensioned by giving both its diameter and length in the rectangular view. A diagonal diameter in the circular view may be used in cases where clearness is gained.

36. Holes to be bored, drilled, reamed, etc., are size-dimensioned by notes in which the leaders point toward the circular view of the holes. Shop processes may be omitted from these notes.

37. Drill sizes are preferably expressed in decimals, especially for drills designated by number or letter.

38. In general, a circle is dimensioned by its diameter, an arc by its radius. Typically, a circular arc goes through an angle less than 180 degrees; a circle goes through an angle of 180 degrees or more.

39. A diameter dimensional value should be preceded by the symbol Ø.

40. The letter R should precede the radius dimensional value.

41. Cylinders should be located by their centerlines.

42. Cylinders should be located in the circular views, if possible.

43. Cylinders should be located by coordinate dimensions in preference to angular dimensions when accuracy is important.

44. When there are several rough non-critical features obviously the same size, such as fillets, rounds, ribs, etc., give only a typical dimension, or use a general note.

45. Mating dimensions should be given on the drawings of mating parts.

46. Decimal dimensions should be used when greater accuracy than 1/64 of an inch is required on a machined dimension. Normally 2-place decimals are preferred over fractional dimensions.

47. Avoid cumulative tolerances, especially in limit dimensioning. This can be done by not "chaining" dimensions, but rather specifying each one from a reference datum.

Creating an A-size Format Sheet

A format sheet is used to create a custom environment for your drawings. It's a layout sheet, a special kind of drawing used as a starting point when creating engineering drawings. Rather than drawing the border and title block on every detailed part or assembly drawing, you create a format sheet/drawing and then reference it when you are creating a new drawing.

All drawings done in this textbook will be made on A-size (8.5″ × 11″) paper, thus we will create only a format drawing containing a border and title block for an A-size page in this section. A similar procedure can be used for other size paper.

Step 1: Start Creo Parametric by *double-clicking* with the LMB on the ***Creo Parametric*** icon on the desktop, or from the Program list: Creo Parametric.

Step 2: Set your working directory, by *selecting* the ***Select Working Directory*** icon in the Home ribbon. Locate your working directory, and then *pick* **OK.**

Step 3: Create a new format by *picking* the ***New*** icon at the top of the screen, or *by selecting* **File>New.**

Step 4: *Select* **Format** in the New window. Type "A-size_BorderTitleBlock" for the name. *Pick* **OK.** See Figure 7-39.

Step 5: When the New Format window appears, be sure **Empty** and **Landscape** are selected. In the drop-down menu beside Standard Size, *select* **A.** *Pick* **OK.** See Figure 7-40.

The edge of an 8.5″ × 11″ page will appear. This is the edge of the paper. Do not assume this is your border; it's the edges of the paper. See Figure 7-41.

Figure 7-39 Open Object Window   **Figure 7-40** New Format Window

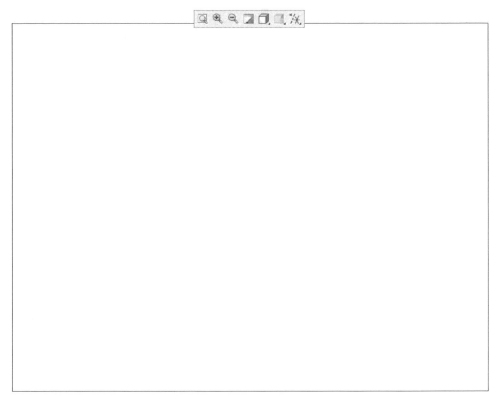

Figure 7-41 8.5" x 11" Paper

Step 6: *Select* the Sketch tab, and then *select* the **Sketcher Preferences** icon. See Figure 7-42.

Step 7: When the Sketch Preferences window appears (Figure 7-43), *select* **Horizontal/Vertical, Grid intersection,** and **Grid angle.** *Check* **Chain sketching** and **Parametric sketching.** Be sure **Vertex, On entity, Angle,** and **Radius** are turned off. *Pick* **Close.**

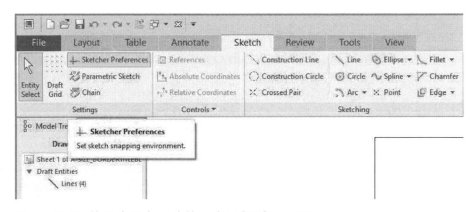

Figure 7-42 Sketch Tab and Sketcher Preferences

Figure 7-43 Sketcher Preferences Set

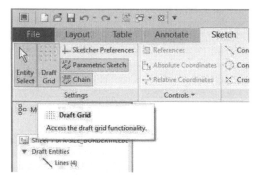

Figure 7-44   Draft Grid

Step 8: *Select* the ***Draft Grid…*** icon. See Figure 7-44.

Step 9: *Select* **Grid Params** in the Menu Manager window. See Figure 7-45.

Step 10: *Select* **X&Y Spacing** (Figure 7-46) from the Menu Manager. *Type* "0.25" <Enter> or ".25" followed by *picking* the ***green checkmark***. *Pick* **Done/Return.**

Step 11: *Pick* **Show Grid** from the Menu Manager window. *Pick* **Done/Return.** See Figure 7-47.

The 8.5″ × 11″ page will have a grid on it with each square equal to ¼ inch by ¼ inch. We will use this grid when we are creating our border and title block. See Figure 7-48.

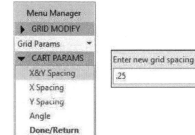

Figure 7-45
Grid Parameters

Figure 7-46   Grid Spacing

Figure 7-47
Show Grid

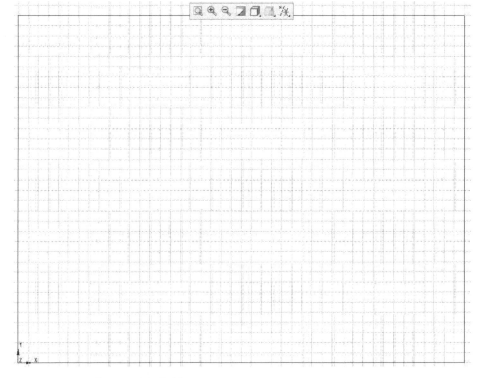

Figure 7-48   Page with Grid

Step 12: *Select* the **Line** icon at the top of the screen (Figure 7-49). Do not close the References window. Each time you draw a line it will be added to the references window. Note that you can only select points on the grid.

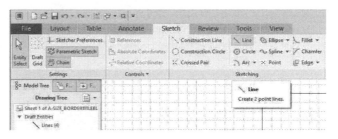

Figure 7-49   Sketch Line

Step 13: Draw a border ¼ inch inside the edges of the page as seen in Figure 7-50. Press the MMB to exit from the chain of lines. *Pick* **Close** to close the References window.

Step 14: *Pick* the Table tab, then *pick* the **Table** icon. Move down so that a 5 × 3 table is highlighted in the table grid. See Figure 7-51.

Step 15: Move the cursor with the 5 × 3 table attached to the lower left corner of the border. <u>Move the lower left point of the table to the lower left corner of the border.</u> See Figure 7-52.

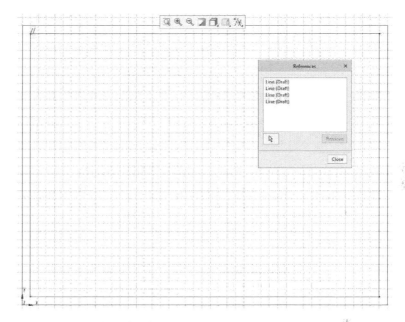

Figure 7-50   Border for Drawing

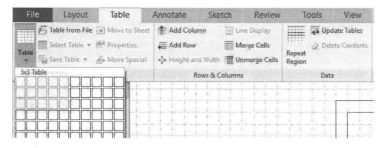

Figure 7-51   Table Tab

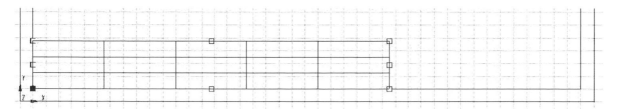

Figure 7-52   Initial Table Placed

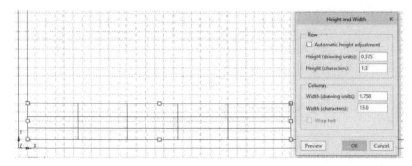

Figure 7-53   Change Column Width

Step 16: 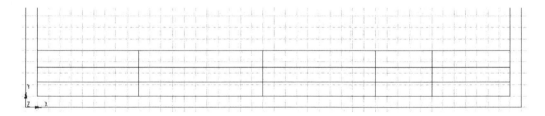 Highlight one of the cells in the right column, then *select* the *Height and Width* icon in the ⬚Table tab. Change the Row Height to 0.375 and the Column Width to 1.75 <Enter>. *Pick* **OK.** *Select* the second column from the right and set the Column Width of 1.25 <Enter>, then *pick* the **Preview** button. See Figure 7-53. *Pick* **OK.**

Step 17: 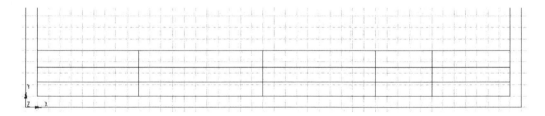 Highlight one of the cells in the far left column of the table, then *select* the *Height and Width* icon in the ⬚Table tab. In the Height and Width window, change the Column Width to 2.25 <Enter>, then *pick* the **Preview** button. *Pick* **OK.**

Step 18: Repeat for the second column from the left. Change the Column Width to 2.75 <Enter>, then *pick* the **Preview** button. *Pick* **OK.**

Step 19: Repeat for the third column from the left. Change the Column Width to 2.50 <Enter>, then *pick* the **Preview** button. The table should extend between the border lines. See Figure 7-54. *Pick* **OK.**

Step 20: Using the <Ctrl> key *select* the four areas in the upper right corner of the table as shown in Figure 7-55.

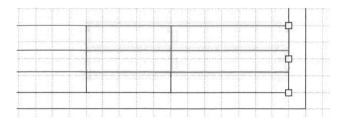

Figure 7-54   Three Left Columns of Table Expanded

Figure 7-55   Combine Cells

Step 21: *Select* the ***Merge Cells...*** icon (Figure 7-56) to make these four cells one cell, as shown in Figure 7-57.

Step 22: Using the <Ctrl> key *select* the three areas in the far left column of the table, then *select* the ***Merge Cells...*** icon again.

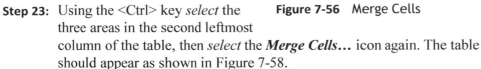
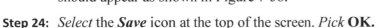

Figure 7-56   Merge Cells

Step 23: Using the <Ctrl> key *select* the three areas in the second leftmost column of the table, then *select* the ***Merge Cells...*** icon again. The table should appear as shown in Figure 7-58.

Step 24: *Select* the ***Save*** icon at the top of the screen. *Pick* **OK.**

Step 25: *Select* the Sketch tab again.

Step 26: *Select* **Sketcher Preferences** again. *Uncheck* **Horizontal/Vertical, Grid intersection,** and **Grid angle.** We do not want to align the text with the corners of the grid. *Pick* **Close** in the Sketch Preferences window.

Step 27: *Select* the Annotate tab.

Step 28: *Select* the ***Note*** icon. See Figure 7-59.

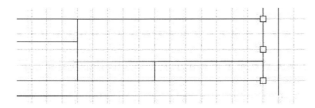

Figure 7-57   Merged Cells in Table

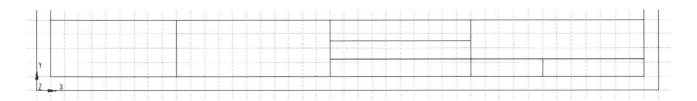

Figure 7-58   Modified Table

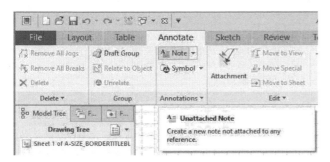

Figure 7-59   Create Note

Step 29: With the first icon selected, move the cursor to any spot on the page, then press LMB. See Figure 7-60.

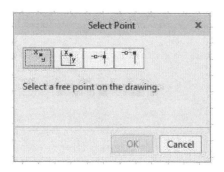

Begin typing the text that goes in the leftmost block. Enter your department, university, city, state, and zip code. The note ends when you pick a point outside the note area. In my case I entered:

"Mechanical Engr Dept."<Enter>
"Ohio Northern University"<Enter>
"Ada, OH 45810"

Step 30: The text on the screen should be highlighted. If not, select it.

Figure 7-60   Place Note

Step 31: *Select* the [Format] tab at the top of the screen. See Figure 7-62. Or press and hold the RMB inside the text box to bring up a pop-up menu. See Figure 7-61.

Step 32: In the Font box pick the down arrow with LMB to bring up a list of possible fonts for this text. I picked SackersEnglishScript for my font. You can pick a different font from the list if you choose. You can change the height of the text by entering a new value for the **Height** followed by <Enter>. I used 0.150 inches for the height. If you like the changes, *pick* any point on the screen outside the text box.

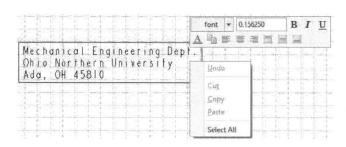

Figure 7-61   Select Text Properties

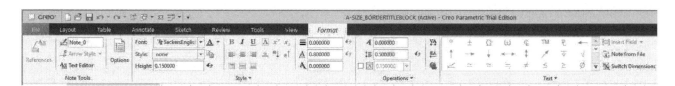

Figure 7-62   Format Tab and Font Options and Special Symbols

Step 33: *Select* the red box around the text, then *move/drag* the text to the correct location on the drawing template to verify that it fits into the first box of the title block. Change the text height as described above if the text doesn't fit into the designated area.

Figure 7-63   Text Symbols with Scroll Bar

Step 34: *Select* the *Note* icon again. *Pick* any location on the screen. Get the ± and ° from the Text Symbol window (Figure 7-63) in the lower right corner of the screen. <u>Place 4 spaces between the tolerances in the third line and 3 spaces before the last 2 lines of this note</u>. Type:

> UNLESS OTHERWISE SPECIFIED, DIMENSIONS \<Enter\>
> ARE IN INCHES. TOLERANCES ARE: \<Enter\>
> .XX ±.005   .XXX ±.001   ANGLE: ±2° \<ENTER\>
> \<ENTER\>
>   DIMENSIONING AND TOLERANCING IN \<ENTER\>
>   ACCORDANCE WITH ASME Y14.5-2009

Step 35: *Pick* the text so that a red box appears around the text area.

Step 36: Set the height of this font to "0.087"\<Enter\>.

Step 37: *Pick* the red box again, then *move/drag* the text to the second area in the title block, as shown in Figure 7-64. Adjust the text if necessary.

Step 38: *Pick* the **Save** icon at the top of the screen.

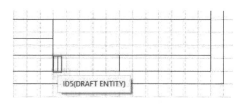

Figure 7-64 Tolerance Block Added

Now we will fill in the table portion of the title block. We want entries in the title block (such as model name, drawing scale, etc.) to be filled in automatically the first time the drawing is created. This can be done by entering the system, drawing, or part parameters in the various table cells. Some common system parameters are shown in Table 7-5.

Table 7-5 Common System Parameters	
Parameter	**Description of parameter**
&model_name	The name of the model
&scale	The view drawing scale
&dwg_name	The name of the drawing file
&todays_date	The current date description
&linear_tol_0_0	The linear tolerance values defined
&angular_tol_0_0	The angular tolerance values defined
&pdmrev:d	The drawing revision number

Step 39: *Select* the **Note** icon again.

Step 40: Place the note in the cell in the lower row fourth column of the table. See Figure 7-65.

Step 41: Set the Font Height to 0.12 \<Enter\>.

Figure 7-65 Select Cell in Table

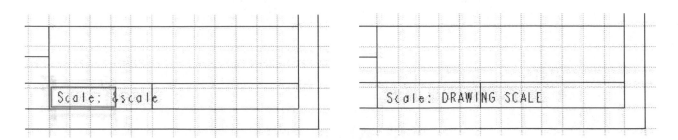

Figure 7-66 Initial Drawing Scale Note   **Figure 7-67** Drawing Scale Expanded

Step 42: Enter the following text: "Scale: &scale". Be sure to add a space after the colon. See Figure 7-66. Note that "&scale" will have changed to "DRAWING SCALE." See Figure 7-67.

Step 43: *Select* the Sketch tab. *Select* the ***Draft Grid*** icon.

Step 44: *Select* **Hide Grid.**

Step 45: *Select* **Done/Return.**

Step 46: *Pick* the ***Refit*** icon.

Step 47: File >Print >Print

Step 48: *Pick* the ***Preview*** icon (Figure 7-68) to see what the border and title block will look like.

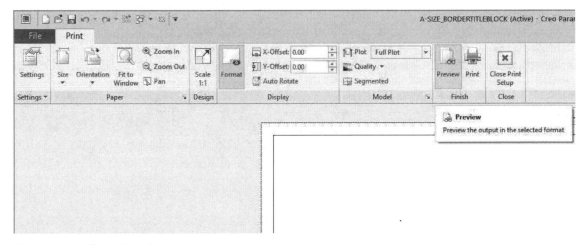

Figure 7-68 Print Preview

Step 49: *Pick* the ***Print*** icon shown in Figure 7-68 and follow the directions to print the border and title block. Without the grid showing, the border and title block should appear similar to Figure 7-69.

Step 50: *Pick* ***Close Print Setup.***

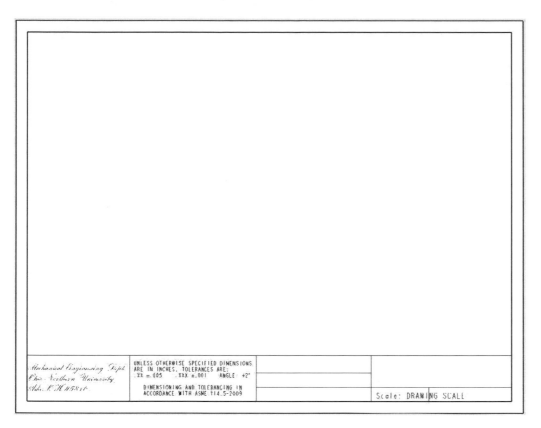

Figure 7-69 Border and Title Block Format Sheet

Step 51: *Select* the *Save* icon at the top of the screen.

Step 52: Assuming everything looks good on the printout, proceed to the next step. If there is a problem, go back to the appropriate step and fix it.

Step 53: *Select* **File> Manage File> Delete Old Versions.** *Pick* the *Yes* button. See Figure 7-70.

Step 54: *Select* **File>Close** or pick the ☒ in the upper left corner of the screen.

Step 55: *Select* **File> Manage Session> Erase Not Displayed** or *pick* the Erase Not Displayed icon.

Step 56: *Pick* **OK** in the **Erase Not Displayed** window.

You have successfully created an A-size formatted sheet with a border and a title block. The text and lines on this format sheet will be exactly the same for every drawing created from now on when this format sheet is referenced.

You can exit from Creo Parametric if you are quitting for now, or you can proceed to the next section.

Figure 7-70 Delete Old Versions

▶ Creating A-size Templates

A template sheet is used to create a custom environment for your drawings. It's a pattern, a special kind of drawing used as the starting point when creating engineering drawings. It contains all information from the format sheet plus information that is specific to the part, the layout of the orthographic views, and the dimensioning and text specifications. Rather than creating a part or assembly drawing from a blank screen, you create a template sheet/drawing and then reference it when you are creating a new engineering drawing for a part or assembly.

In this section we will create a template with two and three orthographic views along with a template containing one 3D view.

Template with Three Orthographic Views

All drawings done in this textbook will be made on A-size (8.5″ × 11″) paper; thus we will create a template drawing containing a border, a filled in title block, and three orthographic views with dimensions for an A-size page in this section.

Step 1: Start Creo Parametric by *double-clicking* with the LMB on the ***Creo Parametric*** icon on the desktop, or from the Program list: Creo Parametric.

Step 2: Set your working directory, by *selecting* the ***Select Working Directory*** icon in the Home ribbon. Locate your working directory, and then *pick* **OK.**

Step 3: Create a new drawing by picking the *New* icon at the top of the screen, or by *selecting* **File>New.**

Step 4: *Select* **Drawing** from the window. Type "A_Template_3-views" for the name. Be sure **Use Default Template** is checked. *Pick* **OK.** See Figure 7-71.

Step 5: Be sure the Default Model is "**none.**" If the Default Model is not "none," highlight the default model name, then type "none." *Pick* **Empty with format.** *Pick* the Format **Browse…** button. *Select*

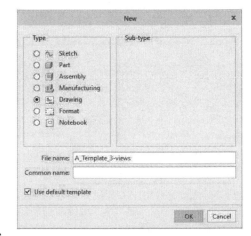

Figure 7-71 New Object Window

Working Directory from the left side of the screen. *Select* **a-size_bordertitleblock.frm.** See Figure 7-72. *Pick* **Open.** *Pick* **OK.**

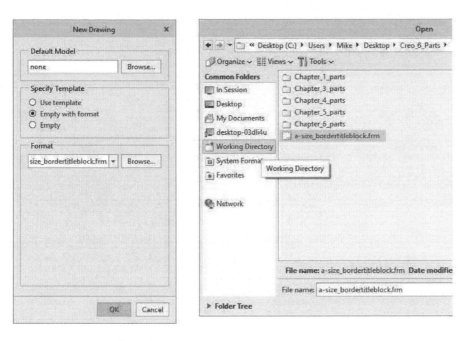

Figure 7-72 New Drawing

The previously created format for an A-size drawing is used to create this drawing template. Remember, format information cannot be changed, whereas template information can be changed. Don't be concerned about the drawing variable, "&scale" (DRAWING SCALE) not fitting into its appropriate table cell. See Figure 7-73.

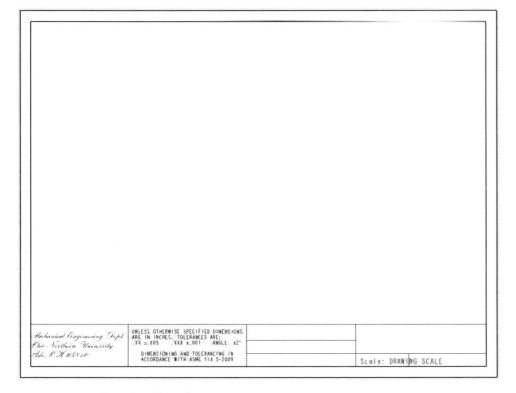

Figure 7-73 Drawing Template

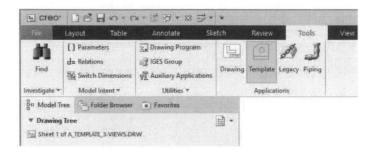

Figure 7-74   Select Template

Step 6: *Select* the ⬚Tools⬚ tab, and then *select* the **Template** icon. See Figure 7-74. Note that some of the available tabs disappear since they are related to the model used in the drawing. We are making a template for a drawing, not a drawing for a part.

Step 7: *Select* the ⬚Layout⬚ tab, then *select* the **Template View** icon.

Step 8: Type "FRONT" <Enter> for the View Name. See Figure 7-75.

Step 9: *Select* the **Model Display** line. *Pick* **Hidden Line.** See Figure 7-76.

Step 10: *Select* **Tan Edge Display.** *Pick* **None.**

Step 11: *Select* **Snap Lines.** *Type* "2" <Enter> for the Number. *Type* "0.375" <Enter> for the Incremental Spacing. *Type* "0.5" <Enter> for the Initial Offset. See Figure 7-77.

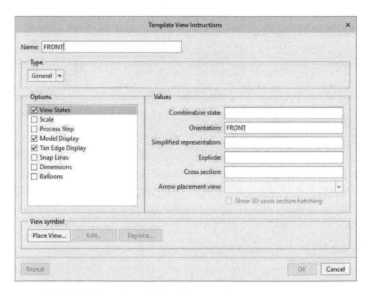

Figure 7-75 Template View Instructions

Figure 7-76   Model Display

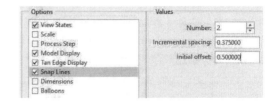

Figure 7-77   Snap Lines

➡ Drag a box keeping the LMB depressed to define a view bounding box or select a point for standard placement.

Figure 7-78 Create Viewing Area Message

Step 12: *Select* **Dimensions.** *Check* the **Create Snap Lines** box.

Step 13: *Pick* the **Place View...** button. The following message (Figure 7-78) appears at the bottom of the screen.

Step 14: Move the mouse cursor to the region in the lower left corner of the drawing area. *Press* LMB once. The view icon will appear at the cursor location as shown in Figure 7-79.

Step 15: *Pick* the OK button to close the Template View Instructions window. Select the **Template View** icon again.

Step 16: Type "TOP" <Enter> for the View Name when a new Template View Instructions window appears. See Figure 7-80. In the pull-down menu in View Type, *select* **Projection.** Repeat steps 9, 10, 11, and 12 for this view as well.

Step 17: *Pick* the **Place View...** button. Move the cursor above the FRONT view, and then *press* the LMB once to place it. Don't worry if the TOP view is not exactly above the FRONT view.

Step 18: *Pick* OK. Select the **Template View** icon again.

Step 19: Type "RIGHT" <Enter> for the View Name when a new Template View Instructions window appears. In the pull-down menu in View Type, *select* **Projection.** Verify projection parent is FRONT view. Repeat steps 9 through 12 for this view also.

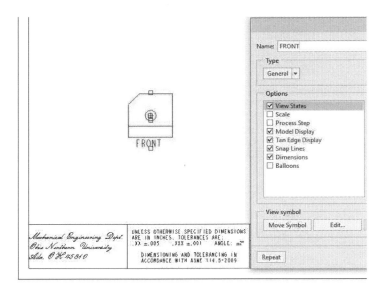

Figure 7-79 Place FRONT View

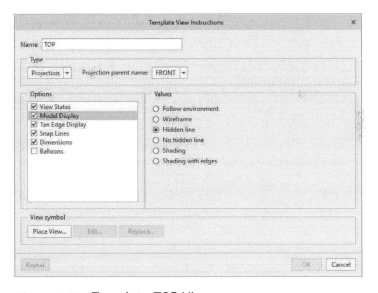

Figure 7-80 Template TOP View

Step 20: *Pick* **Place View...** Move the cursor to the right of the FRONT view, and then *press* the LMB once to place it. Don't worry if the RIGHT view is not exactly horizontal from the FRONT view.

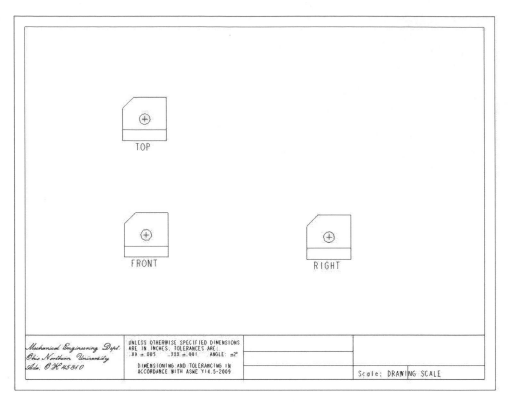

Figure 7-81   Drawing Template with Three Views

Step 21: *Pick* **OK** to close the Template View Instructions window. Your drawing template should appear similar to the one pictured in Figure 7-81.

Step 22: *Pick* the *Save* icon. *Pick* **OK.**

Step 23: *Select* **File>Prepare>Drawing Properties** from the upper right corner of the screen. In the Drawing Properties window, on the line containing Detail Options, *pick* **change.**

Step 24: Type "text_height" <Enter> followed by ".125" <Enter>. See Figure 7-82. Note that when you type "text" the rest of the word appears. You only need to type enough of the parameter word to make it unique from all others.

Step 25: Type "text_width_factor" <Enter> followed by ".75" <Enter>.

Step 26: Type "dim_leader_length" <Enter> followed by ".25" <Enter>.

Step 27: Type "draw_arrow_length" <Enter> followed by ".125" <Enter>.

Step 28: Type "draw_arrow_width" <Enter> followed by ".042" <Enter>.

Step 29: Type "Radial_pattern_axis_circle" <Enter> followed by "Yes" <Enter>

Step 30: Type "tol_display" <Enter> followed by "Yes" <Enter>.

Step 31: Type "gtol_datums" <Enter> followed by "std_asme" <Enter>. *Pick* the **Apply** button. *Pick* **OK.**

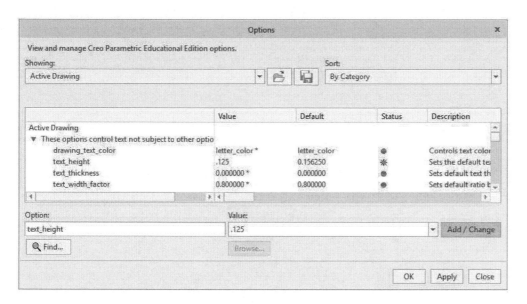

Figure 7-82   Modify Text Height

Step 32: 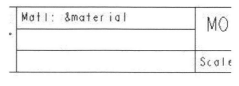 *Pick* the ***Save a Copy of the currently displayed configuration file*** icon at the top of the Options window. The default file name is "Active Drawing"; *pick* **OK** to store this configuration in your working directory.

Step 33: *Pick* **Close** to close the Options window. *Pick* **Close** to close the Drawing Properties window.

Step 34: *Select* the ***Save*** icon at the top of the screen to save the template and all modifications so far.

Now we will add text to the title block area that changes from drawing to drawing. Note that the text on the format page cannot be changed on the drawing. Your template text will initially appear on all new engineering drawings; however, this text can be changed on any drawing. Placing an "&" in front of the text makes it a variable instead of just text. If the variable is already defined in Creo Parametric, then the appropriate value is substituted for the variable on the new engineering drawing. If the variable is not defined, then Creo Parametric prompts the user for the value of the variable on a new drawing. Text or numbers can be entered.

Step 35: *Select* the [Table] tab.

Step 36: *Double-click* on the top cell in the third column of the table to activate the [Format] tab. Set the font size to 0.125 <Enter>, then type "Matl: &material" (Figure 7-83). Next, *Select* the cell below the Material cell.

Figure 7-83   Note Properties

Step 37: *Double-click* on this cell below the Material cell. Set the font height to .125 <Enter>, then type: "Drwn By: &name". Next, Select the cell below the Drwn By cell.

Step 38: *Double-click* on this cell below the Drawn By cell. Set the font height to 0.125 <Enter>, then type "Date: &todays_date Appd:".

Step 39: *Double-click* on this table cell in the lower right corner of the table. Set the font height to 0.125 <Enter>, then type: "Dwg No: &dwg_no". *Highlight* just "&dwg_no", then increase its height to .156 <Enter>.

Step 40: Locate the cell in the upper right corner of the title block. Set the font size to 0.20 <Enter>. *Enter* "&model_name". *Pick* anywhere on the drawing.

Note that the variable "&scale" changed to "DRAWING SCALE" and the variable "&model_name" changed to "MODEL NAME." Don't worry about DRAWING SCALE not fitting into the box. See Figure 7-84.

Step 41: *Select* the **Save** icon to save the template and all modifications.

Step 42: *Select* **File> Manage File> Delete Old Versions.** *Pick* the *Yes* button in the Delete Old Versions window. Your 3-View Engineering Drawing Template should appear similar to Figure 7-85.

Step 43: File> Print> Print.

Step 44: *Pick* the **Preview** icon.

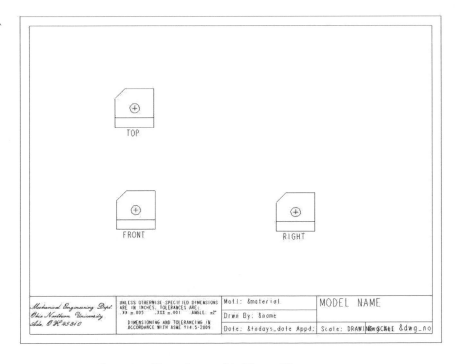

Figure 7-84   Modified Title Block

Figure 7-85   Drawing Template with Three Views

Step 45: If everything looks good, refit it to the screen, then *select* the ***Print*** icon, and then follow the directions for your computing environment.

Step 46: *Pick **Close Print Preview***.

Step 47: Assuming everything looks good on the printout, proceed to the next step. If there is a problem, go back to the appropriate step and fix it.

Step 48: File>Close.

Step 49: File> Manage Session> Erase Not Displayed. *Pick* **OK.**

You have successfully created an A-size template sheet with a border and a title block and three orthographic views in the inch system.

Template with Two Orthographic Views

Some parts only require two orthographic views to completely describe their shape and feature sizes. In this section, we will copy, then modify the template containing three views to only include two orthographic views. If you skipped the previous section on Template with three orthographic views, then go back and do the exercise.

All drawings done in this textbook will be made on A-size paper; thus, we will create a template drawing containing a border, a filled in title block, and two orthographic views with dimensions for an A-size page in this section. Skip steps 1 and 2 if you are continuing from the previous section.

Step 1: If necessary, start Creo Parametric by *double-clicking* with the LMB on ***the Creo Parametric*** icon on the desktop, or from the Program list: Creo Parametric.

Step 2: Set your working directory, by *selecting* the ***Select Working Directory*** icon in the Home ribbon. Locate your working directory, and then *pick* **OK.**

Step 3: **Select** the ***Open*** icon to load a file that already exists. The shortcut keystroke is <Ctrl>-O. You can also use **File>Open.**

Step 4: *Select* "a_template_3-views.drw", and then *pick* **Open.** The previously created three-view template appears on the screen. See Figure 7-86.

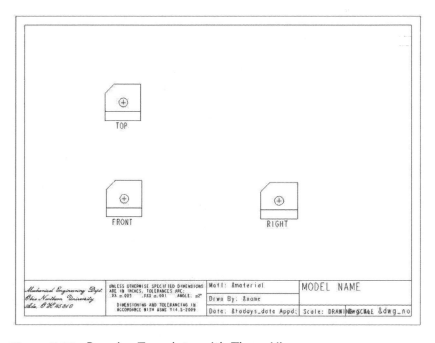

Figure 7-86 Drawing Template with Three Views

Step 5: **File> Save As> Save a Copy.**

Step 6: In the blank area after New
Name, type
"A_TEMPLATE_2-VIEWS".
Do not change the Model
Name from
A_TEMPLATE_3-VIEWS
.DRW. *Pick* **OK.** See Figure
7-87.

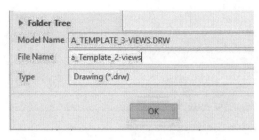

Figure 7-87   Save As New Name

Step 7: **File>Close.**

Step 8: **File> Manage Session> Erase Not Displayed;** *pick* **OK.**

Step 9: *Select* the OPEN icon or **File>Open.**

Step 10: *Select* "a_template_2-views.drw" from the list, and then *pick* **Open.**

Step 11: *Select* the ⌷Tools⌷ tab. *Select the **Template** icon. Pick* the TOP view with
the LMB. It will turn some color. Press <Delete>. The TOP view will
disappear.

Step 12: *Pick,* then *drag* the FRONT and RIGHT side views so they are vertically
in the middle of the drawing area. Move the cursor over the + sign in the
FRONT view, and then *press* the LMB. *Press* the LMB down again and
hold it this time. Move the mouse cursor and the view will follow. The
view will stop moving when you release the LMB. Do the same for the
RIGHT view. Your finished template should appear similar to Figure 7-88.

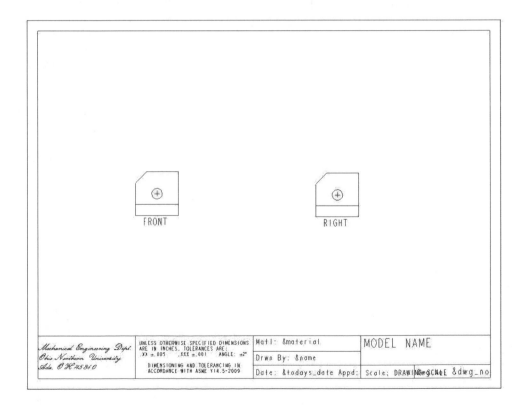

Figure 7-88   Drawing Template with Two Views

Step 13: *Select* the **Save** icon to save the template and all modifications.

Step 14: *Select* **File> Manage File> Delete Old Versions.** *Pick* the **Yes** button in the Delete Old Versions window.

Step 15: **File> Print> Print**.

Step 16: *Pick* the **Preview** icon.

Step 17: If everything looks good, *select* the **Print** icon, and then follow the directions for your computing environment.

Step 18: *Pick* **Close Print Setup**.

Step 19: Assuming everything looks good on the printout, proceed to the next step. If there is a problem, go back to the appropriate step and fix it.

Step 20: **File>Close** or pick ⬚ in the upper left corner of the screen.

Step 21: **File> Manage Session> Erase Not Displayed.** *Pick* **OK.**

You have very quickly created a template containing two orthographic views. Since all the information is the same for 3-views and 2-views, we used the 3-view template to create the 2-view template.

Template with One 3D Projection View

Occasionally we want to simply show a 3D object in one 3D view. In this section, we will copy, then modify the template containing three views to include only one 3D projection view.

All drawings done in this tutorial will be made on A-size paper, thus we will create a template drawing containing a border, a filled in title block, and one 3D projection view for an A-size page in this section. Skip steps 1 and 2 if you are continuing from the previous section.

Step 1: If necessary, start Creo Parametric by *double-clicking* with the LMB on **the Creo Parametric** icon on the desktop, or from the Program list: Creo Parametric.

Step 2: Set your working directory, by *selecting* the **Select Working Directory** icon in the Home ribbon. Locate your working directory, and then *pick* **OK.**

Step 3: *Select* the **Open** icon to load a file that already exists. The shortcut keystroke is <Ctrl>-O. You can also use **File>Open.**

Step 4: *Select* "a_template_3-views.drw", and then *pick* **Open.** The previously created three-view template appears on the screen.

Step 5: **File> Save As> Save a Copy.**

Step 6: In the blank area after New Name, *type* "A_TEMPLATE_3DVIEW". Do not change the Model Name from A_TEMPLATE_3-VIEWS.DRW. *Pick* **OK.**

Step 7: **File>Close.**

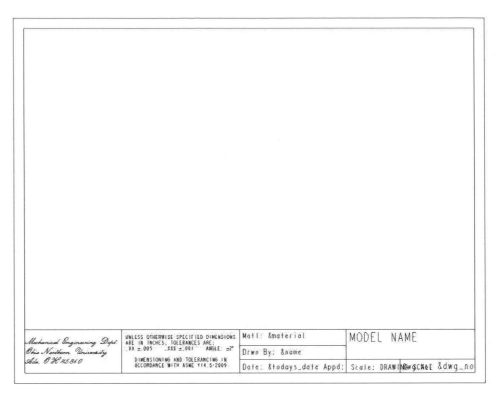

Figure 7-89   Border and Title Block

Step 8:   **File> Manage Session> Erase Not Displayed;** *pick* **OK.**

Step 9:   *Select* the OPEN icon or **File>Open.**

Step 10: *Select* "a_template_3dview.drw" from the list, and then *pick* **Open.**

Step 11: *Select* the ⎡Tools⎤ tab. *Select* the ***Template*** icon.

Step 12: *Pick* the TOP view with the LMB. It will change color. Press <Delete>. The TOP view will disappear. Do the same thing with the RIGHT view, and with the FRONT view. Your template should appear similar to Figure 7-89. We have erased all three views. We will start again placing a single 3D view. (You could draw an imaginary box around all three views to select them, then press the <Delete> key to erase all three at once.)

Step 13: *Pick* the ⎡Layout⎤ tab. *Pick* the ***Template View*** icon.

Step 14: *Type* "3-D_PROJECTION"<Enter> for the View Name. *Type* "DEFAULT ORIENTATION"<Enter> for the Orientation. See Figure 7-90.

Step 15: *Pick* **Tan Edge Display.** *Pick* **Dimmed** for the view option.

Step 16: *Pick* **Place View...** button. Move the mouse cursor to the region in the center of the drawing area. *Press* LMB once. The view icon will appear at the mouse cursor location. Do not make a box in the drawing area, but rather *press* the LMB only once.

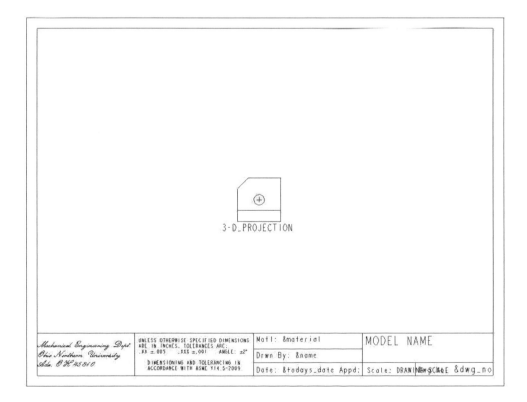

Figure 7-90 Template View Instructions

Step 17: *Pick* **OK** from the Template View Instructions window. The view icon will appear in the middle of the drawing area as shown in Figure 7-91.

Step 18: *Select* the ***Save*** icon to save the template and all modifications.

Step 19: *Select* **File> Manage File> Delete Old Versions.** *Pick* the ***Yes*** button in the Delete Old Versions window.

Figure 7-91 Drawing Template with One View

Step 20: Assuming everything looks good on the screen, proceed to the next step. If there is a problem, go back to the appropriate step and fix

Step 21: File>Close or pick ⊠ in the upper left corner of the screen.

Step 22: File> Manage Session> Erase Not Displayed. *Pick* **OK.**

You have very quickly created a template containing one 3D projection view. Since the border and title block information are the same we used the 3-views template to create the 3D projection view template. You can exit from Creo Parametric if you are quitting for now, or you can proceed to the next chapter.

*** Can you create drawing templates for metric parts now using A4 paper? ***

▶ ## Review Questions

1. What is the difference between a drawing template and a drawing format?

2. Why would you use a drawing template or a drawing format?

3. Why are dimensions placed on a part drawing?

4. What is the smallest fractional dimension that a worker can directly measure?

5. In the two place decimal system, the following measurements round to:

    a. 4.613 inches

    b. 2.418 inches

    c. 3.455 inches

    d. 1.625 inches

6. How would one specify that the part is not to have any sharp edges?

7. What is the contour principle?

8. Cylinders should have their diameters dimensioned in the view where they appear _____.

9. What information is needed to properly dimension a threaded hole?

10. What is the proper way to dimension a keyway in a hub?

11. What does REF stand for, and why would it appear on an engineering drawing?

12. Where should general notes be located on an engineering drawing?

13. How should the dimensioning scheme change if the part is to be created using a numerically-controlled machine?

14. Why should you avoid dimensioning to a hidden line?

15. How close to the part's outline should the first dimension line appear?

16. Can extension lines cross? Dimension lines? Leader lines? Hidden lines?

17. Why should an overall dimension in each of the three major directions be present on an engineering drawing?

18. How do you know if a dimension pointing to a curved section is a radius dimension or a diameter dimension?

ENGINEERING DRAWINGS

▶ Engineering Drawings Explored

Design Intent

Two steel 3.00-inch by 6.00-inch, 1-inch thick support braces need three $3/4$-inch diameter holes plus a $1/4$-inch diameter hole. The leftmost $3/4$-inch hole must be 1.00 inches from the left edge and 2.00 inches from the bottom edge. The rightmost hole must be 5.00 inches from the left edge and 2.00 inches from the bottom edge. The third large hole must be located 1.12 inches below the leftmost hole with its center located 2.75 inches from the leftmost hole's center. The $1/4$-inch diameter hole must be 2.25 inches from the left edge and 2.00 inches from the bottom edge. There must be no sharp edges since the part will be handled by the workers.

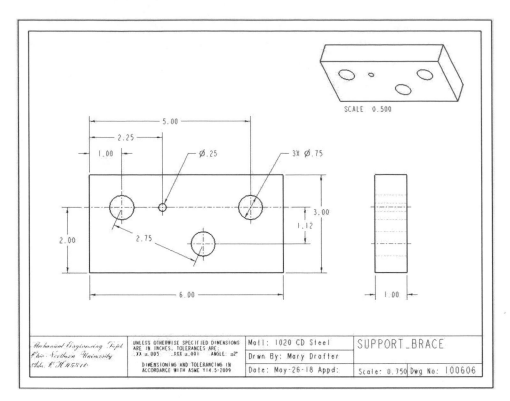

Figure 8-1   Engineering Drawing

The main way of presenting engineering designs is through a 2D engineering drawing (Figure 8-1). This drawing must contain all the information necessary to produce the part such as its geometry, its size, the type of material to be used, its surface finish, any general notes, etc. There is a standard as far as the part layout, dimensions, and notes on the engineering drawing so that anyone familiar with these practices can read the drawing. Creo Parametric has most of these standards built in if you accept the default actions.

Creo Parametric provides a number of additional commands to improve the looks of the drawing along with assisting you in the placement of the orthographic views. After placing the first view, Creo Parametric lets you add projection views that are always correct according to their placement relative to the first view. This does not mean that Creo Parametric does everything perfectly correct and you don't have to think, but rather it gets you close to a valid engineering drawing without a lot of work. For example, Creo Parametric will let you display all the dimensions you used when you created the model so if you paid close attention to design intent when you created the model, then the dimensions that show up should be appropriate. However, because you may have used sketching constraints when you made the model it may be necessary to add some additional dimensions to convey these constraints.

There is a lot of power and options in Creo Parametric's engineering drawing section. This chapter covers just the basic functions necessary to create some typical engineering drawings or detailed drawings as they are sometimes called.

Figure 8-2 is the detailed drawing of the support brace in Creo Parametric. The drawing sheet includes the border and title block that were defined in the drawing format file. Upon creation of the drawing some of the information in the title block

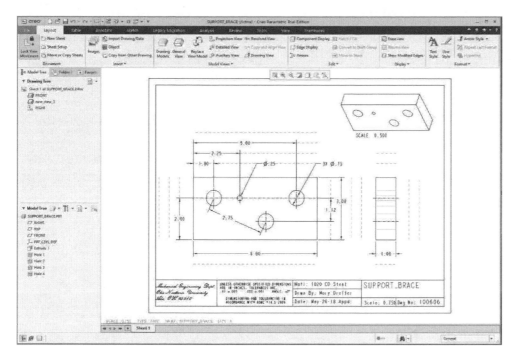

Figure 8-2   Drawing in Creo Parametric

is filled in using values from the 3D model. The drawing shows the standard front and right side views. The top view is omitted because it wouldn't add any new information. A general view of the 3D part is shown in the upper right corner of the sheet. This is becoming more common on computer generated engineering drawings. Dimensions and notes have been added to the drawing. The light gray dashed lines are guidelines for the proper placement of the dimensions and will not show up when a hard copy is made. Most of the shown dimensions come from the model, thus they are referred to as driving dimensions and have a variable name starting with "d." If you change the value of a driving dimension, the model will change size accordingly. The 1.12-inch and the 2.75-inch dimensions are not part of the 3D model, thus they are referred to as driven dimensions and have variable names beginning with "ad." You cannot change the value of driven dimensions; they are sometimes called draft dimensions.

The drawing ribbon has ten tabs: File, Layout, Table, Annotate, Sketch, Review, Legacy Migration, Analysis, Review, Tools, and View. Each tab has an associated set of commands. The ribbon and the drawing tree are linked. The model tree appears below the drawing tree. The search filter in the lower right refers to elements that appear in the drawing. Its default is General.

The middle mouse button (MMB) has changed its function. Instead of controlling spin in part mode, it now controls pan in drawing mode. The scroll wheel still controls zoom and the Refit icon resizes and repositions the drawing in the current window. The left mouse button (LMB) is used to select items. The right mouse button (RMB) is used to bring up the appropriate pop-up menu.

We will discuss the ten drawing ribbons next. The commands available to the user will change depending upon which drawing ribbon tab is selected and what is currently highlighted in the graphics window.

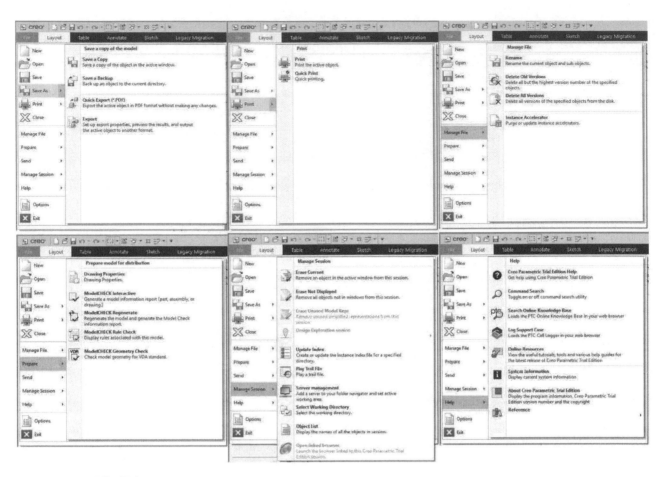

Figure 8-3   File Tab

On the far left in Figure 8-3 is the File tab. This tab is used to create a new file, open an existing file, save the current file, save the current file under a new name, print and print preview, manage files in the working directory, prepare the current working environment, email the active window as an attachment, manage the session, get help, change the options in Creo Parametric, and exit Creo Parametric.

Figure 8-4 shows the Layout ribbon. At its far left are the sheet options. Complicated part drawings tend to have more than one sheet of information. In the middle are the options for creating different types of views. A general view is the first view created and must come before you can create an auxiliary, detailed, or projection view. The right side view is a projection view off the general (front) view. The 3D view is a second general view which can use the default orientation.

Figure 8-4   Layout Tab

Figure 8-5 Table Tab

The right portion of this ribbon allows you to add cutting plane lines, modify edge display, modify a text, arrow, or line style, overlay another drawing, etc.

The Table ribbon (Figure 8-5) is used to create the title block, repeat regions, and the bill of materials table. You can draw/create a table, add columns or rows to an existing table, and format the information inside the table. In our case, some of the information in the title block was automatically filled in using values from the 3D model when the drawing was created.

If this were an assembly drawing, then the bill of materials could be automatically filled in with the part names, descriptions, and quantity from the 3D assembly model.

Annotate is the fourth tab (Figure 8-6) and is where you will probably spend most of your time when creating an engineering drawing. It allows you to see the variable names of the driving and driven dimensions being displayed and to set the number of decimal places for each dimension. It allows you to add driven dimensions in standard form, in ordinate form, or in radial form. You can add general notes, balloon notes, or notes with leaders along with reference dimensions, surface finishes, and geometric tolerances. It provides you with guidelines for the proper spacing of dimensions and a cleanup tool that attempts to properly space the existing dimensions. You can modify existing text or lines and the shape of the dimensional arrowheads.

Figure 8-7 shows the Sketch tab. This is similar to sketcher ribbon in part creation except you are sketching on the drawing sheet and not on the 3D part. If you do not associate a sketched feature with a particular view, then the feature will not move when the view is moved. However, if you tie the feature to the view, then it will move with the view. This tab allows the draftsman to add extra information to the part drawing without affecting the original 3D model.

Figure 8-6 Annotate Tab

Figure 8-7 Sketch Tab

Figure 8-8 shows the Legacy Migration tab. This tab is for adding a model to a drawing, creating model views, copying annotations, and converting 2D annotations to 3D annotations. As of the first release of Creo Parametric, this option must be installed on your system prior to its usage.

Figure 8-9 shows the Analysis tab. This tab allows you to quickly estimate mass properties such as the volume, surface area, mass, center of gravity, mass moments of inertia, etc. if the material density is specified. It allows you to measure distances, lengths, angles, diameters, radii, and surface areas, etc.

Figure 8-10 shows the Review tab. The Review tab is eighth in line (Figure 8-9). These commands let you check the status of your drawing prior to its release. You can regenerate the active model, update the drawing sheets, update the draft dimensions, update the drawing views, update any tables, highlight attributes, etc.

Figure 8-11 shows the Tools tab. This tab allows you to display information about a feature or the model, define parameters and relations, and switch between drawing, template, and piping.

Figure 8-8   Legacy Migration Tab

Figure 8-9   Analysis Tab

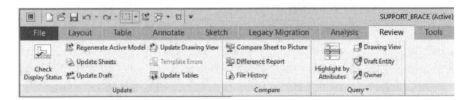

Figure 8-10   Review Tab

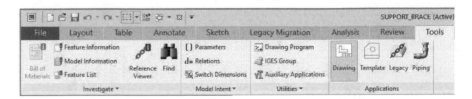

Figure 8-11   Tools Tab

Figure 8-12 shows the View tab. This tab allows you to set or move between layers, refit, pan, zoom, etc., create saved views, set display styles, turn on or off datums, activate or close a window, and switch between windows. Most of the features are available in the graphics toolbar which is always visible.

Figure 8-13 shows the Framework tab and its options.

Figure 8-14 shows the finished engineering drawing. Your drawing should be similar.

Figure 8-12   View Tab

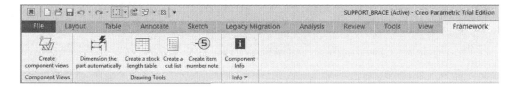

Figure 8-13   Framework Tab

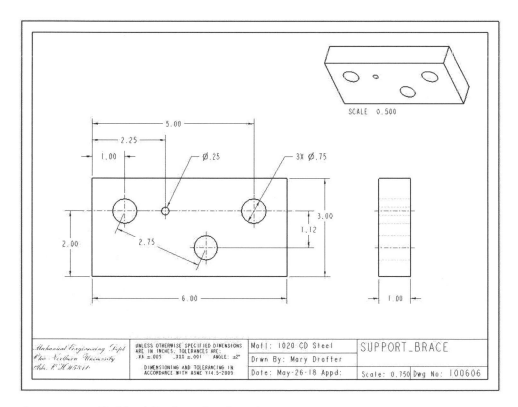

Figure 8-14   Finished Engineering Drawing

▶ **Engineering Drawings Practice**

The primary form of design documentation is 2D detailed engineering drawings. The drawing must contain information about the part's geometry and size, information about the part's material and surface finish, and any necessary manufacturing notes. The placement of the views is by third-angle projection in the United States. That is, the top view is above the front view, and the right side view is to the right of the front view. This placement along with drawing practice standards makes it easy for anyone to read a drawing that is familiar with the standard. Creo Parametric has these standards built in if you accept the default action. This assures you that the placement of the views is correct so you can concentrate on the cosmetics of the drawing. Creo Parametric lets you focus on what to show in the drawing instead of how to show it. From the 3D model Creo Parametric will draw all views correctly including all hidden lines in the proper location. In a 2D drafting system, the user is responsible for placing all of the solid and hidden lines in the correct position for each view, thus it is possible to draw a view incorrectly in the 2D drafting system.

Creating detailed engineering drawings in Creo Parametric is not automatic. Although the views will be correct, the placement and addition of dimensions will need to occur. Creo Parametric uses a set of internal rules for its initial placement of the dimensions. In general, this placement will not meet all of the standard drawing practices. The user will need to move dimensions to meet the standards. Also, sketcher constraints will not be reflected in the driving dimensions created by Creo Parametric. It will be necessary to add driven dimensions to the drawing to reflect sketcher constraint information from the model. Manufacturing notes and surface finishes will also have to be added to the drawing.

This practice session assumes that you have completed the previous chapter and have created the necessary A-format and A-size templates. If you skipped the previous chapter, please go back and do it before continuing.

Design Intent

Create a detailed engineering drawing of the upper plate for a take-up frame subassembly. The 1020 CD steel bar is 1.00 inches wide by 2.50 inches long by 0.268 inches thick. A ½-inch hole is located in its geometric center. The long edges of the bar are rounded using a 0.02-inch radius. Create the part symmetric about the right plane in Creo Parametric. See Figure 8-15.

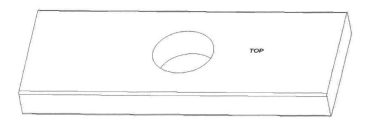

Figure 8-15 Upper Plate

Step 1:   Start Creo Parametric by *double-clicking* with the LMB on the ***Creo Parametric*** icon on the desktop, or from the Program list: Creo Parametric.

Step 2:   Set your working directory, by *selecting* the ***Select Working Directory*** icon in the Home ribbon. Locate your working directory, and then *pick* **OK.**

Step 3:   Create a new part by *picking* the *New* icon at the top of the screen, or by *selecting* **File>New.**

Step 4:   *Select* **Part** from the window and name it "upper_plate". The sub-type should be **Solid.** Make sure the default template is checked, and then *pick* **OK.**

Step 5:   We need to set the units for the upper plate. Let's use the IPS (Inch-Pound-Second) system of units for practice. Use **File>Prepare>Model Properties.** When the Model Properties window appears, *pick* the word **change** on the right end of the Units row. When the Units Manager window appears, if the system of units, **IPS,** is not selected, *pick* it from the list, then *pick* the ⟦ → Set... ⟧ button. *Select* Convert dimensions (for example 1″ becomes 25.4mm), then *pick* **OK.** *Pick* **Close** on the Units Manager window. *Pick* **Close** on the Model Properties window.

Step 6:   *Select* the ***Extrude*** tool. *Select* the TOP plane. *Select* ***Sketch Setup*** and orient the RIGHT plane toward the Right. *Pick* **Sketch.**

Step 7:   ⟦icon⟧ If necessary, *pick* the ***Sketch View*** icon to orient the sketch plane so that it is parallel with the display screen.

Step 8:   ⟦icon⟧ Create a 1.00-inch by 2.50-inch rectangle symmetric about the origin. See Figure 8-16.

Step 9:   Exit sketcher, then extrude the plate 0.268 inches thick. *Pick* the ***green checkmark.***

Step 10: Use the ***Hole*** tool to create a ½-inch ***through*** hole in the center of the plate. Align it with the RIGHT and FRONT datum planes.

Step 11: Use the ***Round*** tool to place a 0.02-inch radius on the four long edges of the bar. Remember to hold down the <Ctrl> key when *picking* the second, third, and fourth edges of the bar.

Step 12: ***Save*** the part. *Pick* **OK** in the Save Object window.

Step 13: **File>Close Window** or pick the ⟦✕⟧ in the upper left corner of the screen**.**

Now we will create the detailed engineering drawing for this part using the A-size 3-views template we created in the previous chapter.

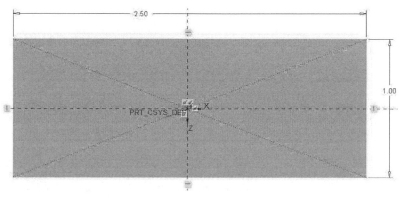

Figure 8-16 Sketch of Symmetric Rectangle

Figure 8-17   New Drawing Window

Step 14: *Select* the *New* icon to create a new object. The shortcut keystroke is <Ctrl>-N.

Step 15: *Select* **Drawing** from the window and name it "upper_plate". Make sure the default template is checked, and then *pick* **OK.**

Step 16: Make sure "upper_plate" is the default model. If it is not, *pick* the **Browse...** button, then *select* the "upper_plate".

Step 17: *Pick* the **Browse...** button inside the Template area. *Pick* Working Directory from the left side of the open window. *Select* "a_template_3-views.drw" from the directory. *Pick* the **Open** button. *Pick* the **OK** button in the New Drawing window, Figure 8-17.

Step 18: At the top of the window, enter "1020CD STEEL" for the material, then *pick* the *green checkmark.* Enter your name, then press the <Enter> key. (This is the same as *picking* the *green checkmark.*) Enter the drawing number, "111001"<Enter>.

Step 19: Turn off *Plane display, Axis display, Point display,* and *Csys display. Pick* the *Regenerate Active Model* icon to update the display. Your drawing should look similar to Figure 8-18. Note that the .268-inch dimension may show up as .27 inches if the default was 2 decimal places on all dimensions.

The hole must be dimensioned in the view where it appears to be round. The .02-inch radius needs to be shown in the right side view.

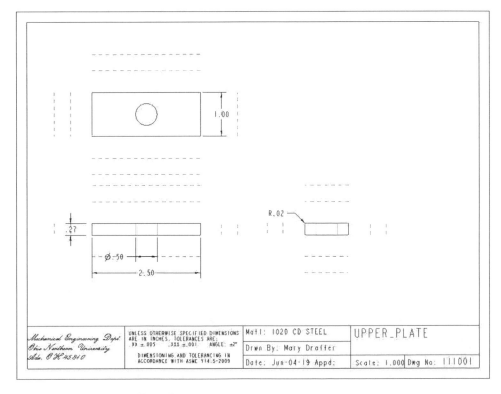

Figure 8-18 Current Drawing

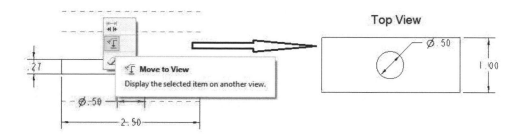

Figure 8-19   Move Item to View

Step 20: *Select* the Annotate tab.

Step 21: *Pick* the ½-inch diameter dimension using the LMB. Press and hold the RMB until a pop-up menu appears. *Select Move to View* icon. *Pick* the TOP view using the LMB to move the dimension to the top view. See Figure 8-19.

Step 22: *Pick* the "R.02" dimension, then select the **Dimension Text** icon. Add "4X " in front of the **R**.

Step 23: *Pick* the ½-inch diameter dimension in the top view and drag it out to the first guideline below the top view. *Pick* the "4X R.02" dimension in the right side view and drag it out to the first guideline above the right side view. Move the 2.50 dimension above the front view. Move the .27 dimension to the right side of the front view. See Figure 8-20.

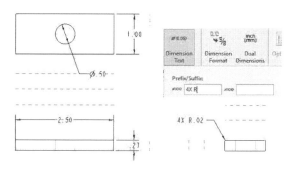

Figure 8-20   Reposition Dimensions

Step 24: *Pick* the TOP view. Hold down the <Ctrl> key and *pick* the FRONT and RIGHT SIDE views. *Select* the Annotate tab. *Pick* the **Show Model Annotations** icon. *Select* the last tab in the Show Model Annotations window. *Check* the three boxes so the centerlines for the hole will show up in all three views. *Pick* **OK**. See Figure 8-21.

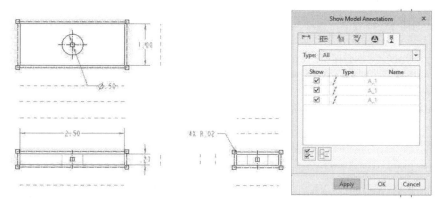

Figure 8-21   Add Centerlines

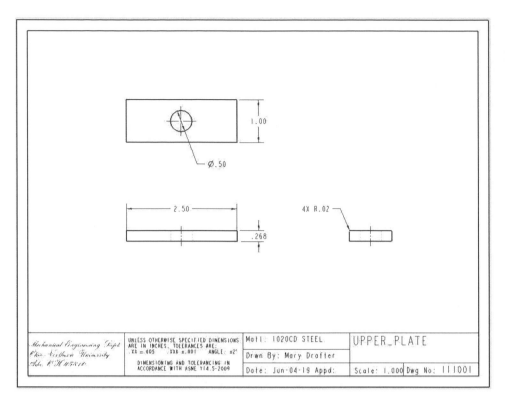

Figure 8-22   Preview Drawing

Step 25: *Pick* the .27-inch dimension. In the Precision box at the top of the screen, use the pull-down menu to change 0.12 to 0.123.

Step 26: File>Print>Print. *Pick* the ***Preview*** icon. Your drawing should look similar to Figure 8-22. *Pick* ***Close Print Setup***.

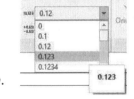

Design Intent

The design engineer has just sent you two changes for this part. The new length is 2.57 inches instead of 2.50 inches. The thickness of the bar stock is 0.250 inches instead of 0.268 inches.

You could go back to the model and change its dimensions, then bring the detailed engineering drawing back up on the screen to reflect the changes. However, since the model is tied directly to the drawing, changing the driving dimensions of the drawing will change the 3D part model.

Step 27: *Pick* the Annotate tab again.

Step 28: *Pick* the 2.50-inch dimension using the LMB. In the upper left corner of the screen, change the value from 2.50 to 2.57, then *press* <Enter>.

Step 29: *Pick* the .268-inch dimension using the LMB. In the upper left corner of the screen, change the value from 0.268 to 0.250, then *press* <Enter>.

Step 30: 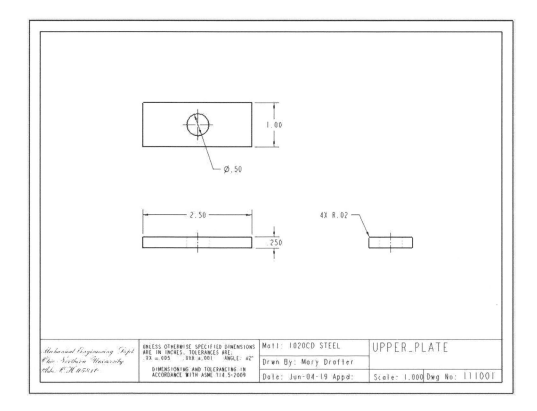 *Select* the ***Regenerate Active Model*** icon in the upper left corner of the screen, or press <Ctrl>+G. The part's shape will change to reflect the change in its dimensions.

Step 31: Move the views so they are in the center of the drawing area. *Pick* the FRONT view, then press down and hold the RMB until a pop-up menu appears. *Pick **Lock View Movement*** to unlock the view. Move the views. Repeat the RMB procedure to lock the views again.

Step 32: *Save* the drawing. *Pick* **OK** in the Save Object window. This also saves the 3D model and its changes.

Step 33: **File>Print>Print.** *Pick* the ***Preview*** icon. Your drawing should look similar to Figure 8-23. *Pick **Close Print Setup***.

Can you make this part from the drawing below? The answer is NO. Why not? The location for the ½-inch diameter hole is missing so the machinist will not know where to place the hole. There were no driving dimensions for the placement of the hole; thus, no dimensions showed up on the drawing. We need to add driven dimensions to locate the hole's center.

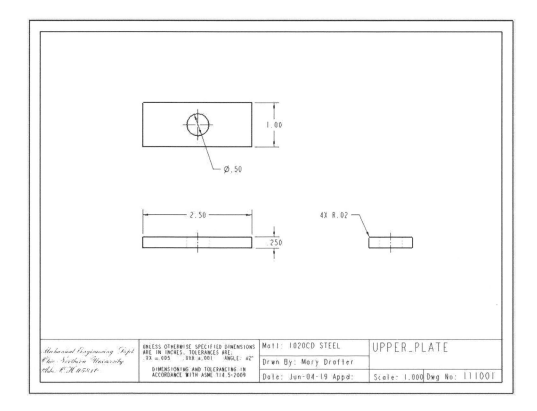

Figure 8-23 Modified Drawing

Step 34: *Pick* the [Annotate] tab again.

Step 35: *Select* the **Dimension** icon.

Step 36: Make sure **Select an Entity** is selected. *Pick* the vertical centerline in the top view with the LMB. Hold down <Ctrl> and *pick* the left edge of the top view with the LMB. Move to the guideline below the top view, then press the MMB to place the dimension.

Step 37: *Pick* the horizontal centerline in the top view with the LMB. Hold down <Ctrl> and pick the bottom edge of the top view with the LMB. Move to the guideline on the left edge of the top view, then press the MMB to place the dimension.

Step 38: Press the MMB again to exit from this mode.

Step 39: If the horizontal dimension shows up as 1.29-inch dimension, change it to 3-decimal places. (Note that the actual dimension is 1.285 inches.)

Step 40: *Select* the 1.00-inch dimension, then drag it to the second dimension guideline on the right side of the top view. *Select* the .50-inch dimension, then drag it to the first dimension guideline on the right side of the top view. Flip the arrows on the 0.50 dimension.

Step 41: You can move the 2.57-inch dimension to the top view or leave it in the front view.

Step 42: *Save* the drawing. *Refit* the drawing to the screen.

Step 43: Your drawing should look similar to Figure 8-24. Print the drawing to prove that you have completed this practice. **File>Print>Print,** then follow your environment instructions. *Pick **Close Print Setup***.

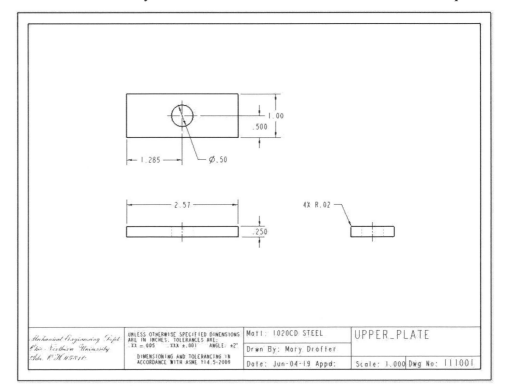

Figure 8-24 Preview Final Drawing

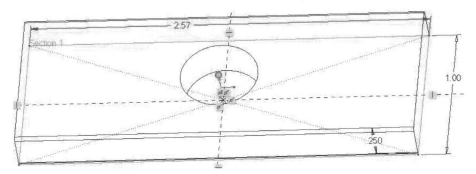

Figure 8-25 Modified Part

Step 44: **File> Manage File> Delete Old Versions,** then *pick* the **Yes** button. **File>Close.**

Step 45: **File> Manage File> Delete Old Versions**, with the upper_plate.prt on the screen. *Pick* the **Yes** button to delete any old versions of the 3D model.

Step 46: **File>Close.**

Step 47: **File> Manage Session> Erase Not Displayed…** *Pick* **OK.**

If you open up the part file "upper_plate.prt" you will see that the changes made in the drawing are truly reflected in the 3D model, as shown in Figure 8-25.

▶ Sectioned View

Design Intent
Create a detailed engineering drawing of a four-hole hub, as shown in Figure 8-26. Because of the complexity of the hub's shape, make the front view a sectioned view. This part was created in the Revolves Explored section and is called "Four-hole _hub.prt" or "4-hole_hub.prt".

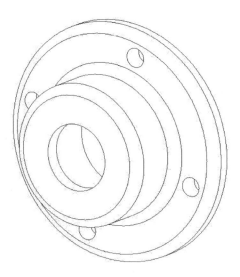

Figure 8-26 Four-Hole Hub

Step 1: *Select* the **New** icon to create a new drawing. The shortcut keystroke is <Ctrl>-N.

Step 2: *Select* **Drawing** from the window and name it "four-hole_hub" or "4-hole_hub.drw". Make sure the default template is checked, and then *pick* **OK.**

Figure 8-27 New Drawing Window

Step 3: Make sure "four-hole_hub" or "4-hole_hub.prt" is the default model. If it is not, *pick* the **Browse...** button, then *select* the "four-hole_hub" or "4-hole_hub.prt".

Step 4: *Pick* the **Browse...** button inside the Template area. *Pick* Working Directory from the left side of the open window. *Select* "a_template_2-views.drw" from the directory. *Pick* the **Open** button. *Pick* the **OK** button in the New Drawing window, Figure 8-27.

Step 5: Enter STEEL for the material, your name for name, and 112000 for the drawing number.

Step 6: *Select* the Annotate tab.

Step 7: *Double-click* on the actual SCALE value in the lower left corner of the screen to bring up a new window that will allow you to change the scale. Type .500 <Enter> or *pick* the **checkmark**. See Figure 8-28.

Step 8: *Pick* the FRONT view with the LMB. Press and hold the RMB until a pop-up menu appears. *Select* **Lock View Movement** to unlock the view. Place the cursor on the small hollow square in the middle of the front view. *Press and hold down* the LMB, then drag the cursor and front view to the left so that the two views and their dimensions do not touch. **Lock View Movement** again. See Figure 8-29.

Figure 8-28   Modifying the Drawing Scale

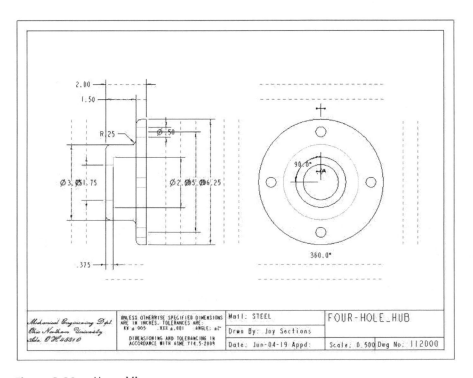

Figure 8-29   Move Views

Step 9: *Select* the ***Plane Display*** icon to turn datum planes on.

Step 10: *Select* the ☐Layout☐ tab.

Step 11: *Double-click* on the FRONT view using the LMB to bring up the Drawing View window. *Select* the Sections category. *Pick* the 2D cross-section button. *Pick* the ***plus sign*** icon. With Planar and single selected in the menu manager window, *pick* **Done.** In the Enter NAME for cross- section [QUIT]: window, type "A" <Enter>. Using the mouse, and the LMB *pick* the FRONT datum plane in the right side view. *Pick* the **Apply** button. The front view will change to a sectioned view with the cutting plane through the FRONT plane. *Pick* **OK.** See Figure 8-30.

Step 12: *Pick* the ☐Annotate☐ tab again.

Step 13: *Pick* the "ϕ.50" dimension with the LMB. *Select* ***Move to View*** icon. *Pick* the RIGHT view to move the dimension. *Press and hold* the RMB again, then *select* **Flip Arrows** from the pop-up menu. Repeat the **Flip Arrows** option again to get the arrowhead outside the hole. *Select* the **Dimension Text** icon, then modify the dimension to read "4X ϕ.5 0 EQUALLY SPACED". Delete the 360.0, 90.0, and .0 degree dimensions in the RIGHT view. See Figure 8-31.

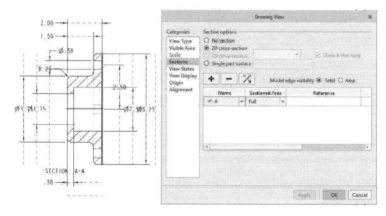

Figure 8-30 Create Sectioned View

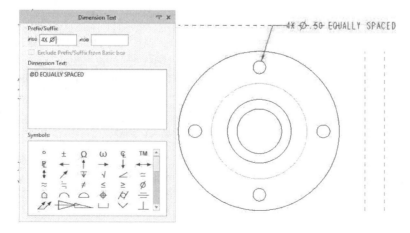

Figure 8-31 Dimension Text Window

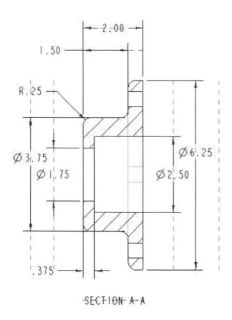

Figure 8-32   Front View

Step 14: *Move* the 5.00-inch diameter dimension to the right side view. *Select* the **Dimension Text** icon in the Dimension tab. Add "BC" for bolt circle to the end of the current text, then *pick OK.*

Step 15: *Move* the dimensions in the front view to reflect Figure 8-32.

Step 16: *Pick* the FRONT view, and then *pick* the **Show Model Annotations** icon. Pick the far right tab in the Show Model Annotations window. Check the box for the middle, top, and bottom centerlines in the hub. See Figure 8-33. Pick **OK.**

Step 17: *Pick* the RIGHT view, and then *pick* the **Show Model Annotations** icon. *Pick* the far right tab in the Show Model Annotations window. *Check* all the boxes for this view. Pick **OK.**

Step 18: If there are extra dimensions on the drawing, *select* them, then press <Delete>.

Step 19: *Save* the drawing.

Step 20: *Refit* the drawing to the screen. *Select* **File>Print>Print**. *Pick* the **Preview** icon. Your drawing should look similar to Figure 8-34. *Pick **Close Print Setup**.*

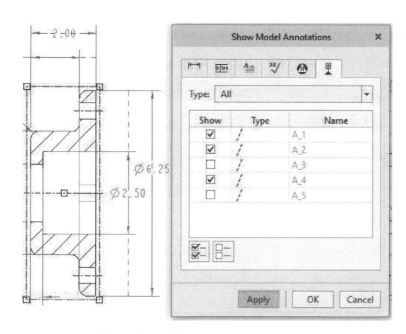

Figure 8-33   Show Centerlines

For the drawing to be correct, a cutting plane line must be shown in the right side view to show where the part was cut to create the front sectioned view.

Step 21: *Select* the Layout tab again.

Step 22: *Select* the **Arrows** icon at the top of the screen.

Step 23: *Pick* the sectioned FRONT view, and then *pick* the circular RIGHT view. (The cutting plane line will probably pass through the 4X Ø.50-inch diameter equally spaced hole note.)

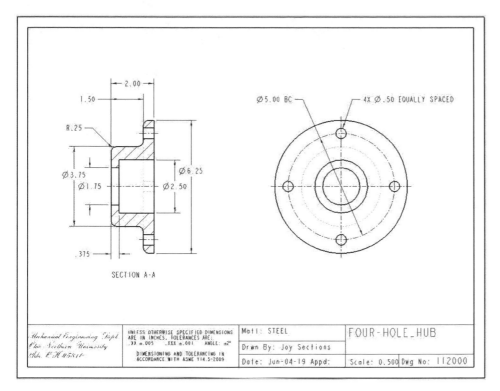

Figure 8-34 Preview Drawing

Step 24: *Pick* the ⃞Annotate⃞ tab.

Step 25: *Highlight* the note, then drag it to the other side of center, as shown in Figure 8-35.

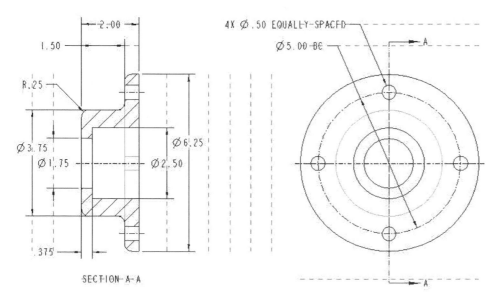

Figure 8-35 Adjust Note

Design Intent

The design engineer stops by and notices that the 1.50-inch dimension is not important. Rather the ½-inch thickness of the outer hub is important. Modify the detailed drawing to show this.

Step 26: *Pick* the ⬚Annotate⬚ tab.

Step 27: *Pick* the 1.50-inch dimension in the front view, and then press the <Delete> key.

Step 28: *Select* the **Dimension** icon. Add the dimension for the thickness of the outer hub and position it at the first upper guideline in the front view.

Step 29: *Save* the drawing. *Refit* the drawing to the screen.

Step 30: **File>Print>Print.** *Pick* the **Preview** icon. Your drawing should look similar to Figure 8-36. *Pick* **Print** to prove you have completed the Sectioned View portion of this chapter. *Pick* **Close Print Setup.**

Step 31: **File> Manage File> Delete Old Versions,** then *pick* the **Yes** button. **File>Close.** *Open* four_hole_hub.prt.

Step 32: **File> Manage File> Delete Old Versions.** Pick the Yes button to delete any old versions of the 3D model.

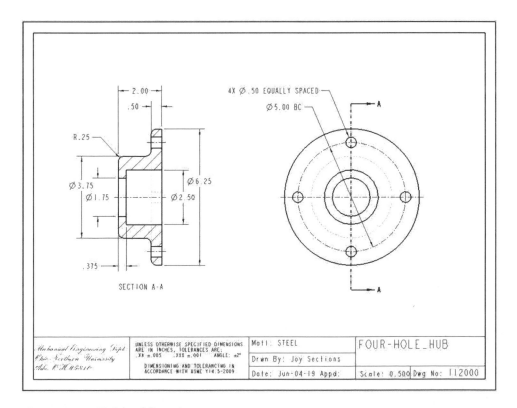

Figure 8-36 Finished Drawing

Auxiliary Views

Many parts have shapes where their principal faces are not aligned with the front, top, or right side views. If we draw the three orthographic views of the "rod guide" (Figure 8-37), the upper protrusion does not appear true size or shape in any of the three views. In order to show the true circular shape of the upper protrusion, it is necessary to assume a direction of sight perpendicular to the plane of its

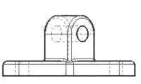

Figure 8-37 Rod Guide

curves. The resultant view is referred to as an auxiliary view. The top view and this auxiliary view will completely describe the part. In this case, the front and right side views are not needed.

Step1: Create a new part and name it "rod_guide". Use the default template.

Step 2: Create the base sketch shown in Figure 8-38 in the TOP datum plane, and then extrude it upward 0.69 inches.

Step 3: 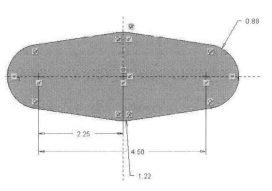 Use the **Datum Axis** tool under the Model tab to create an axis through the intersection of the RIGHT and FRONT planes, as shown in Figure 8-39. *Pick* the RIGHT plane, then hold down the <Ctrl> key and *pick* the FRONT plane. *Pick* **OK.**

Figure 8-38 Base Sketched

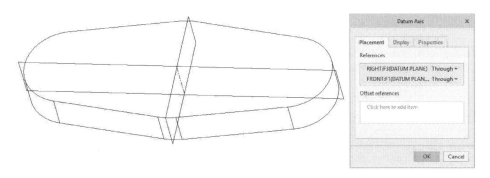

Figure 8-39 Add Datum Axis

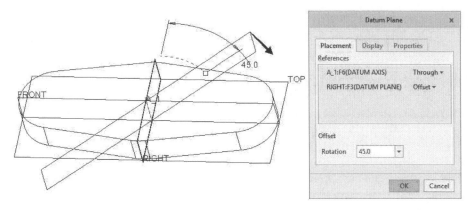

Figure 8-40   Add Datum Plane

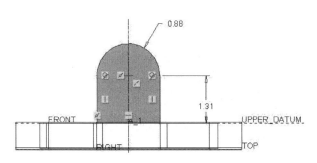

Figure 8-41   Sketched Upper Protrusion

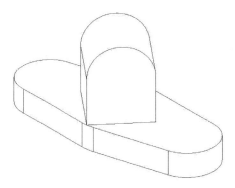

Figure 8-42   Upper Protrusion

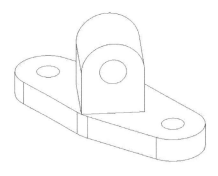

Figure 8-43   Holes Added

Step 4: ▱ Plane Use the ***Datum Plane*** tool to create a datum plane that goes through the datum axis and is 45 degrees from the RIGHT datum plane, as shown in Figure 8-40. *Pick* the Properties tab in the Datum plane window and name it "UPPER_DATUM." *Pick* **OK.**

Step 5: With the UPPER_DATUM highlighted, *pick* the ***Extrude*** tool. *Select* the ***Sketcher Setup*** icon, then orient the TOP datum plane toward the top. *Pick* **Sketch**. *Pick* the ***Reference*** icon in the upper left corner, then *select* the top surface of the base and the datum axis as sketcher references. *Pick* **Close**. Create a vertical construction centerline on top of the datum axis. *Pick* the ***Sketch View*** icon. Sketch the shape shown in Figure 8-41. Exit sketcher by *picking* the ***green checkmark***.

Step 6: Extrude the shape symmetric about the datum plane to an overall length of 1.70 inches. *Pick* the ***green checkmark***. See Figure 8-42.

Step 7: Use the ***Hole*** tool to place a 0.750-inch diameter through hole 1.31 inches above the base at the center of the rounded top. *Pick* the ***green checkmark***.

Step 8: Use the ***Hole*** tool to place a 0.625-inch diameter through hole 2.25 inches from the center of the base. Use the ***Pattern*** tool to create a second hole in the base 4.50 inches from the previous hole. See Figure 8-43.

Step 9: Use the *Round* tool to create 0.12-inch radius rounds on the two ends of the upper protrusion (Figure 8-44).

Step 10: Use the *Round* tool to create a 0.12-inch radius round on the top edge of the base protrusion (Figure 8-44).

Step 11: Use the *Round* tool to create a 0.12-inch radius round where the top protrusion meets the base protrusion. See Figure 8-45.

Step 12: *Save* the part. *Pick* **OK.**

Step 13: **File> Manage File> Delete Old Versions,** then *pick* the *Yes* button.

Step 14: **File>Close.**

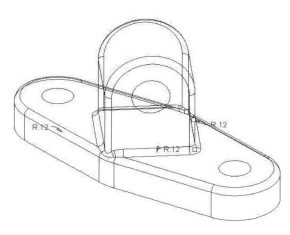

Figure 8-44　　Rounds Added

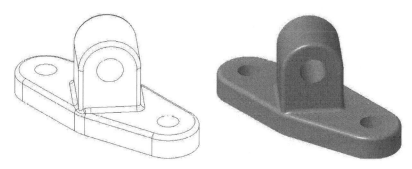

Figure 8-45　Completed Part

Design Intent

Create a detailed engineering drawing of "rod guide" and, because of the rotated upper protrusion, create an auxiliary view. This part can also be found on the publisher's website.

Now we will create the detailed engineering drawing for this part starting with the A-size 2-views template we created in the previous chapter.

Step 15: *Select* the *New* icon to create a new object. The shortcut keystroke is <Ctrl>-N.

Step 16: *Select* **Drawing** from the window and name it "rod_guide". Make sure the default template is checked, and then *pick* **OK.**

Step 17: Make sure "rod_guide" is the default model. If it is not, *pick* the **Browse...** button, then *select* the "rod_guide".

Step 18: *Pick* the **Browse...** button inside the Template area. *Pick* **Working Directory** from the left side of the open window. *Select* "a_template_2-views.drw" from the directory. *Pick* the **Open** button. *Pick* the **OK** button in the New Drawing window.

Step 19: At the top of the window, type "1018 CD STEEL" for the material, then *pick* the **green checkmark**. Enter your name, then press the <Enter> key. (This is the same as *picking* the **green checkmark**.) Enter the drawing number, "123001" <Enter>.

Step 20: After the drawing appears (Figure 8-46) *pick* the right side view, then press the <Delete> key.

Step 21: *Double-click* on the front view to bring up the Drawing View window. Change the view name to "TOP". *Pick* TOP from the list in the Model view names area (Figure 8-47).

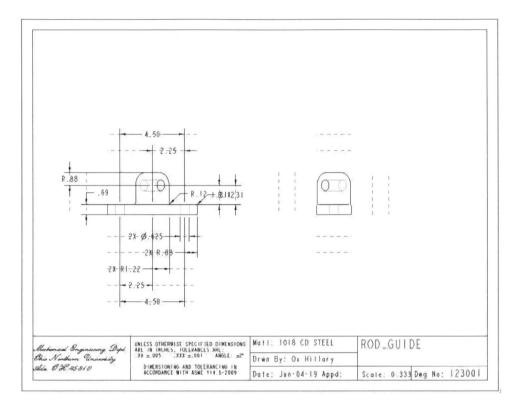

Figure 8-46 Initial Drawing

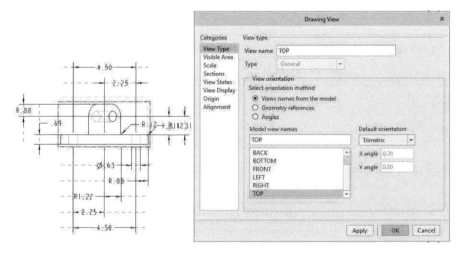

Figure 8-47 Change Views

Step 22: *Pick* the **Apply** button. When the confirmation window appears, *pick* **Yes** to modify the view and **Yes** to delete all highlighted dimensions. See Figure 8-48.

Step 23: *Pick* **OK** to close the Drawing View window.

Step 24: *Select* the TOP view and drag it into the upper left corner of the drawing. Unlock the view if necessary.

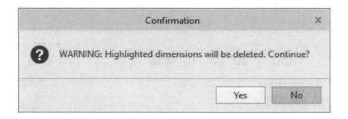

Figure 8-48   Confirmation Window

Step 25: 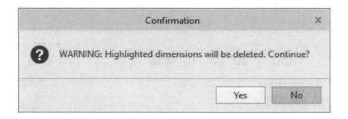 *Select* the *Auxiliary view* icon in the Layout tab.

Step 26: *Pick* the front edge of the upper rotated protrusion. When you do, a rectangular box appears on the screen. You can only move this box perpendicular to the selected line. *Drag* this box into the lower area of the drawing. *Press* the LMB to place it.

Step 27: *Double-click* on this new view to bring up the Drawing View window. *Select* the **View Display** category. *Select* **Hidden** from the pull-down menu for the display style. *Pick* **Apply.** *Pick* **OK.** See Figure 8-49.

Step 28: *Select* the Annotate tab. *Double-click* on the actual scale value in the lower left corner of the screen. Type "0.50" <Enter> to change the drawing scale to half.

Step 29: *Pick* the word **Edit▼** at the bottom of the Annotate ribbon, to display a pull-down menu. *Select* the **Create Snap Lines.** In the Menu Manager window, *select* **Offset Object.** You will add 2 or 3 snap lines to the auxiliary view now.

Step 30: *Pick* the bottom line of the "guide_rod" in the auxiliary view. *Press* the MMB. Type "0.5" <Enter> for the answer to the question, "Enter the distance of the first snap line from the reference point. Type "3" <Enter> for the answer to the question, "Enter the number of snap lines to create." Type "0.375" <Enter> for the answer to the question, "Enter the distance between snap lines."

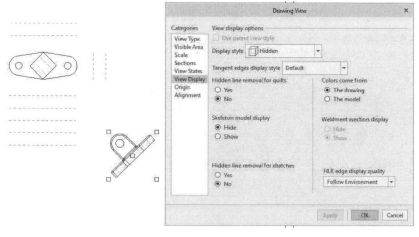

Figure 8-49   Auxiliary View

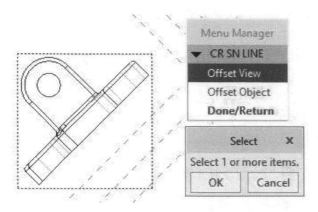

Figure 8-50   Add Snap Lines

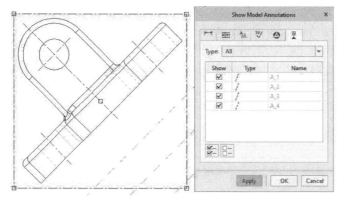

Figure 8-51   Add Centerlines

Step 31: *Pick* the right edge line of the upper protrusion of the "guide_rod" in the auxiliary view with LMB. *Press* the MMB. Type "0.5" <Enter> for the answer to the question, "Enter the distance of the first snap line from the reference point." Type "2" <Enter> for the answer to the question, "Enter the number of snap lines to create." Type "0.375" <Enter> for the answer to the question, "Enter the distance between snap lines."

Step 32: *Select* **Done/Return** from the Menu Manager window. See Figure 8-50.

Step 33: *Pick* the auxiliary view, and then *select* the **Show Model Annotations** tool. *Select* the far right tab in the Show Model Annotations window. *Check* each of the centerline checkboxes, then *pick* **OK.** See Figure 8-51.

Step 34: *Pick* the TOP view, and then *select* the **Show Model Annotations** tool from the top of the screen. *Select* the far right tab in the Show Model Annotations window. *Check* each of the centerline checkboxes, then *pick* **OK.**

Step 35: Zoom In on the view so you can *select* the small centerlines at the middle of the view. Extend the horizontal centerline so that it meets the two holes' centerlines. Extend the vertical centerline so it goes through the entire part. See Figure 8-52.

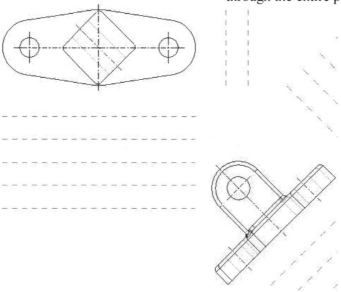

Figure 8-52   Extend Centerlines

Step 36: *Pick* the auxiliary view, and then *select* the ***Show Model Annotations*** tool again. *Select* the far left tab in the Show Model Annotations window. *Pick* the appropriate dimensions on the view, then *pick* **OK.** See Figure 8-53.

Step 37: *Pick* the TOP view, and then *select* the ***Show Model Annotations*** tool again. *Select* the far left tab in the Show Model Annotations window. *Pick* the appropriate dimensions on the view, then *pick* **OK.** See Figure 8-53.

Step 38: *Select* the ***Dimension*** tool. Add the 45-degree angle between the upper protrusion's centerline and the horizontal centerline of the base. Press the MMB to exit from dimensioning mode. Add "2X " to the "R.88", "R1.22" and .625 diameter dimensions in the top view. See Figure 8-53.

Step 39: $\boxed{\text{A}_{\equiv} \text{ Note}}$ *Select* the ***Note*** tool. From the pull-down menu, *pick* **Unattached Note.** With ***Select A Free Point On The Drawing*** highlighted, use the LMB and *pick* a point in the lower left corner of the drawing area. Type, "ALL FILLETS AND ROUNDS R.12" *Press* the MMB to terminate the note. Your drawing should appear similar to Figure 8-53. If it doesn't move the appropriate views, dimensions, or notes.

Step 40: *Pick* the ***Save*** icon. *Pick* **OK.**

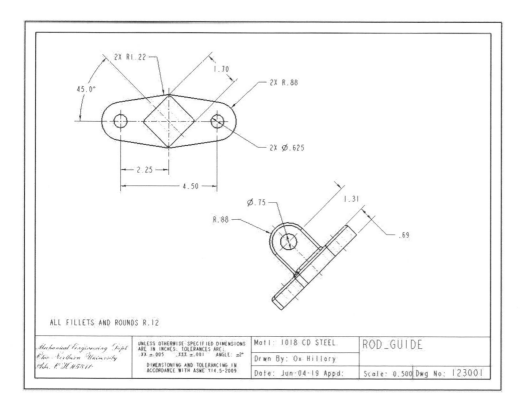

Figure 8-53 Finished Drawing

Step 41: File> Manage File> Delete Old Versions. *Pick* the *Yes* button.

Step 42: File>Print>Print. *Pick* the *Preview* icon. Your drawing should look similar to Figure 8-53. *Pick **Print*** to make a hard copy of the detailed engineering drawing to prove you completed the Auxiliary View portion of this chapter. *Pick **Close Print Setup**.*

▶ Engineering Drawings Exercise

Design Intent
Create three orthographic views with the appropriate dimensions of the clamp base shown in Figure 8-54. Section the front view by cutting through the three holes in the top view. The part file is called, "Clamp base." Create this part or get this part from the publisher's website under the same name. The origin is at the intersection of the middle of the lower left undercut and the bottom plane. (Be sure clamp_base.prt is in your working directory.)

Step 1: Start Creo Parametric by *double-clicking* with the LMB on the ***Creo Parametric*** icon on the desktop, or from the Program list: Creo Parametric.

Step 2: Set your working directory, by *selecting* the ***Select Working Directory*** icon in the Home ribbon. Locate your working directory, and then *pick* **OK.**

Step 3: Create a new drawing by *picking* the ***New*** icon at the top of the screen, *or by selecting* **File>New.**

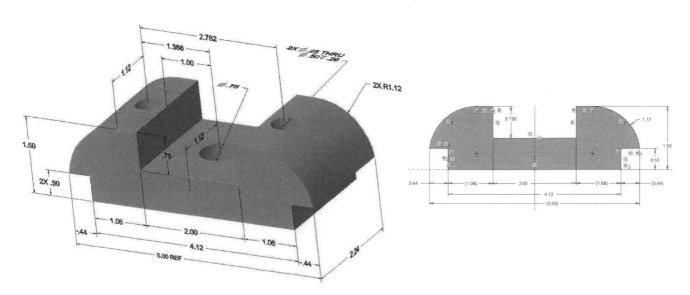

Figure 8-54   Clamp Base

Step 4: *Select* **Drawing** from the window and name it "clamp_base." Make sure the default template is checked, and then *pick* **OK.**

Step 5: Make sure "clamp_base" is the default model. If it is not, *pick* the **Browse...** button, then *select* the "clamp_base".

Step 6: *Pick* the **Browse...** button inside the Template area. *Pick* Working Directory from the left side of the open window. *Select* "a_template_ 3-views.drw" from the directory. *Pick* the **Open** button. *Pick* the **OK** button in the New Drawing window.

Step 7: At the top of the window, type "1018 CD STEEL" for the material, then *pick* the **green checkmark.** Type your name, then press the <Enter> key. (This is the same as *picking* the **green checkmark.**) Type the drawing number, "197000" <Enter>.

Step 8: Turn off Plane display, Axis display, Point display, and Csys display.

Step 9: *Pick* the **Repaint** icon to update the display. Your drawing may look similar to Figure 8-55.

Because the drawing has so many dimensions on it, it is hard to decide what dimensions are needed and where they should go. In this case, it would be better to start with no dimensions showing on the drawing.

Step 10: *Select* the Annotate tab. In the lower right corner of the page, change the search filter to Annotation. Draw an imaginary box around the drawing area to select most of the dimensions, then delete them by pressing

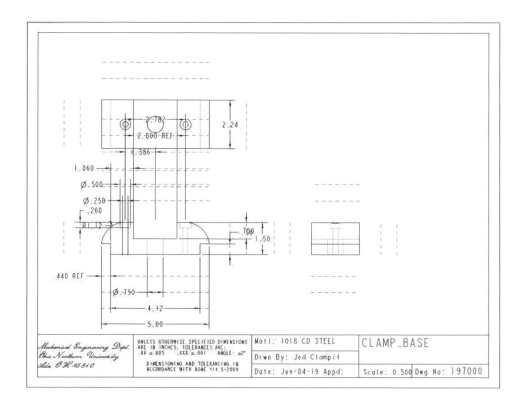

Figure 8-55 Initial Drawing

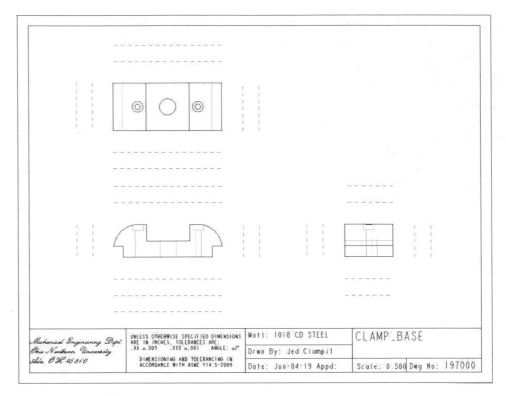

Figure 8-56   Add Snap Lines

<Delete>. Move any dimensions that are in the title block area up into the drawing area, then delete them. Change the search filter back to **General**.

Step 11: Highlight and delete fourth, fifth, etc., snap lines because they are not needed. Position the 3 views similar to Figure 8-56.

Step 12: *Select* the FRONT view so there is a dashed box around it. *Select* the ***Show Model Annotations*** icon from the top of the screen. *Select* the far right tab in the Show Model Annotations window. *Pick* one at a time the three vertical centerlines in the front view. They will change color and their appropriate checkbox will be checked in the Show Model Annotations window. *Pick* **OK** to save the changes.

Step 13: Repeat this procedure for the TOP view and its three centerlines.

Step 14: Repeat this procedure for the RIGHT view. *Select* just the vertical centerline that goes all the way through the right side view. Leave the other centerlines unchecked.

Step 15: In the top view extend the center horizontal centerline so it touches the centerlines of the two outer circles. Extend the vertical centerline so it goes through the entire top view. See Figure 8-57.

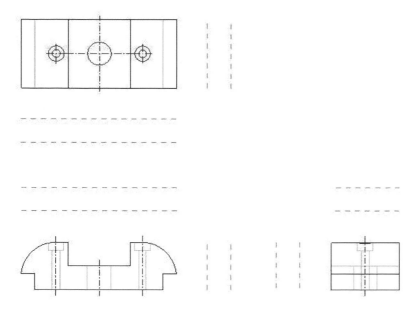

Figure 8-57 Add Centerlines

Step 16: *Select* the ⌐Layout¬ tab. *Double-click* on the front view to bring up the Drawing View window. *Select* the **Sections** category. *Pick* 2D cross-section. *Pick* the ***plus sign***. Turn on the **Plane Display** by *picking* its icon in the graphics toolbar.

Step 17: With **Planar** and **single** highlighted in the Menu Manager window, *pick* **Done.**

Step 18: Type "A" <Enter> for the name of the cross-section.

Step 19: With **Plane** highlighted in the menu manager, *pick* the FRONT datum plane in the TOP view or from the drawing tree, then *pick* the **Apply** button in the Drawing View window. See Figure 8-58. *Pick* **OK.**

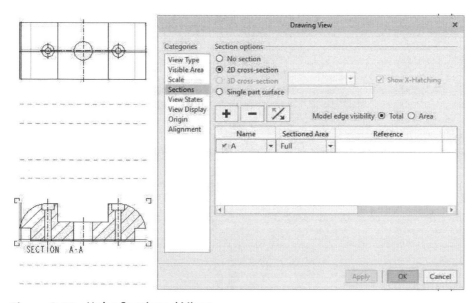

Figure 8-58 Make Sectioned View

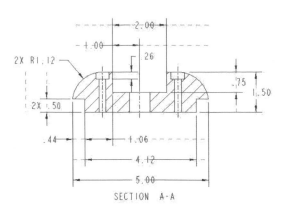

Figure 8-59   Sectioned View

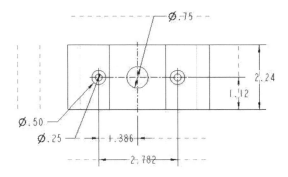

Figure 8-60   Cleanup Dimensions

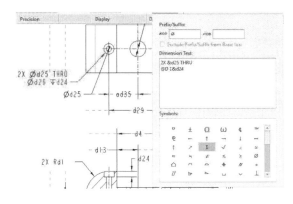

Figure 8-61   Counterbore Note

Step 20: *Select* the ⬚Annotate⬚ tab. *Drag* the words SECTION A-A down away from the front view for now. *Pick* the ***Plane Display*** icon again to turn off datum planes.

Step 21: *Pick* the FRONT view. *Pick* the ***Show Model Annotations*** icon. *Select* the first tab in the Show Model Annotations window. Using the LMB *pick* the dimensions that you want to show up in the front view. *Pick* **OK.**

Step 22: Rearrange the dimensions in the front view similar to Figure 8-59.

Step 23: Rearrange the dimensions in the top view similar to Figure 8-60. Flip the arrowheads on the counterbore dimension in the top view. (Don't be concerned about the arrowheads for the 0.25-inch diameter hole at this time.) Move the FRONT or TOP view if you need more room between the views. If the views are locked, *pick* a view, then press RMB and hold until a pop-up menu appears. *Select* **Lock View Movement** to unlock the views. Repeat this procedure to lock the views again.

Step 24: If necessary, *pick* the 1.39-inch dimension and change the number of decimal places to 3 so it reads 1.386. Do the same with the 2.78-inch dimension (2.782). Flip the arrowheads on the 1.386 dimension if necessary.

Step 25: *Pick* the 0.50 diameter counterbore dimension in the top view. Under the ⬚Annotate⬚ tab, *select* the word **Format ▼** with the LMB to show a pull-down menu. *Select* **Switch Dimensions.** Look at the names for small diameter hole and the depth of the counterbore. In my case, d25 and d24. Your variable names may be different. Modify the text accordingly as shown in Figure 8-61. (Placing an "&" in front of text makes the text a variable name.)

Step 26: Under the ⬚Annotate⬚ tab, *select* the word **Format ▼** with the LMB to show a pull-down menu again. *Select* the **Switch Dimensions** to switch the dimension symbol back to their values.

Step 27: Highlight the dimension for the 0.25-inch diameter hole (d25) in the top view and the 0.26-inch depth (d24) of the counterbore hole in the front view. Press <Delete> to remove them from the drawing. (We used these variable names in the previous step, but no longer need them on the drawing.)

Step 28: Select the Layout tab. *Pick* the **Arrows** icon. *Pick* the sectioned FRONT view with the LMB. *Pick* the RIGHT view with the LMB to display the cutting plane line in the RIGHT view.

Step 29: Unlock the views, and then reposition them similar to Figure 8-62. Lock the views again.

Step 30: Save the drawing. *Pick* **OK**.

Step 31: File>Print>Print. Pick the Preview icon. Your drawing should look similar to Figure 8-63. *Pick **Close Print Setup***.

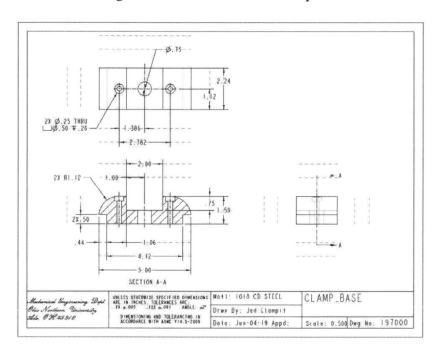

Figure 8-62   Current Drawing

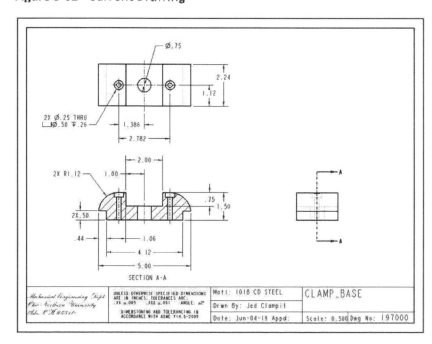

Figure 8-63   Final Drawing

Let's add a 3D view of the part in the upper right corner of the drawing.

Step 32: *Select* the ⎡Layout⎤ tab. *Select* the **General** view icon. *Pick* **OK** in the Select Combined State window. Using the mouse, *pick* a point in the upper right corner of the drawing. *Select* **3D_VIEW** (or **Default Orientation** if 3D_VIEW is not shown) for the Model view name. *Pick* **Apply.** *Select* **View Display** from the categories list. *Select* **No Hidden** for the display style. *Select* **None** for the Tangent edges display style. *Pick* **Apply.** *Pick* **OK** to close the Drawing View window.

Step 33: *Save* the drawing.

Step 34: **File> Manage File> Delete Old Versions,** then *pick* the *Yes* button.

Step 35: **File>Print>Print.** *Pick* the **Preview** icon. Your drawing should look similar to Figure 8-64. *Pick* **Print** to prove you have completed this exercise. *Pick* **Close Print Setup**.

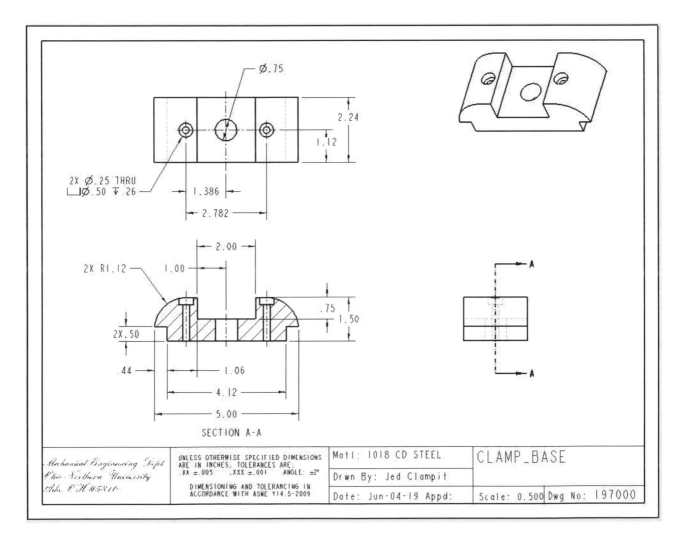

Figure 8-64   Final Drawing with Added 3D View

▶ Review Questions

1. Is it possible to create a 2D drawing of a part without creating the part file first?

2. The first view added to an engineering drawing is called what?

3. What is the easiest way to move a view on the drawing sheet?

4. When you move a view on a drawing sheet, does everything move with it?

5. Is it possible to delete a view created by a drawing template without affecting the drawing template?

6. How do you edit the text in a general note?

7. Describe two ways to move a dimension in one view to another view.

8. What happens if you change the value of a dimension on a part drawing?

9. What is the difference between a shown dimension and a driven dimension?

10. Is it possible to add new features when you are in drawing mode?

11. Is it possible to change the size of features when you are in drawing mode?

12. How would you turn off hidden lines in a section view?

13. How is design intent shown in an engineering drawing? How does design intent relate back to the part?

14. What symbol is used in a note to tell Creo Parametric to display the value of a variable?

15. How do you change drawing options?

16. How do you add drawing symbols to a general note or a dimension?

Engineering Drawings Problems

8.1   Create 3 orthographic views of the "**lower_plate**" shown in Figure 8-65. The 0.56-inch diameter hole is located 1.40 inches from the bottom of the part. All rounds are 0.04 inches in radius. The part is symmetric about the center hole. Part made from 1020 CD steel. The drawing number is 100081. All dimensions are in inches.

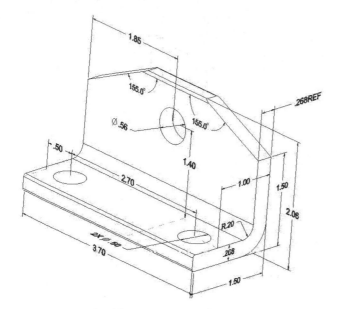

Figure 8-65   Problem 1—Lower Plate

8.2   Create the part, then create 2 orthographic views of the "**front_plate**" shown in Figure 8-66. Part made from 1018 CD steel. The drawing number is 100082. There is a 1.75-inch diameter cutout 5/16-inch deep on the back side. Use a bolt circle dimension to locate the small holes. All dimensions are in inches.

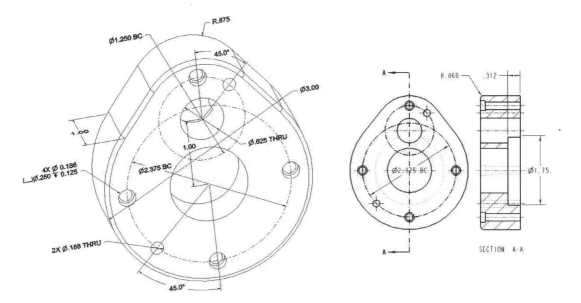

Figure 8-66   Problem 2—Front Plate

8.3 Determine the number of orthographic views necessary for the **"step_shaft"** shown in Figure 8-67 and then create the appropriate drawing. Part made from 1040 CD steel. The drawing number is 100083. All dimensions are in inches.

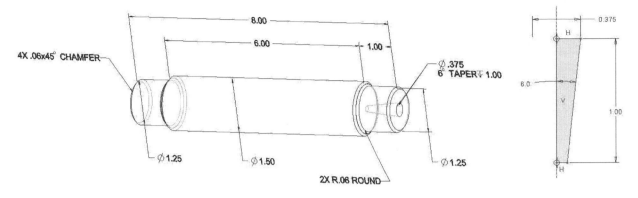

Figure 8-67   Problem 3—Step Shaft

8.4 Create 3 orthographic views of the **"small_cone_clutch"** shown in Figure 8-68 with the front view being a section view. Part made from 1020 CD steel. The drawing number is 100084. All dimensions are in inches.

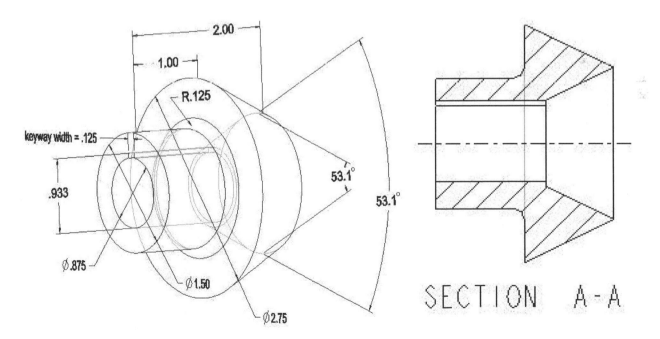

Figure 8-68   Problem 4—Small Cone Clutch

8.5 Determine the number of orthographic views necessary for "**cutoff_holder**" shown in Figure 8-69 and then create the appropriate drawing. The horizontal hole must be 2.094 inches above the top of the flat cutout and in line with the vertical hole. Part made from cast iron. The drawing number is 100085. All dimensions are in inches.

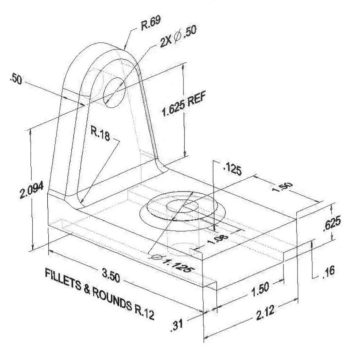

Figure 8-69   Problem 5—Cutoff Holder

8.6 Determine the number of orthographic views necessary for "**anchor_bracket**" shown in Figure 8-70 and then create the appropriate drawing. Part made from 1030 CD steel. The drawing number is 100086. All dimensions are in inches.

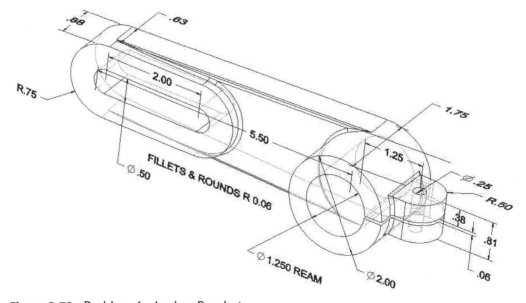

Figure 8-70   Problem 6—Anchor Bracket

8.7 Determine the number of orthographic views necessary for the **"Three-hole_support"** shown in Figure 8-71 and then create the appropriate drawing. All dimensions are in inches. Add a general note to break all sharp edges. Part made from 1018 HR steel. The drawing number is 100087. Note that this part is similar to the "Bearing_Holder", problem 4.8 so you could start with that part, modify it to create this part, then save it as "Three-hole_support". All dimensions are in inches.

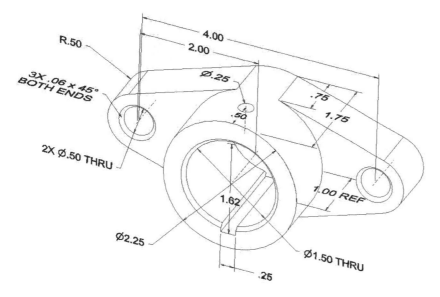

Figure 8-71 Problem 7—Three-Hole Support

8.8 Determine the number of orthographic views necessary for "**index_feed**" shown in Figure 8-72 and then create the appropriate drawing. Part made from 1020 CD steel. The drawing number is 100088. **Design change:** 0.750-inch protrusion must be 0.744 inches. All dimensions are in inches.

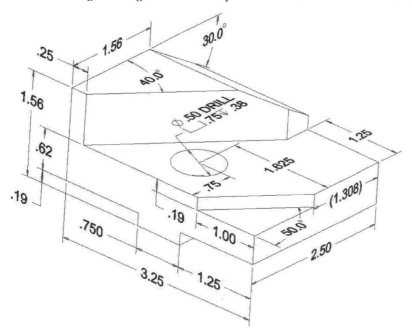

Figure 8-72 Problem 8—Index Feed

8.9 Determine the number of orthographic views necessary for the **"step_shaft_mm"** shown in Figure 8-73 and then create the appropriate drawing. Part made from 1040 CD steel. The drawing number is 100890. All dimensions are in millimeters.

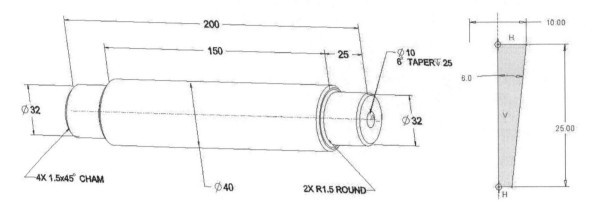

Figure 8-73 Problem 9—Step Shaft (metric)

8.10 Determine the number of orthographic views necessary for the **"Three-hole_brace_mm"** shown in Figure 8-74 and then create the appropriate drawing. Add a general note to break all sharp edges. Part made from 1020 HR steel. The drawing number is 100810. All dimensions are in millimeters.

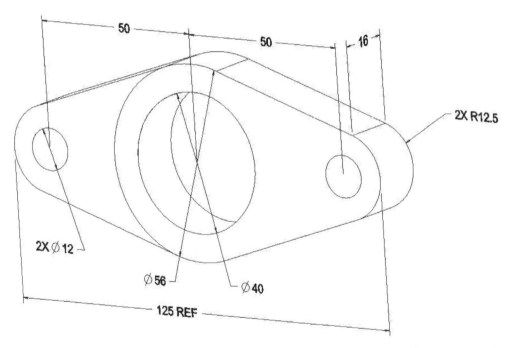

Figure 8-74 Problem 10—Three-Hole Brace (metric)

8.11 Determine the number of orthographic views necessary for **"index feed_mm"** shown in Figure 8-75 and then create the appropriate drawing. Part made from 1020 CD steel. The drawing number is 100811. All dimensions are in millimeters.

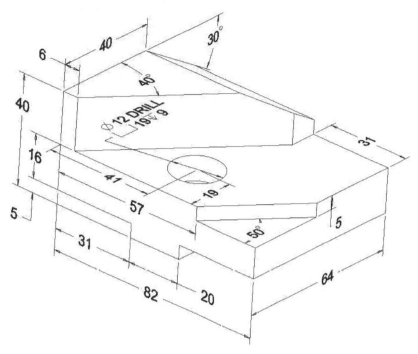

Figure 8-75   Problem 11—Index Feed (metric)

8.12 Create an engineering drawing with the appropriate auxiliary view for the **"control_block_mm"** shown in Figure 8-76. Part made from 1018 CD steel. The drawing number is 100812. All dimensions are in millimeters.

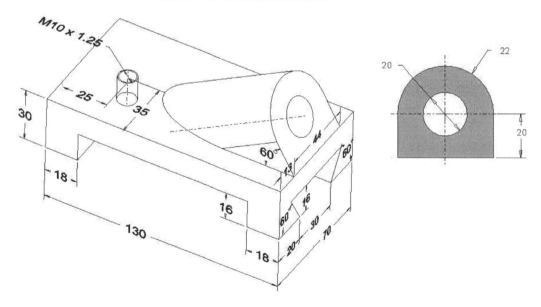

Figure 8-76 Problem 12—Control Block (metric)

8.13 **Challenge problem** - Create an engineering drawing of the **"tension_bracket"** shown in Figure 8-77. The part is symmetrical about its centerline. Part made from 1018 CD steel. The drawing number is 100813. Convert all fractional dimensions to 2-place decimals. After testing the design the engineer noted that a small crack had formed in the sharp underneath corner so he/she has decided to place a 1/4-inch round in this corner (not shown). Modify your part and drawing accordingly. After further testing, the engineer has decided to increase the horizontal plate thickness from 7/16-inch to 1/2-inch. Also, add a general note to break all sharp edges. All dimensions are in inches.

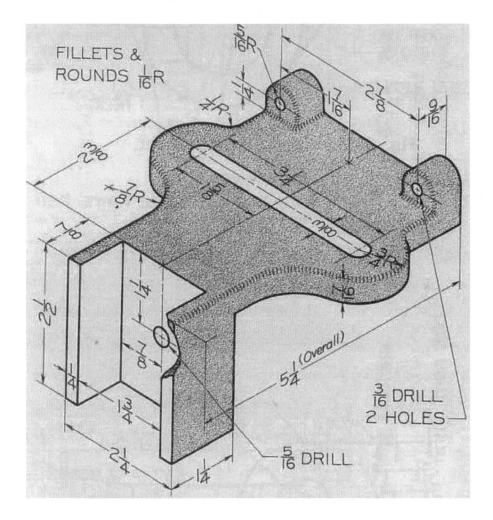

Figure 8-77 Problem 13 and 14—Tension Bracket 207

8.14 **Challenge problem** - Create an engineering drawing of the **"tension_bracket_mm"** shown in Figure 8-77 above. The part is symmetrical about its centerline. Part made from 1018 CD steel. The drawing number is 100814. Convert all fractional inch dimensions to 1-place millimeter decimals. After testing the design engineer noted that a small crack had formed in the sharp underneath corner so he/she has decided to place a 6-mm round in this corner (not shown). Modify your part and drawing accordingly. After further testing, the engineer has decided to increase the horizontal plate thickness from 7/16-inch to 12 mm. Also, add a general note to break all sharp edges. All dimensions were in inches.

ASSEMBLIES

Objectives

- ▶ Create subassemblies
- ▶ Create fixed assemblies
- ▶ Create exploded assemblies
- ▶ Add and reposition parts in an assembly

- ▶ Understand and use assembly constraints
- ▶ Create moveable assemblies
- ▶ Understand the bidirectional associative features of parts

▶ Assemblies Explored

An assembly drawing shows the assembled structure, mechanism, or machine with all the parts in their functional locations. A drawing showing only part of a larger machine is referred to as a subassembly drawing. The views selected for an assembly drawing must keep in mind the purpose of the drawing. That is, to show how the parts fit together and to suggest how the device functions. Its purpose is not to describe the shape and size of the parts—the detailed part drawings are used for that; thus, assembly drawings show the relationship among the parts. Since parts can fit into each other or overlap sectioned views are encouraged, it is necessary not only to show the cut surfaces, but there must be a distinction between the different parts. Cast iron general-purpose section lines are recommended for assembly drawings. It is acceptable to use the actual material's symbolic section lines. Thin parts such as gaskets should be filled in solid instead of drawn with section lines. Solid round parts in the cutting plane should be shown not sectioned, or "in the round." These parts include screws, nuts, pins, keys, ball bearings, roller bearings, gears, and shafts. Hidden lines are typically omitted since they tend to confuse rather than clarify. Dimensions are not given on assembly drawings unless they are added to show the overall height or width, or the maximum movement between two moveable parts.

Parts are identified on an assembly drawing using circles containing part numbers and leader lines that connect the circle to the actual part. The circled part numbers should be arranged in a neat clockwise or counter-clockwise fashion, not scattered all over the drawing. Leader lines should never cross each other and adjacent leader lines should be approximately parallel. A bill of materials must be provided with the assembly drawing that shows the part number, the material specified, the number of items required, and a description that may include its stock or pattern number. The bill of materials may be on a separate sheet.

The previous chapters have covered the fundamentals of creating standard parts and detailed drawings. This chapter discusses how to create and modify assembly drawings. The assembly will reflect any changes made to the assembled parts automatically whether the changes are made in the 3D assembly, the detailed part drawing, or the 3D part model. The bidirectional associative functionality in Creo Parametric allows for very flexible model modifications. Many parallels exist between part design and assembly creation.

The relationships between parts in an assembly need to be discussed next. The best way to assemble a machine or structure is the same way it would be assembled in the real world. A large assembly should be made up of subassemblies and joining parts. Subassemblies (or groups of parts) are treated the same way as individual parts in Creo Parametric. Placement constraints create parent/child relationships that allow us to capture the design intent. The component that is placed becomes a child of the already existing assembled parts.

A spur gear is to be pressed onto a step shaft. User-defined constraints are shown in Figure 9-1. They are Distance, Angle Offset, Parallel, Coincident, Normal, Coplanar, Centered, Tangent, Fix, and Default. Automatic mode selects the appropriate constraint based upon the selection of the component and assembly picked entities.

Besides the user-defined placement constraints, there are constraints that reflect real-world operations. See Figure 9-2. These user-defined constraints include Rigid, Pin, Slider, Cylinder, Planar, Ball, Weld, Bearing, General, 6DOF (6 degrees of freedom), Gimbal, and Slot.

Figure 9-1   User-Defined Placement Constraints

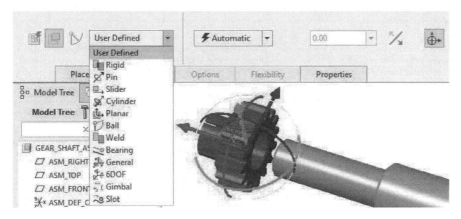

Figure 9-2   Other Placement Constraints

User-Defined Constraints

Coincident—Replaces insert, align, and mate from previous versions of Pro/ENGINEER Wildfire. It is the most common type of constraint. Selected datum axes become coaxial. For example, a spur gear is inserted onto a step shaft. See Figure 9-3. If the spur gear's hole is smaller than the shaft diameter, it would be a press fit. If

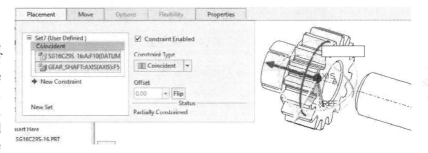

Figure 9-3   Coincident Axes

the hole is larger than the shaft diameter, it would be a running or loose fit. The mating surfaces do not need to touch and the cylindrical surfaces do not have to be complete 360° cylinders. With just this constraint the inserted part is only partially constrained. It still has two degrees of freedom, rotation about the selected axis and translation along the axis.

Coincident—Selected surfaces are made coplanar. Mating the left wall of the step shaft at the change in diameter with the right surface of the spur gear places the spur gear in the appropriate position. See Figure 9-4. This constraint removes the translation of the spur gear along the step shaft's axis, but still allows the spur gear to rotate about its axis as indicated by the red line circle.

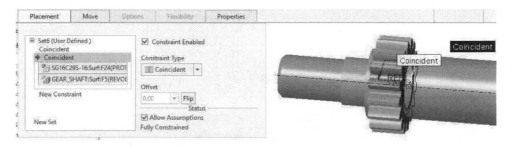

Figure 9-4   Coincident Surfaces

Figure 9-5   Distance

Distance—Selected surfaces are offset by a specified plus or minus distance. The offset distance can be modified at any time. If the design intent was for the spur gear to be located 0.75 inches from the step shaft's change in diameter, this could be done by using the distance constraint (Figure 9-5) and a value of 0.75 inches. A negative value would move the spur gear toward the right and into the large diameter of the step shaft.

Note that *picking* the **Flip** button flips the orientation of the spur gear on the shaft. In this case, the small hub of the spur gear is now toward the large diameter of the step shaft. See Figure 9-6. If the distance was less than 0.75 inches the small hub of the spur gear would be inside the large diameter of the step shaft.

Coincident—Selected datum planes are made coplanar. Aligning the TOP datum plane of the spur gear with the TOP assembly datum plane stops the spur gear from rotating about the step shaft, and fully constrains the spur gear relative to the step shaft. See Figure 9-7.

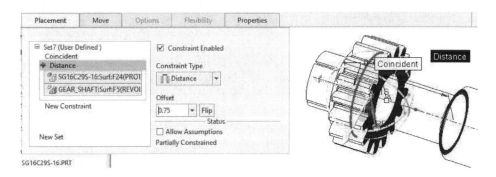

Figure 9-6   Distance Flip

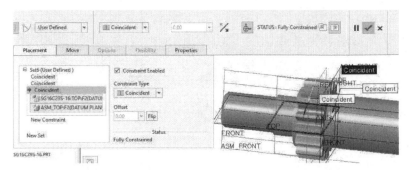

Figure 9-7   Coincident Datum Panes

The other user-defined placement constraints are briefly listed here.

▶ **Angle Offset**—Offsets the component reference from the assembly reference by the angle value entered in the Offset Input box.

▶ **Parallel**—Makes component reference oriented parallel to the assembly reference, but not coplanar.

▶ **Normal**—Offsets the component reference from the assembly reference by 90 degrees so the two references are perpendicular to each other.

▶ **Coplanar**—Makes component reference oriented on the same plane and parallel to the assembly reference.

▶ **Centered**—Makes component reference centered to assembly reference.

▶ **Tangent**—Selected cylindrical surface is made tangent to another surface.

▶ **Fix**—Selected part is grounded at its current location. Degrees of freedom are zero.

▶ **Default**—Selected part is placed with its existing placement constraints.

Predefined Constraints

▶ **Rigid**—Selected part is connected so that it does not move relative to another part. They are constrained with any valid set of constraints. Components so connected become a single body. The rigid connection constraint is similar to a user-defined constraint.

▶ **Pin**—Selected part is connected to a referenced axis so that the component rotates or moves along this axis with one degree of freedom. Select an axis, edge, curve, or surface as an axis reference. Select a datum point, vertex, or surface as a translation reference. A Pin connection set has two constraints: axis alignment and planar mate or align or a point alignment.

▶ **Slider**—Selected part is connected to a referenced axis so that the component moves along the axis with one degree of freedom. Select edges or aligning axes as alignment references. Choose surfaces as rotation references. A Slider connection set has two constraints: axis alignment and planar mate/align to restrict rotation along the axis.

▶ **Cylinder**—Selected part is connected so that it moves along and rotates about a specific axis with two degrees of freedom. Select axes, edges, or curves as axis alignment references. A Cylinder connection set has one constraint.

▶ **Planar**—Selected part is connected so that they move in a plane relative to each other with two degrees of freedom in the plane and one degree of freedom around an axis perpendicular to it. Select Mate or Align surface references. A Planar connection set has a single planar mate or align constraint. The mate or align constraint may be flipped or offset.

▶ **Ball**—Selected part is connected so that it can rotate in any direction with three degrees of freedom (360° rotation). Select points, vertex, or curve ends for alignment references. A Ball connection set has one point-to-point alignment constraint.

▶ **Weld**—Selected part is connected to another so that it does not move relative to another. The component is placed in the assembly by aligning the coordinate system of the component with a coordinate system in the assembly. The component can be adjusted using the open degrees of freedom in the assembly. A Weld connection has one coordinate system alignment constraint.

▶ **Bearing**—A combination of the Ball and Slider connections with four degrees of freedom. There are three degrees of freedom (for 360° rotation) and movement along a referenced axis. For the first reference choose a point on the component or the assembly. For the second reference choose an edge, axis, or curve on the assembly or the component. The point reference can rotate freely about the edge and move along its length. A Bearing connection has one point-on-edge alignment constraint.

▶ **General**—Selected part is connected using one or two configurable constraints that are identical to those in a user-defined set. Tangent, point on curve, and point on nonplanar surface cannot be used for a General connection.

▶ **6DOF**—Does not affect the motion of the component in relation to the assembly because no constraints are applied. The coordinate system of the component is aligned to a coordinate system in the assembly. The X, Y, and Z assembly axes are motion axes allowing rotation and translation.

▶ **Gimbal**—A pivot joint. The coordinate system's centers are aligned, but not the axes to allow free rotation.

▶ **Slot**—This connection has four degrees of freedom, where the point follows the trajectory in three directions. Choose a point on the component or the assembly for the first reference. The referenced point follows the curved reference trajectory. The trajectory has endpoints that are set when the connection is configured. A Slot connection has a single point alignment to multiple edges or curves constraint.

Using Extra Constraints

You can add more constraints than are necessary to place the component in an assembly. This is called overconstraining. Even when the position of a component is completely constrained mathematically, you may want to specify additional constraints to ensure that the assembly follows your design intent. Although you can specify up to 50 constraints, PTC recommends a limit of 10.

▶ Assembly Practice

This practice session uses several of the different assembly connections to put together a simple swinging link assembly.

Design Intent

Assemble the swinging link assembly. The link must be able to rotate about the clevis pin and be held on by the retaining ring. See Figure 9-8.

Before beginning, be sure the following four parts are present in your working directory. You can create the parts or get them from your instructor. See Figures 9-9 through 9-13.

Figure 9-8   Complete Swinging Link Assembly

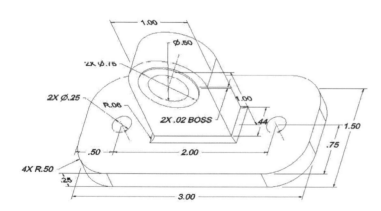

Figure 9-9   Base (base.prt)

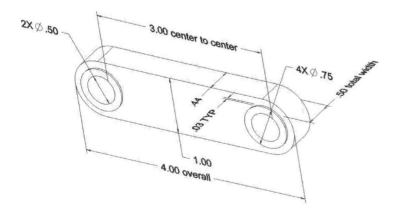

Figure 9-10   Link (link.prt)

Clip Included	A	B	C	E	F	G	H
No.	0.5000	0.625	0.110	0.046	0.388	1.218	1.005

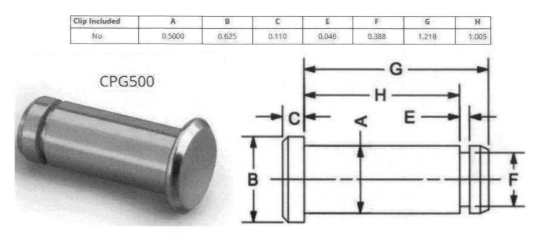

Figure 9-11   Manufacturer's Information for Clevis Pin

We need a 0.50-inch diameter clevis pin to connect the link to the base. Rather than manufacture one, we are going to search the Internet for an acceptable part. After doing an Internet search, we found this website:

Clevis Pin, Grooved—Midwest Control Products Corp.

Related Words: clevis pins, clevis pins for safety clips, grooved clevis pins, *http://midwestcontrol.com/part.php?id=4618*

On their website, we obtained the following information (Figure 9-11).

We will use this information to create a 3D model of the needed Grooved clevis pin (part number CPG500) which we will buy from Midwest Control Products. Note that the base and link width of this assembly totals 0.98 inches. This clevis pin has a usable length of 1.005 inches so it will work just fine. See Figures 9-11 and 9-12.

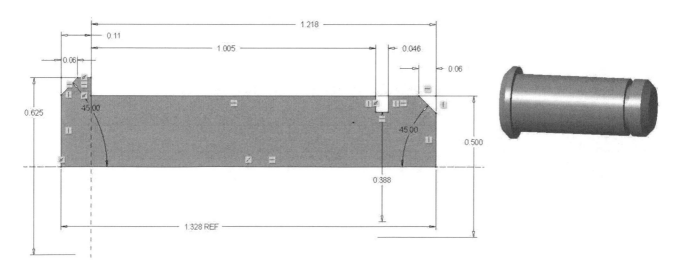

Figure 9-12   Clevis Pin (cpg500-clevispin.prt)

After selecting the clevis pin, we need to find a retaining ring that will slip into the clevis pin's groove. Research leads us to the e-rings from Fabory. Instead of creating this part in Creo Parametric, we are going to download the 3D model from the manufacturer's website and use it. See Figure 9-13. The procedure goes something like this:

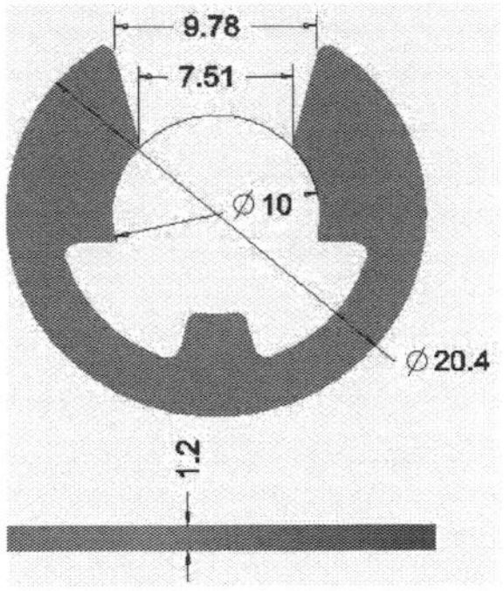

Step 1: Start Creo Parametric and set your working directory.

Step 2: *Pick* the **PARTcommunity 3D CAD manufacturer catalog** tab. See Figure 9-15.

Figure 9-13 Metric E-Style Retaining Ring (sdin6799.prt)

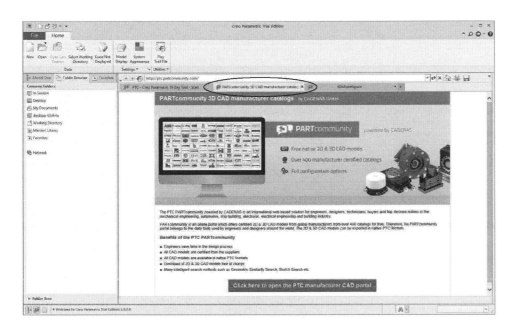

Figure 9-15 Access PARTcommunity 3D CAD manufacturer catalog

Figure 9-14 Retaining Ring on the Web Page

Step 3: *Pick* the red **Click here to open the PTC manufacturer CAD portal** button. See Figure 9-16.

Click here to open the PTC manufacturer CAD portal

Figure 9-16 Access manufacturer 's CAD Portal

Step 4: *Type* "e-clip" <Enter> in the search dialog box.

Step 5: *Pick* "DIN 6799 - Retaining washers for shafts, by Fabory" from the list of parts.

Step 6: Set the Short Description to "10 MM" using the pull-down list. *Pick* the red CAD Download button. Enter a valid email address and password after registering. *Check* the Accept license agreement box if asked. See Figure 9-17.

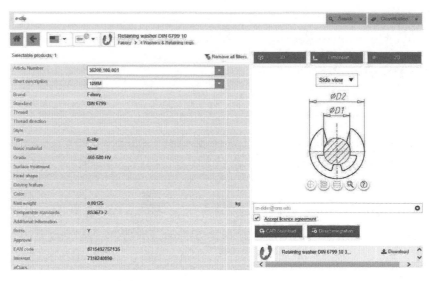

Figure 9-17   Set the E-ring diameter, D1, to 10 MM

Step 7: *Pick* the red **CAD download** button.

Figure 9-18   Select CAD download button

Step 8: Check the "Creo Parametric 2.0 (3D)" box in the list of formats.

Step 9: *Pick* the **Start Download** button.

Step 10: Scroll down and *pick* the ***Download*** icon. See Figure 9-19.

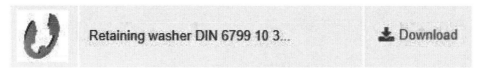

Figure 9-19   Select **CAD** Model Type

Step 11: Pick Save. See Figure 9-20.

Step 12: Pick Save again.

Step 13: Pick Close. Close all windows.

Step 14: Exit from Creo Parametric.

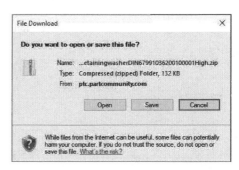

Figure 9-20   Save zipped part file

Step 15: Move the zipped folder to your working directory.

Step 16: Select the zipped folder. Press and hold the RMB until a pop-up menu appears. *Select* **Extract All...** Follow the procedure on your computer to extract the file.

Open the folder, rename the part file "sdin6799.prt", then move the part file to your working directory. See Figure 9-21. Next, verify your download by opening this file in Creo Parametric.

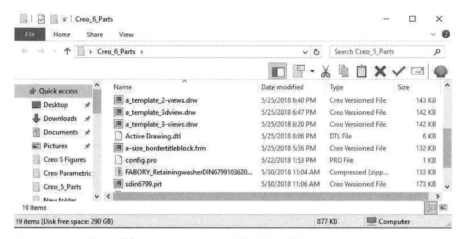

Figure 9-21   Part File moved to your Working Directory

Step 17:   Start Creo Parametric.

Step 18: Set the working directory to the directory where your part files are located.

Step 19: **File>Open.** *Select* the file "sdin6799.prt", and then *pick* the **Open** button to verify that the part file was downloaded correctly. See Figure 9-22.

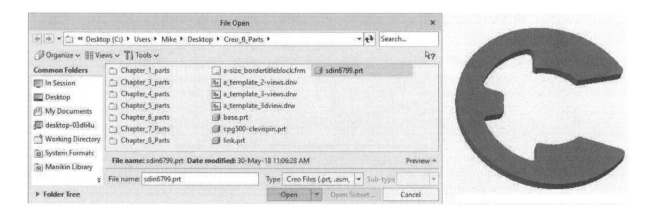

Figure 9-22   Open Downloaded E-Ring Part File

Figure 9-23   New Assembly Created

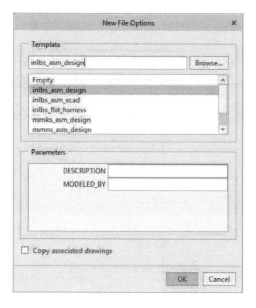

Figure 9-24   New File Options if Use Default Template is not checked

Step 20: File>Close.

Step 21: File> Manage Session> Erase Not Displayed. *Pick* **OK.**

Now let's begin creating the swinging link assembly.

Step 1: Create a new object by *selecting* the *New* icon with the LMB. Typing Ctrl-N, or **File>New** will do the same thing.

Step 2: Be sure **Assembly** and **Design** options are selected. Name this part "swinging_link". Be sure the Use default template box is checked, then *pick* **OK.** See Figure 9-23.

Step 3: If Use Default template is unchecked, then in the Template window, *select* **inlbs_asm_design**, then *pick* **OK.** See Figure 9-24.

Step 4: *Pick* the *Assemble* tool in the Model tab.

Step 5: *Select* "base.prt" from the Open window. *Pick* the Open button. *Set* the relation type to **Fix**, then *press* the MMB. The base is located in the assembly at its current position. See Figure 9-25. If **Empty** was selected, then no default coordinate system is present, and the first part's coordinate system becomes the default coordinate system for the assembly.

Step 6: *Pick* the Assemble tool in the Model tab again.

Step 7: *Select* "cpg500-clevispin.prt" from the Open window. *Pick* the **Open** button.

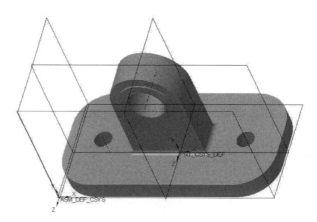

Figure 9-25   Base Part Placed

Step 8: Note the three colored circles and three colored arrows at the origin of the clevis pin. Rotate and translate to position the new part similar to Figure 9-26. *Pick* the red circle with the LMB, then *press down and hold* the LMB to rotate the clevis pin so its axis lines up with the base's big hole axis. *Pick* the ***red arrow***, then move the clevis pin over in line with the base's axis. *Pick* the ***green arrow***, then move the clevis pin up in line with the base's axis. *Pick* the ***blue arrow***, then move the clevis pin back behind the hole.

Step 9: *Select* the Placement tab near the top of the screen. *Pick* the axis of the clevis pin and *pick* the axis of the base hole. See Figure 9-26. Four of the six degrees of freedom will be removed.

Step 10: *Pick* **New Constraint** under the Placement tab. *Select* the underside of the clevis pin's head. *Pick* the raised surface near the base's hole. The constraint type will be distance. Change the constraint type from **Distance** to **Coincident.** See Figure 9-27.

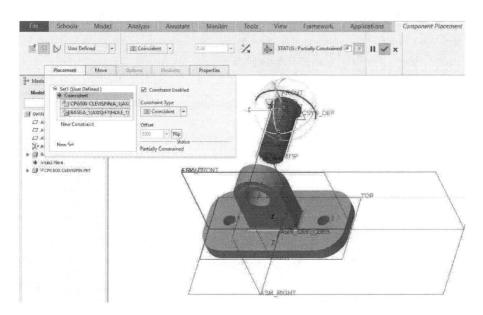

Figure 9-26 Coincident Axes of Base and Clevis Pin

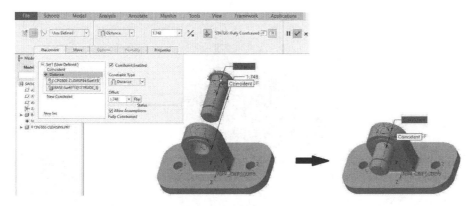

Figure 9-27 Coincident Clevis Pin with Base

With Allow Assumptions checked, the clevis pin is fully constrained. If we uncheck this assumption, then the clevis pin would be free to rotate and we would need to add one more constraint. <Ctrl><Alt> MMB allows you to move the pin without unchecking the Allow Assumptions box.

Step 11: *Pick* the ***green checkmark*** to accept the clevis pin placement.

Step 12: Highlight the clevis pin, then *select* **Appearance Gallery** in the Model tab. *Select* a yellowish shade from the appearance gallery at the top of the screen to color the clevis pin dark grey. See Figure 9-28.

Step 13: *Pick* the ***Assemble*** tool in the Model tab again.

Step 14: *Select* "link.prt" from the Open window. *Pick* the **Open** button.

Step 15: Above the Placement tab, change the connection type to **Pin.** If you do not use the pin connection, the link will not be able to rotate around the clevis pin. (Do not use Coincident or Automatic.)

Step 16: Use the red, green, and blue arrows to position the link similar to Figure 9-29.

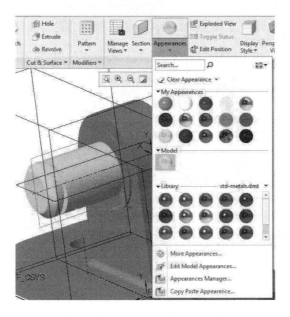

Figure 9-28 Color the Clevis Pin

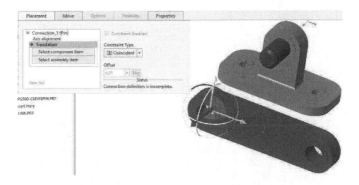

Figure 9-29 Align Axes of Clevis Pin and Link for Pin Connection

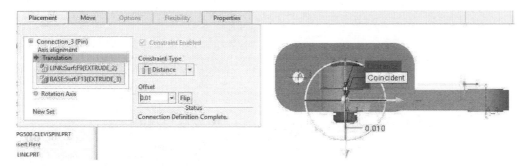

Figure 9-30 Distance Translation of Link for Pin Connection

Step 17: *Select* the [Placement] tab near the top of the screen.

Step 18: For Axis alignment, *pick* the A_1 axis of the clevis pin, then *pick* the A_1 axis of the hole in the link. See Figure 9-29. The Axis alignment will become coincident.

Step 19: For Translation, *pick* the raised surface on the back side of the link's hole, then *pick* the raised front surface of the base's hole. Change the constraint type to Distance. Enter a value of 0.01 inches for the offset. See Figure 9-30. Note that the connection definition is complete.

Step 20: *Pick* the **green checkmark** to accept the link placement.

Step 21: Highlight the link, then *select* a color from the appearance gallery at the top of the screen to color the link. See Figure 9-31.

Step 22: Select the **Drag Component** icon from the top of the screen.

Step 23: *Pick* the link with the LMB. Move the cursor. The link should follow the cursor as it rotates around the clevis pin. See Figure 9-32. *Press* LMB, then Close the Drag window.

Figure 9-31   Color the Link

Step 24: *Pick* the **Assemble** tool in [Model] tab again.

Step 25: Select "sdin6799.prt" from the Open window. *Pick* the **Open** button. Be sure the **Manual** placement icon is selected.

Figure 9-32 Move the Link into Different Positions

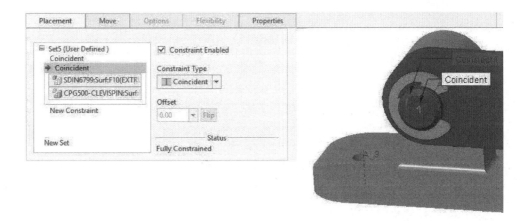

Figure 9-33   Coincident the E-ring to Clevis Pin Groove's Sidewall

Figure 9-34   E-ring placed in Clevis Pin Groove

Step 26:   Rotate and translate to position the e-ring similar to Figure 9-33.

Step 27:   *Select* the Placement tab near the top of the screen. *Pick* the outer surface of the retaining ring. *Pick* the sidewall of the clevis pin's groove closest to you. See Figure 9-33.

Step 28:   *Select* **New Constraint.** *Pick* the inner surface of the retaining ring and the base diameter of the clevis pin's groove. See Figure 9-34.

Step 29:   *Pick* the **green checkmark** to accept the clevis pin placement. See Figure 9-35.

Step 30:   [icon] *Select* the **Drag Component** icon from the top of the screen.

Step 31:   *Pick* the link with the LMB. Move the cursor. The link should follow the cursor as it rotates around the clevis pin. Press the MMB, then close the Drag window. Type <Ctrl> G to regenerate the model.

Step 32:   **File>Save.** *Pick* **OK.**

Figure 9-35   Complete
Swinging Link Assembly

Step 33:   **File>Manage File>Delete Old Versions.** *Pick* **Yes** button.

End of Assembly Practice, part 1.

Design Intent

Create an exploded assembly view of the Swinging Link Assembly to show how it goes together. See Figure 9-36. Start with the completed assembly view shown in Figure 9-35.

Step 1: 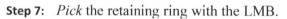 Select **Exploded View** in the Model tab. Parts may move apart similar to Figure 9-37. Don't worry if your initial exploded view looks different. This icon is also located in the View tab.

Step 2: Select **Edit Position** in the Model tab. (Also, under the View tab.)

Figure 9-36 Exploded Assembly View

Step 3: Select the References tab in the upper left corner of the screen so you can see what's going on.

Step 4: Pick the clevis pin with the LMB.

Step 5: Pick the x-axis arrow (Figure 9-38).

Step 6: Press down and hold the LMB, then move the mouse cursor upward. The clevis pin will move along the x-axis as the cursor is moved. When the clevis appears to be in the approximate location, release the LMB.

Step 7: Pick the retaining ring with the LMB.

Figure 9-37 Initial Exploded View

Step 8: Pick the x-axis arrow.

Step 9: Press down and hold the LMB, then move the cursor downward. The retaining ring will follow. When the retaining ring appears to be in the approximate location, release the LMB. See Figure 9-39.

Figure 9-38 Move Clevis Pin along X-axis

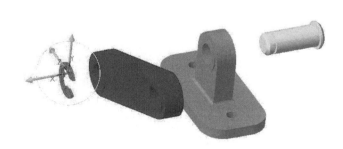

Figure 9-39 Move Retaining Ring along X-axis

Figure 9-40   Move Retaining Ring
along Z-axis

Note: use the MMB to move through
the model until you locate the proper
axis, then press the LMB.

Step 10:   *Pick* the z-axis of the retaining ring.

Step 11:   *Press down and hold* the LMB, then move the cursor
toward the lower left corner of the screen. The
retaining ring will follow. When the retaining ring
appears to be in the approximate location, release the
LMB. See Figure 9-40.

Step 12:   *Select* the **Axis Display** icon in the graphics toolbar to
turn on datum axes.

Step 13:   *Select* the **Create Cosmetic Offset Lines** icon. Or
select it under the Explode Lines tab. See Figure 9-41.

Step 14:   *Pick* the center tab of the retaining ring with the LMB
while still in Edit Position mode, then *pick* the A_1 axis
of the link. *Pick* the **Apply** button in the Cosmetic Offset
Line window.

Step 15:   *Pick* the A_1 axis of the link with the LMB, then *pick*
the A_1 axis of the base. *Pick* the **Apply** button in the
Cosmetic Offset Line window.

Step 16:   *Pick* the A_1 axis of the base with the LMB, then *pick*
the A_1 axis of the clevis pin. *Pick* the **Apply** button in
the Cosmetic Offset Line window. See Figure 9-42.
Pick **Close.**

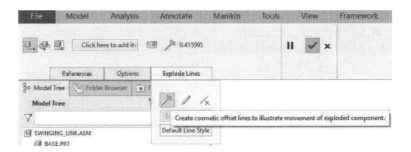

Figure 9-41   Explode Lines Tab

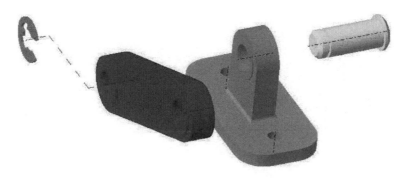

Figure 9-42   Add Offset Lines

Step 17: *Pick* the offset cosmetic line between the retaining ring and the link, then under the $\boxed{\text{Explode Lines}}$ tab, *pick the* Edit ***the selected explode line icon***. See Figure 9-43. Little circles will appear at the two bends in the offset line.

Step 18: *Pick* the leftmost circle with the LMB, then *press and hold* the LMB and drag the circle along the line toward the retaining ring. See Figure 9-44.

Step 19: Add a cosmetic offset line between the link's centerline and the base's upper hole centerline. *Pick* **Apply**. Also, add a cosmetic offset line between the base's centerline and the clevis pin's centerline. *Pick* **Apply**.

Step 20: *Select* the ***Refit*** icon at the top of the screen.

Step 21: *Select* the ***Axis Display*** icon to turn the datum axes off.

Step 22: *Pick* the ***green checkmark*** to accept the exploded view.

Step 23:  *Select* ***Exploded View*** in the $\boxed{\text{Model}}$ or $\boxed{\text{View}}$ tab to unexplode the view. See Figure 9-45.

Step 24:  *Select* ***Exploded View*** in the $\boxed{\text{Model}}$ or $\boxed{\text{View}}$ tab to explode the view again. See Figure 9-46.

Now we need to update the Default Exploded View so it reflects the changes we just made.

Step 25: *Select* the $\boxed{\text{View}}$ tab.

Step 26: *Select* the ***View Manager*** from $\boxed{\text{View}}$ tab. (Also, available in the $\boxed{\text{Model}}$ tab: **Manage View > View Manager.**)

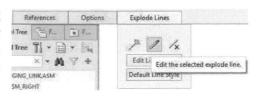

Figure 9-43 Edit Offset Lines

Figure 9-44 Modify Offset Lines

Figure 9-45 Unexploded View

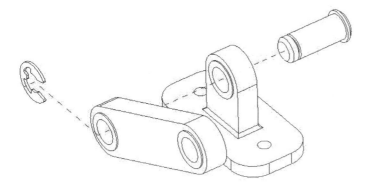

Figure 9-46 Exploded View

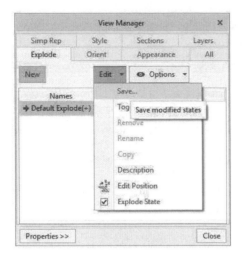

Figure 9-47   Save Modified State   **Figure 9-48**   Update Default State

Step 27: *Select* the ⬚Explode⬚ tab in the View Manager window.

Step 28: *Select* the **Edit** button with the LMB to bring up a pop-up menu. *Select* **Save…** from this menu. See Figure 9-47.

Step 29: When the Save Display Elements window appears, *pick* the **OK** button.

Step 30: When the Update Default State window appears, *pick* the **Update Default** button. See Figure 9-48.

Step 31: *Pick* t h e ⬚Orient⬚ tab. Orient the assembly as desired. *Pick* the **New** button. *Type* "3DVIEW" <Enter> for the new name. *Pick* the **Close** button to close the View Manager window.

Step 32: **File>Save.** (If you get a message "Model not regenerated," *pick* **OK**.

   Pick 🗒 in the upper left corner of the screen to regenerate the model, then *select* **File>Save** again.)

Step 33: **File>Manage File>Delete Old Versions.** *Pick* **Yes**.

Step 34: **File>Close.**

Step 35: **File>Manage Session>Erase Not Displayed.** *Pick* **OK.**

Step 36: **File>Open.**

Step 37: *Pick* "swinging_link.asm" from the list, then *pick* **Open.**

Step 38: With a fresh copy of the assembly back on the screen, verify that you can change back and forth from the exploded view to the unexploded view. If not, then go back and correct your mistake. (You probably left out a step or two.)

Step 39: With the assembly exploded and NO Hidden Line mode selected, print the image to prove you have completed this practice.
   >File>Print>Print.

End of Practice session, part 2.

Assembly Exercise

Design Intent

Assemble a side-mounting take-up frame made up of four different steel parts as shown in Figure 9-49. The actual assembly will be welded.

Before you begin a new assembly you should consider how the individual parts fit together. In the picture below part 2 is joined to part 1, part 3 is joined to part 2, and part 4 is joined to part 1. The procedure of creating the parts, then assembling them is called "Bottom-Up Design." The reverse procedure is "Top-Down Design" where the parts are created within the assembly as needed. The two methods are not mutually exclusive, thus you can switch back and forth during the assembly process.

For this practice session, we will use the Bottom-Up Design process. This means that the parts are created and numbered, then assembled. Part 1 is usually the main part of the assembly. We will call the lower plate part 1 (Figure 9-50).

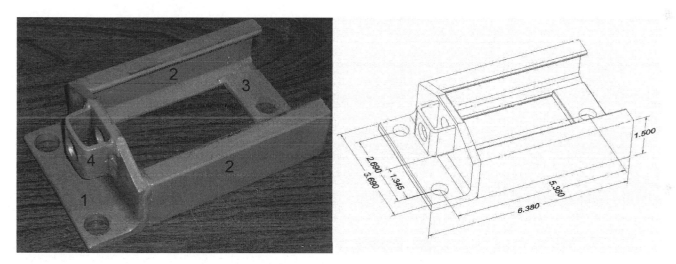

Figure 9-49 Assembled Take-Up Frame

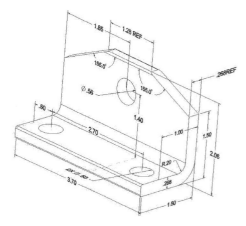

Figure 9-50 Lower Plate
(lower_plate.prt)

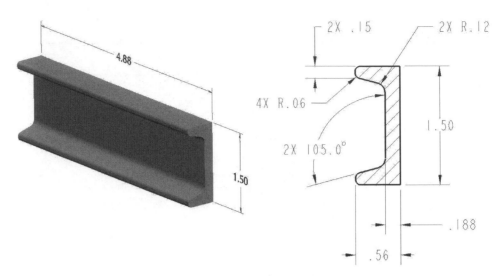

Figure 9-51   C15 x 14 x 488 Side Slide (c15x14x488.prt)

Part 2 is the side slide, which is made from a C15 × 14 × 488 steel C-channel. See Figure 9-51.

The upper plate, part 3, is made from ¼-inch thick by 1-inch wide steel, bar stock (Figure 9-52).

The steel nut holder (Figure 9-53) is part number 4. It is made from bar stock that is bent in two places.

These parts should be created in the IPS unit system. In order to create this assembly, all four parts must be in the same working directory. Be sure they are before continuing.

Step 1: Start Creo Parametric.

Step 2: Set Working Directory to the directory containing the four different parts.

Step 3: **File>New.**

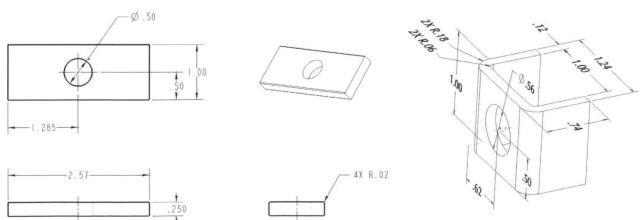

Figure 9-52   Upper Plate (upper_plate.prt)

Figure 9-53   Nut Holder (nut_holder.prt)

Step 4: *Select* **Assembly** as the file type and **Design** as the sub-type. Name the file "take-up_frame." Be sure "Use default template" is checked. *Pick* **OK.** See Figure 9-54.

Step 5: If Use default template is unchecked, select "**inlbs_asm_design**" in the New File Options window, then *pick* **OK**. See Figure 9-55. Otherwise, skip this step.

Step 6: Select the **Assemble** tool in the Model tab.

Step 7: In the Open window, *select* the **lower_plate.prt.** *Pick* the **Open** button. Set the relation type to **Fix**, then *press* the MMB again. The lower plate is located in the assembly at its current position.

Step 8: *Select* the View tab.

Step 9: Orient the part similar to Figure 9-56 below, then *select* the **View Manager** icon from the top of the screen.

Step 10: *Select* the Orient tab in the View Manager window. *Pick* the **New** button. *Type* "3DVIEW" <Enter>. *Pick* the **Close** button. The current view orientation is saved under the name "3DVIEW".

Step 11: Select the **Saved Orientations** icon from the top of the screen. *Pick* the **Standard Orientation** from the list. Repeat the procedure picking **3DVIEW** the second time. The part should return to the orientation you just created.

Step 12: *Select* the Model tab.

Step 13: Select the **Assemble** tool in the Model tab.

Step 14: *Select* **c15x14x488.prt** from the list, then *pick* the **Open** button.

Step 15: Use the colored circles and arrows icons to position the part similar to Figure 9-57.

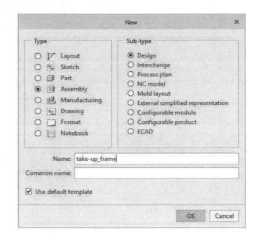

Figure 9-54   New Object Window

Figure 9-55   Empty Template

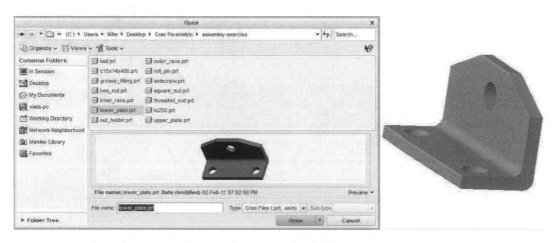

Figure 9-56   Part 1 – Lower Plate and its default orientation

Step 16: *Select* the Placement tab near the top of the screen. *Pick* the back of the lower_plate and the left end of part c15x14x488 as shown in Figure 9-57. The constraint type should be Coincident.

Step 17: *Select New Constraint* under the Placement tab, then *pick* the outside surface of the c15x14x488 part and the end surface of the lower_plate. See Figure 9-58. The constraint type should be Coincident again.

If you accidentally exit from the placement feature in Creo Parametric, you can select part c15x14x488, and a pop-up menu will appear. *Select* the **Edit Definition** icon to enter part placement again.

Step 18: *Select New Constraint* under the Placement tab, then *pick* the bottom surface of the c15x14x488 part and the bottom surface of the lower_plate. See Figure 9-59. The constraint type should be Coincident. After this constraint has been added, the status should show as Fully Constrained.

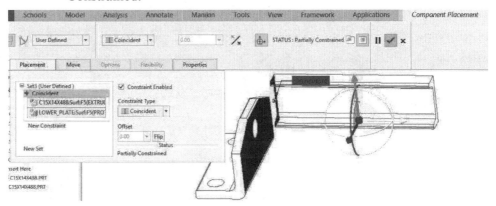

Figure 9-57   First Coincident Constraint

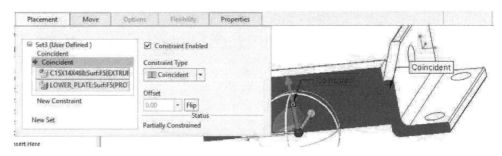

Figure 9-58   Second Coincident Constraint

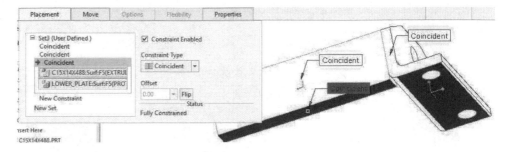

Figure 9-59   Third Coincident Constraint

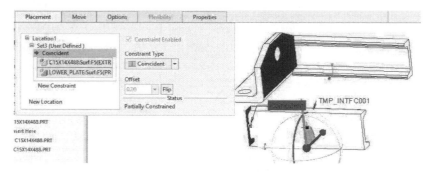

Figure 9-60   First Coincident Constraint of Second Slide

Step 19: *Pick* the ***green checkmark*** to accept the part placement since it is fully constrained.

Step 20: *Select* the ***Save*** icon to save the assembly to disk. *Pick* **OK.**

Step 21: *Select* the ***Assemble*** tool in the Model tab.

Step 22: *Select* **c15x14x488.prt** from the list again, then *pick* the **Open** button.

Step 23: Use the colored circles and arrows icons to position the part similar to Figure 9-60.

Step 24: *Select* the Placement tab near the top of the screen and *delete* the constraints shown. *Pick* the back of the lower_plate and the left end of the rotated part, c15x14x488, as shown in Figure 9-60. The constraint type should be Coincident.

Step 25: *Select* ***New Constraint*** under the Placement tab, then *pick* the outside surface of the c15x14x488 part and the end surface of the lower_plate. See Figure 9-61. The constraint type should be Coincident.

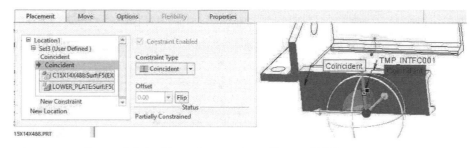

Figure 9-61   Second Coincident Constraint of Second Slide

Figure 9-62   Third Coincident Constraint of Second Slide

Step 26: *Select* *New Constraint* under the Placement tab, then *pick* the bottom surface of the c15x14x488 part and the bottom surface of the lower_plate. See Figure 9-62. The constraint type should be Coincident. After this constraint has been added, the status should show as Fully Constrained.

Step 27: *Pick* the **green checkmark** to accept the part placement since it is fully constrained.

Step 28: *Select* the *Save* icon to save the assembly to disk.

Step 29: *Select* the *Assemble* tool in the Model tab.

Step 30: *Select* **upper_plate.prt** from the list, then *pick* the **Open** button.

Step 31: Use the colored circles and arrows icons to position the part similar to Figure 9-63.

Step 32: *Select* the Placement tab near the top of the screen. *Pick* the bottom of the upper_plate and the bottom of the left part, c15x14x488, as shown in Figure 9-63. The constraint type should be Coincident.

Step 33: *Select* *New Constraint* under the Placement tab, then *pick* the right end of the left part, c15x14x488, and the right edge surface of the upper_plate. See Figure 9-64. The constraint type should be Coincident.

Step 34: *Select* *New Constraint* under the Placement tab, then *pick* the bottom inside surface of the c15x14x488 part and the left end of the upper_plate. See Figure 9-65. The constraint type should be Coincident. After this constraint has been added, the status should show as Fully Constrained.

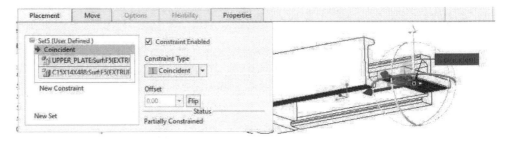

Figure 9-63   First Coincident Constraint for Third Part

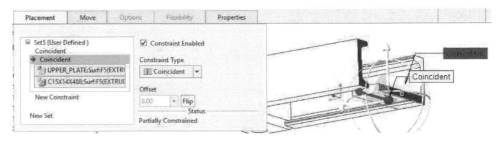

Figure 9-64   Second Coincident Constraint for Third Part

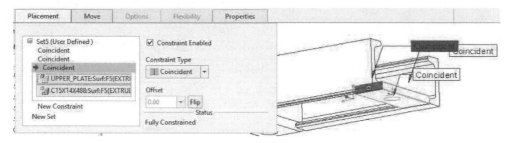

Figure 9-65 Third Coincident Constraint for Third Part

Step 35: *Pick* the **green checkmark** to accept the part placement as defined.

Step 36: *Select* the **Save** icon to save the assembly to disk.

Step 37: *Select* the **Assemble** tool in the ⃞Model⃞ tab.

Step 38: *Select* **nut_holder.prt** from the list, then *pick* the **Open** button.

Step 39: Use the colored circles and arrows icons to position the part similar to Figure 9-66.

Step 40: *Select* the ⃞Placement⃞ tab near the top of the screen. *Pick* the front surface of the Lower_plate and the right end of the nut_holder as shown in Figure 9-66. The constraint type should be Coincident.

Step 41: Turn on **Axis Display** from the graphics toolbar.

Step 42: *Select* **New Constraint** under the ⃞Placement⃞ tab, then *pick* the axis, A_3, of the lower_plate and the axis, A_1, of the nut_holder. See Figure 9-67. With Allow Assumptions checked, the status is Fully Constrained. However, the nut holder is not in the desired orientation so we need to add a third constraint.

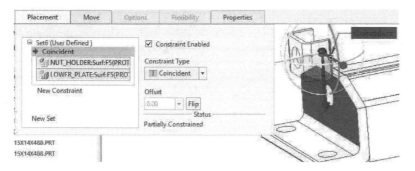

Figure 9-66 First Coincident Constraint for Fourth Part

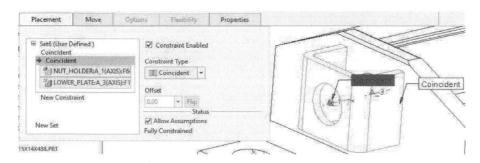

Figure 9-67 Second Coincident Constraint for Fourth Part

Design Intent

We want the top of the nut holder to be parallel with the top of the lower plate.

Step 43: *Select* ***New Constraint*** under the ☐Placement☐ tab, then *pick* the top surface of the nut_holder and the top surface of the lower_plate. See Figure 9-68. Change the constraint type to Parallel. After this constraint has been added, the status should show as Fully Constrained (without assumptions).

Step 44: *Pick* the ***green checkmark*** to accept the part placement as defined. The assembled take-up frame should appear as shown in Figure 9-69.

Step 45: *Select* the ***Save*** icon to save the assembly to disk.

Step 46: **File>Manage File>Delete Old Versions.** *Pick* the ***Yes***.

Step 47: Orient the assembly similar to Figure 9-69.

Step 48: 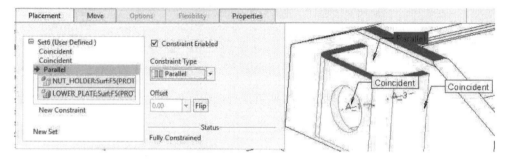 *Select* the ***Refit*** icon from the graphics toolbar.

Step 49: *Select* the ***No Hidden*** icon from the graphics toolbar.

Figure 9-68 Third Coincident Constraint for Fourth Part

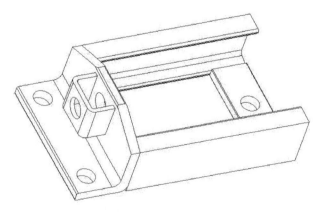

Figure 9-69 Assembled Take-up Frame
(take-up_frame.asm)

Step 50: **File>Print>Print** to prove that you have completed this portion of the assembly exercise.. *Pick* **OK.**

Step 51: File>Close.

Step 52: File>Manage Session>Erase Not Displayed. *Pick* **OK.**

You can stop at this point or go on with this exercise. If you choose to continue, be sure that the following four parts are located in your working directory. See Figures 9-70 through 9-73. These parts should be created in the IPS unit system. In order to create this assembly, all four parts must be in the same working directory. Be sure they are before continuing.

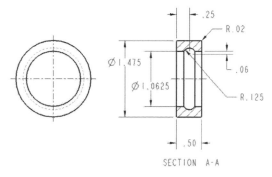

Figure 9-70 Part 1, Outer Race of Ball Bearing (outer_race.prt)

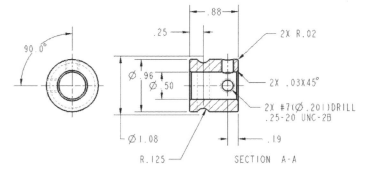

Figure 9-71 Part 2, Inner Race of Ball Bearing (inner_race.prt)

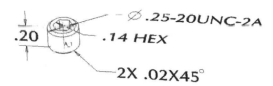

Figure 9-72 Part 3, ¼-inch Set Screw (setscrew.prt)

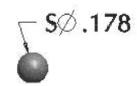

Figure 9-73 Part 4, Bearing Ball (ball.prt)

Design Intent

Assemble the ball bearing made up of four different steel parts as shown in Figure 9-74.

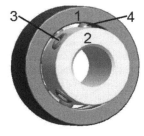

Figure 9-74 Ball Bearing Assembly

The first step is to add the set screws, part 3, to the inner race, part 2.

Step 1: **File>New.**

Step 2: *Select* Assembly as the file type and Design as the sub-type. Name the file "inner_race". Be sure to *check* "Use default template." *Pick* **OK.**

Step 3: If Use Default Template is unchecked, *select* **inlbs_asm_design** in the New File Options window. *Pick* **OK.** Otherwise, skip this step.

Step 4: *Select* the ***Assemble*** tool in the Model tab.

Figure 9-75 Inner Race of Ball Bearing

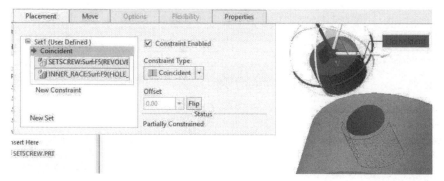

Figure 9-76   Insert Setscrew in the Inner Race

Step 5: In the Open window, *select* the **inner_race.prt.** *Pick* the **Open** button. Set the relation type to Fix, then press the MMB again. The inner race is located in the assembly. See Figure 9-75.

Step 6: *Select* the *Assemble* tool in the Model tab.

Step 7: *Select* **setscrew.prt** from the list, then *pick* the **Open** button.

Step 8: Use the colored circles and arrows icons to position the part similar to Figure 9-76.

Step 9: *Select* the Placement tab near the top of the screen. *Pick* the outside surface of the setscrew and the inner surface of the inner_race's threaded hole as shown in Figure 9-76. The constraint type should be Coincident.

Step 10: *Select New Constraint* under the Placement tab, then *pick* the top surface of the setscrew. Turn on Datum Planes. *Pick* the TOP datum plane of the inner_race. See Figure 9-77. The constraint type should be Distance with a 0.50-inch offset.

The setscrew will still rotate about the axis of the threaded hole if the Allow Assumptions box is unchecked. With Allow Assumptions checked, the setscrew is fully constrained. We will use the assumption made by Creo Parametric. Leave the Allow Assumptions checked.

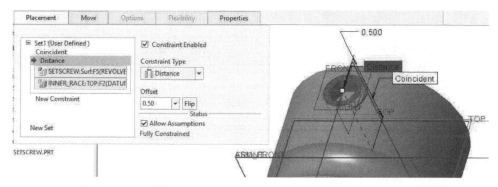

Figure 9-77 Setscrew with Distance Offset

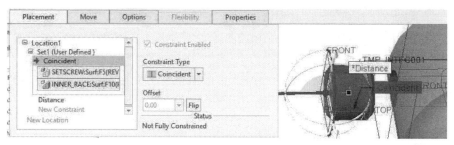

Figure 9-78 Insert Second Setscrew

Step 11: *Pick* the ***green checkmark*** to accept the placement of the setscrew.

Step 12: *Select* the ***Assemble*** tool in the $\boxed{\text{Model}}$ tab.

Step 13: *Select* **setscrew.prt** from the list again and *pick* the **Open** button.

Step 14: Use the colored circles and arrows icons to position the second setscrew above the second threaded hole. See Figure 9-78.

Step 15: *Select* the $\boxed{\text{Placement}}$ tab near the top of the screen. The outside surface of the setscrew has already been selected. *Pick* the inner surface of the inner_race's threaded hole as shown in Figure 9-78. The constraint type should be Coincident.

Step 16: *Select **Distance*** constraint under the $\boxed{\text{Placement}}$ tab. The top surface of the setscrew has already been selected. *Pick* the RIGHT datum plane of the inner_race. See Figure 9-79. The constraint type should be Distance with a 0.50-inch offset. Enter -0.50 if the setscrew moved the wrong direction.

The setscrew will still rotate about the axis of the threaded hole if the Allow Assumptions box is unchecked. With Allow Assumptions checked, the setscrew is fully constrained. We will use this assumption again.

Step 17: *Pick* the ***green checkmark*** to accept the placement of the setscrew.

Step 18: *Select* the ***Save*** icon to save the assembly to disk. *Pick* **OK.**

Step 19: File>Close.

Step 20: File>New.

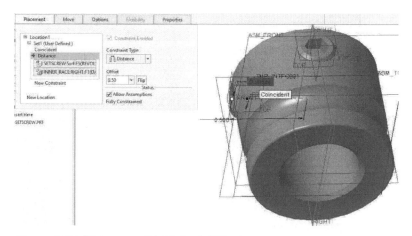

Figure 9-79 Distance with ¹/₂-inch Offset

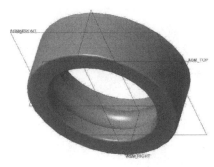

Figure 9-80   Outer Race of Ball
Bearing

Step 21: *Select* **Assembly** as the file type and **Design** as the
sub-type. Name the file "ball_bearing". *Check* Use
default template. *Pick* **OK.**

Step 22: If Use Default Template is unchecked, select **inlbs_asm_
design** in the New File Options window. *Pick* **OK.**

Step 23: [icon] *Select* the *Assemble* tool in the [Model] tab.

Step 24: In the Open window, *select* the **outer_race.prt.** *Pick* the
Open button. *Set* the relation type to **Fix**, then *press* the
MMB again. The outer race is located in the assembly. See
Figure 9-80.

Step 25: *Select* the *Assemble* tool in the [Model] tab.

Step 26: *Select* **ball.prt** from the list and *pick* the **Open** button.

Step 27: Use the colored circles and arrows icons to position the part similar to
Figure 9-81.

Step 28: *Select* the [Placement] tab near the top of the screen. Set the constraint
type to **Tangent.** *Pick* the outer surface of the ball and the inner
indented surface of the outer_race as shown in Figure 9-81. The
constraint type must be Tangent.

Step 29: Turn on Datum *Planes Display* if they are off.

Step 30: [icon] *Select* the *Show component in separate window* icon to make it
easier to pick items on the ball.

Step 31: *Select* **New Constraint** under the [Placement] tab, then *pick* the RIGHT
datum plane of the ball and the RIGHT datum plane of the outer_race.
See Figure 9-82. The constraint type should be Coincident.

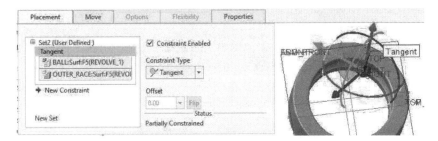

Figure 9-81 Add One Ball to the Outer Race

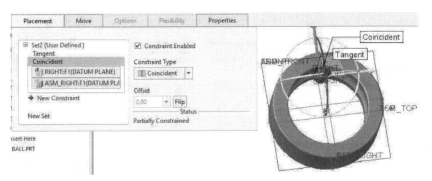

Figure 9-82 Coincident Constraint

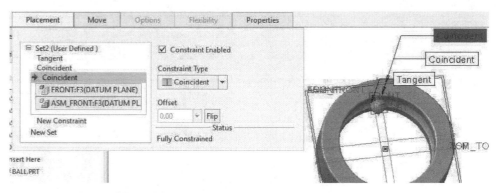

Figure 9-83 Second Coincident Constraint

Step 32: *Select* **New Constraint** under the Placement tab, then *pick* the FRONT datum plane of the ball and the FRONT datum plane of the outer_race. See Figure 9-83. The constraint type should be coincident. The ball is now fully constrained.

Step 33: *Pick* the **green checkmark** to accept the placement of the bearing ball.

Step 34: *Select* the **Save** icon to save the assembly to disk. *Pick* **OK.**

Step 35: Turn Datum **Planes Display** off. Turn Datum **Axis Display** on. Turn off **Show components in Separate window** by *picking* the icon again.

Step 36: Highlight **ball.prt** in the model tree, then *select* the **Pattern** icon from the Model tab.

Step 37: Set the pattern type (first box) to **Axis,** then *pick* the A_1 axis of the outer_race. Set the number of patterns to 8. Set the pattern angle to 45 degrees. See Figure 9-84.

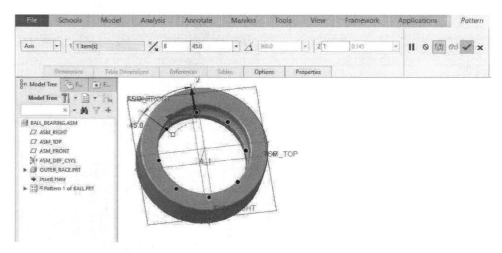

Figure 9-84 Outer Race with Ball Pattern

Step 38: *Pick* the **green checkmark** to accept the pattern. See Figure 9-85.

Step 39: *Select* the **Save** icon to save the assembly to disk. *Pick* **OK**.

Step 40: *Select* the **Assemble** tool in the Model tab.

Step 41: *Select* the sub-assembly **inner_race.asm** (not inner_race.prt) from the list and *pick* the **Open** button.

Step 42: Use the colored circles and arrows icons to position the part similar to Figure 9-86.

Figure 9-85 Outer Race with Eight Balls

Step 43: *Select* the Placement tab near the top of the screen. Change the User-Defined connection to Pin connection in the menu area. *Pick* the A_1 axis of the outer_race and the A_1 axis of the inner_race as shown in Figure 9-86.

Step 44: *Select* the word **Translation** in the Placement window, then *pick* the leftmost surface of the outer_race and the leftmost surface of the inner_race. See Figure 9-87. (An alternative placement would use the FRONT datum plane of outer_race.prt and the FRONT datum plane of inner_race.asm if Datum planes are turned on.)

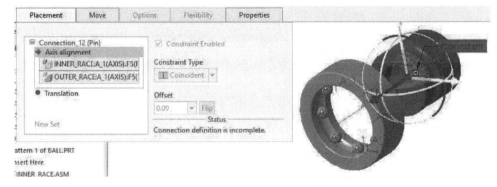

Figure 9-86 Axis Alignment for Inner and Outer Races

Figure 9-87 Translation for Inner and Outer Races

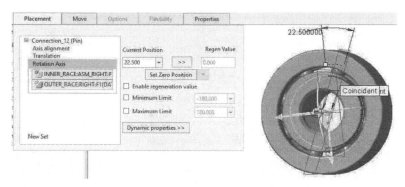

Figure 9-88 Rotation Axis for Inner and Outer Races

Step 45: *Select* the words **Rotation 1** in the Placement window. Turn Datum Planes on at the top of the screen if they are off. *Pick* the RIGHT datum plane of the inner_race and the RIGHT datum plane of the outer_race. Set the current position to 22.5 degrees. Be careful when selecting the datum planes since there are many datum planes visible. See Figure 9-88. *Zoom In* if necessary.

Step 46: *Pick* the ***green checkmark*** to accept the placement.

Step 47: Turn off Datum Planes and Datum Axis.

Step 48: *Select* the ***Save*** icon to save the assembly to disk.

Step 49: *Select* the ***Hidden Line*** icon from the top of the screen.

Step 50: Orient the ball bearing assembly similar to Figure 9-89.

Step 51: *Select* the ***Refit*** icon.

Step 52: **File>Print>Print.** *Pick* **OK.**

Step 53: *Select* the ***Shading*** icon to make the ball bearing assembly appear as a solid.

Step 54: Highlight **outer_race.prt** in the model tree. *Select* the Appearance Gallery at the top of the screen, then *pick* standard steel "ptc-std-steel" from the list. See Figure 9-90.

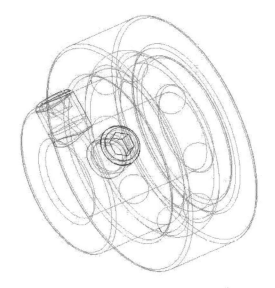

Figure 9-89 Assembled Ball Bearing

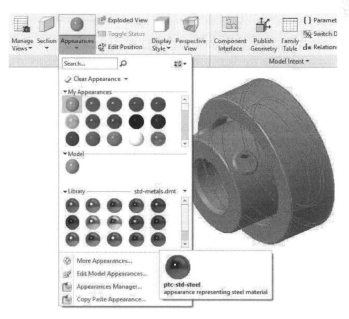

Figure 9-90 Color Outer Race of Bearing

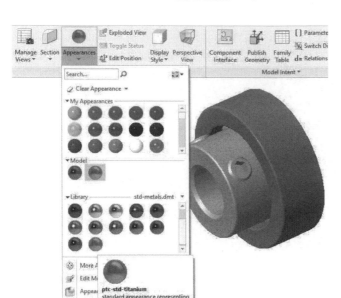

Figure 9-91   Color Inner Race of Bearing
(ball_bearing.asm)

Step 55: Highlight **inner_race.asm** in the model
tree. *Select* the Appearance Gallery at the
top of the screen again, then *pick* another
color from the list. See Figure 9-91.

Step 56: *Regenerate* the model, then *pick* the *Save*
icon to save the assembly to disk.

Step 57: *Select* the *Drag Components* icon
in the Model tab. *Pick* an edge of the
inner race. Move the mouse cursor in a
circle. The inner race should rotate in
place if you successfully implemented
the pin joint. If the inner race will not
rotate, go back and fix it.

Step 58: *Pick* **Close** in the Drag window.

Step 59: **File>Manage File>Delete Old
Versions.** *Pick Yes*.

Step 60: File>Close.

**Step 61: File>Manage Session>Erase Not
Displayed.** *Pick* **OK.**

Before you continue, be sure that the following parts are located in your
working directory. Parts **TU250.prt** and **grease_fitting.prt** should be
created in the IPS unit system. See Figure 9-91 above and Figures 9-93 and 9-94
following.

Design Intent

Assemble the Take-up Piece by inserting the ball bearing assembly and the
grease fitting. See Figure 9-92.

Figure 9-92 Take-up Piece Assembly

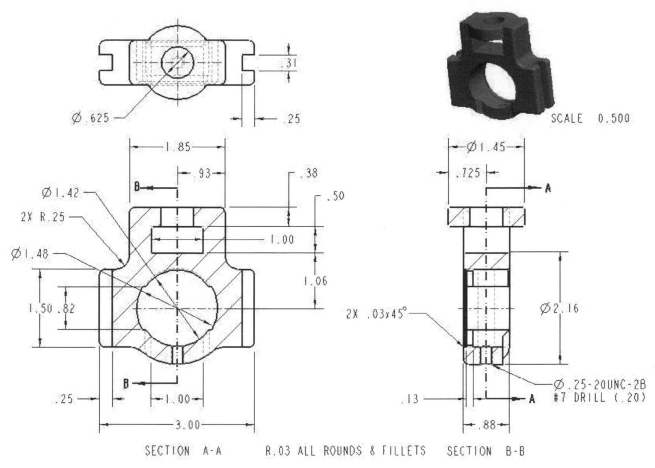

Figure 9-93 TU250—Take-up Piece (TU250.prt)

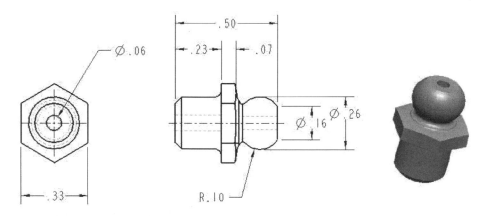

Figure 9-94 Grease Fitting (grease_fitting.prt)

Step 1: **File>New.**

Step 2: *Select* Assembly as the file type and Design as the sub-type. Name the file "Take-up_piece."

Step 3: Be sure to *check* Use default template. *Pick* **OK.**

Step 4: *Select* the **Assemble** tool in the Model tab.

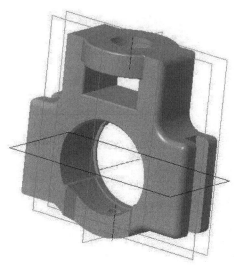

Figure 9-95   Take-up Piece

Step 5: In the Open window, *select* the **TU250.prt.** *Pick* the **Open** button. Set the relation type to Fix, then press the MMB again. The take-up piece is located in the assembly. See Figure 9-95.

Step 6: *Select* the *Assemble* tool in the ⎵Model⎵ tab.

Step 7: *Select* **ball_bearing.asm** from the list and *pick* the **Open** button.

Step 8: Use the colored circles and arrows icons to position the part similar to Figure 9-96.

Step 9: *Select* the ⎵Placement⎵ tab near the top of the screen. *Pick* the outer surface of the ball bearing assembly and the inner surface of the take-up piece (TU250.prt) as shown in Figure 9-96. The constraint type must be Coincident.

Step 10: Turn Datum Planes on if they are not on.

Step 11: *Select New Constraint* under the ⎵Placement⎵ tab, then *pick* the front surface of the outer race and the front lower flat surface of the take-up piece. See Figure 9-97. The constraint type should be Distance with an offset value of 0.10 inches. If the bearing protrudes out of the take-up piece, enter -0.10 for the offset distance.

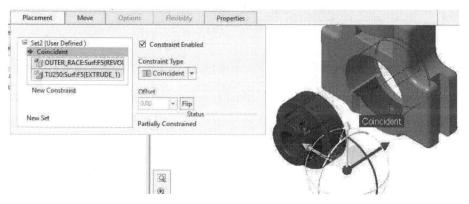

Figure 9-96   Add Ball Bearing Assembly to the Take-up Piece

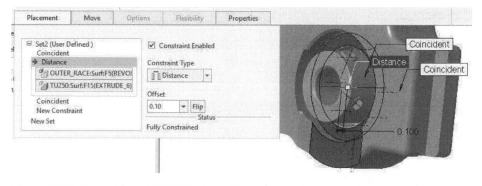

Figure 9-97   Coincident FRONT Datum Planes

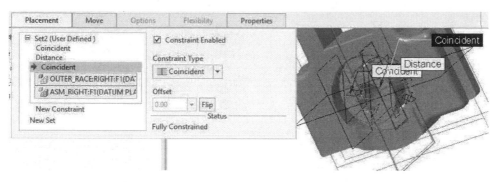

Figure 9-98 Coincident RIGHT Datum Planes

Step 12: *Select New Constraint* under the Placement tab, then *pick* the RIGHT datum plane of the outer race and the RIGHT datum plane of the take-up piece (or RIGHT_ASM). See Figure 9-98. The constraint type should be Coincident.

Step 13: *Pick* the **green checkmark** to accept the placement of the bearing ball.

Step 14: *Select* the **Save** icon to save the assembly to disk. *Pick* **OK.**

Step 15: Turn Datum **Planes Display** off again.

Step 16: *Select* the **Drag Component** icon from the top of the screen.

Step 17: Using the LMB *pick* the inner race of the ball bearing assembly. Move the mouse cursor around and watch the inner race rotate. If it doesn't rotate, then go back and verify that the ball bearing assembly was created using a pin connection and you used the outer race datum planes, not the inner race datum planes.

Step 18: *Pick* **Close** in the Drag window.

Step 19: *Select* the **Assemble** tool in the Model tab.

Step 20: Select grease_fitting.prt from the list, then pick the Open button.

Step 21: Use the colored circles and arrows icons to position the part similar to Figure 9-99.

Step 22: *Select* the Placement tab near the top of the screen. *Pick* the outer round surface of the grease fitting and the inner surface of the small hole of the take-up piece (TU250.prt) as shown in Figure 9-99. The constraint type must be Coincident.

Figure 9-99 Insert Constraint for Grease Fitting

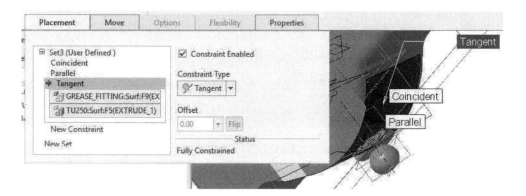

Figure 9-100 Align Edge of Fitting with Take-up Piece

Step 23: *Select **New Constraint*** under the ⏍Placement⏎ tab, then *pick* a flat surface of the grease fitting and the corresponding flat surface of the take-up piece. See Figure 9-100. The constraint type should be Parallel.

Step 24: *Select **New Constraint*** under the ⏍Placement⏎ tab, then *pick* the underside of the hex nut of the grease fitting and the bottom surface of the take-up piece. See Figure 9-101. The constraint type should be Tangent.

Step 25: *Pick* the **green checkmark** to accept the placement of the grease fitting. See Figure 9-102.

Figure 9-101   Tangent to Surface

Figure 9-102 Completed Take-up Piece Assembly (take-up_piece.asm)

Step 26: *Select* the ***Save*** icon to save the assembly to disk. **Step 27:**

File>Manage File>Delete Old Versions. *Pick Yes*.

Step 27:   File>Close.

Step 28:   File>Manage Session>Erase Not Displayed. *Pick* **OK.**

Before you continue, be sure that the following assemblies and their corresponding parts are located in your working directory. See Figure 9-69 (take-up_frame.asm) and Figure 9-102 (take-up_piece.asm).

Design Intent

Assemble the entire take-up assembly with ball bearing present. See if you can make the assembly operate as it would in the real world. That is, turning the threaded rod moves the take-up bearing relative to the frame. See Figure 9-103. This exercise will not perform this task. The take-up bearing will be fixed in position as shown.

Four new parts, Figures 9-104 through 9-107, are shown below. These parts should be created in the IPS unit system.

Step 1:   File>New.

Step 2: *Select* Assembly as the file type and Design as sub-type. Name the file "take-up_assembly."

Step 3: Be sure to *check* Use default template. *Pick* **OK.**

Step 4: *Select* the ***Assemble*** tool in the Model tab.

Figure 9-103 Completed Take-up Assembly (take-up.asm)

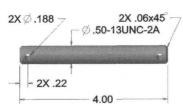

Figure 9-104 Threaded Rod (threaded_rod.prt)

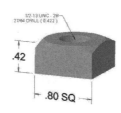

Figure 9-105 Square Nut (square_nut.prt)

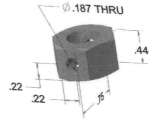

Figure 9-106 Hex Nut (hex_nut.prt)

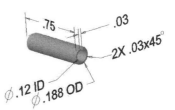

Figure 9-107 Roll Pin (roll_pin.prt)

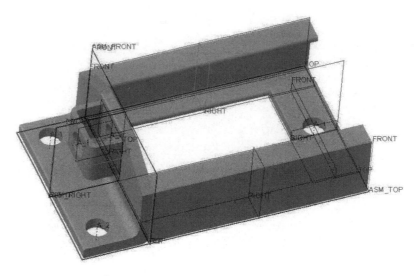

Figure 9-108 Take-up Frame

Step 5: In the Open window, *select* the **take-up_frame.asm.** *Pick* the **Open** button. Set the relation type to Fix, then press the MMB again. The take-up frame is located in the assembly. See Figure 9-108.

Step 6: *Select* the *Assemble* tool in the Model tab.

Step 7: *Select* **square_nut.prt** from the list, then *pick* the **Open** button.

Step 8: Use the colored circles and arrows icons to position the part similar to Figure 9-109. *Turn* off Plane and Csys Displays.

Step 9: *Select* the Placement tab near the top of the screen. *Pick* the A_1 axis of the threaded hole in the square nut and the A_3 axis of the hole in the lower plate.

Step 10: *Select New Constraint* under the Placement tab. *Pick* the back surface of the square nut and the left outside surface of the lower plate as shown in Figure 9-109. The constraint type must be Coincident. Make sure Allow Assumptions is checked to fully constrain the part.

Step 11: *Pick* the **green checkmark** to accept the placement of the square nut.

Step 12: *Select* the *Assemble* tool in the Model tab.

Step 13: *Select* **threaded_rod.prt** from the list, then *pick* the **Open** button.

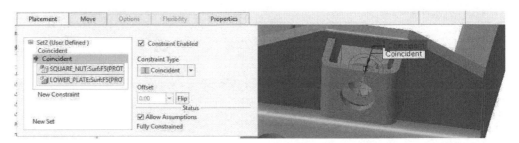

Figure 9-109 Add Square Nut to the Take-up Frame

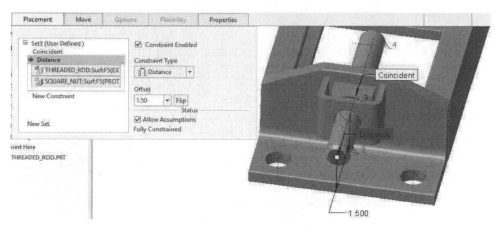

Figure 9-110 Add Threaded Rod to the Take-up Frame

Step 14: Use the colored circles and arrows icons to position the part similar to Figure 9-110.

Step 15: *Select* the Placement tab near the top of the screen. *Pick* the A_1 axis of the square nut and the A_1 axis of the threaded rod. (You may need to make this a cylinder connection if you want it to rotate and move as a threaded rod would.)

Step 16: *Select New Constraint* under the Placement tab. *Pick* the left end of the threaded rod and the left surface of the square nut as shown in Figure 9-110. The constraint type must be distance with a 1.50-inch offset.

Step 17: *Uncheck* Allow Assumptions so the threaded rod is only partially constrained.

Step 18: *Select New Constraint* under the Placement tab.

Step 19: Turn on datum planes if they are off.

Step 20: *Pick* the FRONT datum plane of the threaded rod and the FRONT datum plane of the square nut. Set the constraint type to Angle Offset with an angle of 0 degrees. See Figure 9-111. The hole in the rod should be straight up.

Step 21: *Pick* the **green checkmark** to accept the placement of the threaded rod.

Step 22: *Select* the **Assemble** tool in the Model tab.

Step 23: *Select* **hex_nut.prt** from the list, then *pick* the **Open** button.

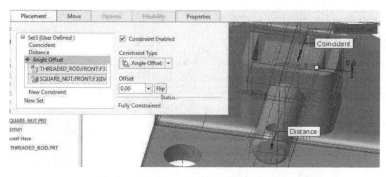

Figure 9-111 Add the Third Constraint for Threaded Rod

Step 24: Use the colored circles and arrows icons to position the part similar to Figure 9-112. Turn off Plane Display.

Step 25: *Select* the $\boxed{\text{Placement}}$ tab near the top of the screen. *Pick* the A_1 axis of the threaded rod and the A_1 axis of the threaded hole in the hex nut. Set the constraint type to Coincident.

Step 26: *Select New Constraint* under the $\boxed{\text{Placement}}$ tab. *Pick* the A_3 or A_4 axis of the small hole in the threaded rod and the A_3 axis of the small hole in the hex nut as shown in Figure 9-112. The constraint type can be Oriented or Coincident.

Step 27: *Pick* the **green checkmark** to accept the placement of the hex nut.

Step 28: *Select* the **Assemble** tool in the $\boxed{\text{Model}}$ tab.

Step 29: *Select* **hex_nut.prt** from the list again, then *pick* the **Open** button. Place this hex nut on the other end of the threaded rod in a similar fashion by picking the threaded rod's axis A_1 and the hole's axes A_3 or A_4. See Figure 9-113. You may have to zoom in to select the hex nut's axis.

Step 30: *Pick* the **green checkmark** to accept the placement of the second hex nut.

Step 31: *Select* the **Assemble** tool in the $\boxed{\text{Model}}$ tab.

Step 32: *Select* **roll_pin.prt** from the list, then *pick* the **Open** button. Insert the roll pin in the small hole in the hex nut by selecting the two mating surfaces. Align the end of the roll pin with the top surface of the hex nut. See Figure 9-114. Make sure **Allow Assumptions** is checked.

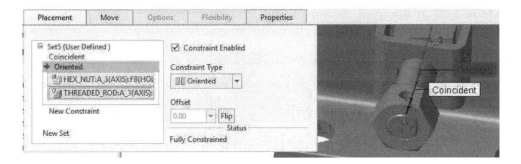

Figure 9-112   Add One Hex Nut to the Threaded Rod

Figure 9-113   Add Second Hex Nut to the Threaded Rod

Step 33: *Pick* the **green checkmark** to accept the placement of the roll pin.

Step 34: Repeat the procedure for the second roll pin inserted into the second hex nut.

Step 35: *Select* the **Assemble** tool in the [Model] tab.

Step 36: *Select* **take-up_piece.asm** from the list and *pick* the **Open** button. (If you want to make the **take-up_piece.asm** move, *select* a **Slider** connection.)

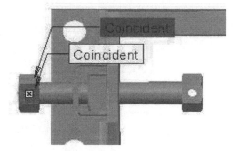

Figure 9-114 Add Both Roll Pins to the Two Hex Nuts

Step 37: *Pick* the A_2 axis of part TU250A and the A_3 axis of the lower plate as Coincident.

Step 38: *Pick* the top surface of the take-up frame and the top surface of TU250A's cutout where it touches the take-up frame and make them a Parallel constraint.

Step 39: *Pick* the left inside surface of the lower plate and the outer top surface of part TU250A. Set the distance to 0.03 inches. See Figure 9-115.

Step 40: *Pick* the **green checkmark** to accept the placement of the take-up piece assembly.

Step 41: *Select* the **Save** icon to save the assembly to disk.

Step 42: Select the left hex nut, then select the Appearance Gallery at the top of the screen. Locate the **ptc-std-steel-polished** icon and pick it.

Step 43: Repeat this procedure for the other two nuts. See Figure 9-116.

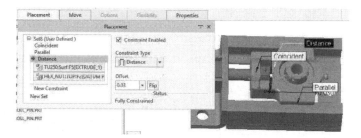

Figure 9-115 Add Take-up Piece to Take-up Frame

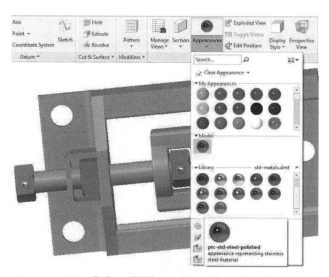

Figure 9-116 Color All Three Nuts Metallic Gold

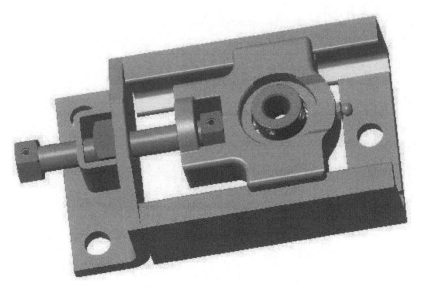

Figure 9-117   Color Threaded Rod Metallic Steel Light

Step 44: *Select* the threaded rod, then *select* the Appearance Gallery at the top of the screen. Locate the ptc-std-titanium color and *pick* it. See Figure 9-117.

Step 45: *Select* the ***Save*** icon to save the assembly to disk.

Step 46: With No Hidden line mode selected, **File>Print>Print** to prove that you have completed this entire assembly exercise. *Pick* **OK**.

Step 47: **File>Manage File>Delete Old Versions.** *Pick Yes*.

Step 48: **File>Close.**

Step 49: **File>Manage Session>Erase Not Displayed.** *Pick* **OK.**

Step 50: Exit from Creo Parametric.

End of Assembly Exercise

▶ Review Questions

1. Name five joint constraints used to connect parts or subassemblies.

2. How do you determine the types of constraints used to join two parts in an assembly?

3. What is the difference between using an empty assembly template and the default assembly template?

4. Can you modify dimensions of a part while in assembly mode?

5. How do you explode an assembly?

6. How can you make some of the assembly parts transparent?

7. How do you determine which part is active in an assembly?

8. What is the difference between Hide and Suppress?

9. How would you change the assembly reference for a part in an assembly?

10. Name at least three reasons for creating an assembly in Creo Parametric.

11. What is the difference between a cylinder and a pin joint?

12. How many degrees of freedom does a slot joint have?

13. Can you download parts from the internet, then assemble them into your assembly? If so, why would you want to do this?

14. Can you assemble parts created in different measurement systems (inches, feet, millimeters, etc.) in the same assembly?

15. Can you create a datum plane in an assembly?

16. How often should you save your work when creating an assembly?

17. How do you get rid of old copies of your assembly file?

18. Is it possible to create a subassembly, and then treat it as a part in an assembly?

Assembly Problems

9.1   Assemble the pulley (Figure 9-121) on the step shaft shown in Figure 9-118. Use the square key (Figure 9-119) to allow the pulley to transmit torque to the shaft. The shaft and key are made from 1020 CD steel. The 4.5-inch diameter pulley is a zinc die-cast. You can create these parts from Figures 9-118 through 9-121 or get them from your instructor. Start with the step shaft, then add the key and the pulley. Insert the set screw (Figure 9-120) into the threaded hole of the pulley. Name the assembly **"step-shaft_assembly"**. All dimensions are in inches.

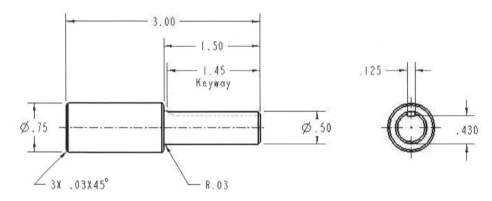

Figure 9-118   Step Shaft

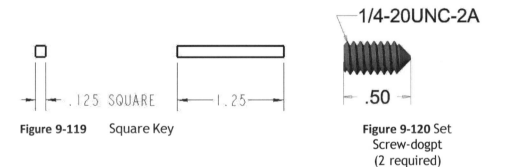

Figure 9-119   Square Key

Figure 9-120 Set
Screw-dogpt
(2 required)

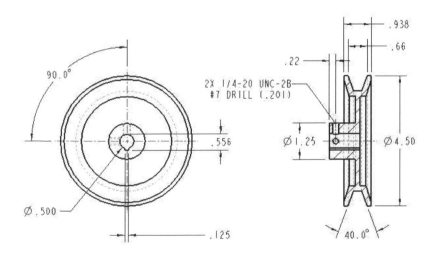

Figure 9-121   Pulley

9.2 Create an exploded assembly for the "**step-shaft_assembly**" found in problem 1. Use **View>Explode> Explode View** and **View>Explode>Unexplode View** to toggle between the two different views shown in Figure 9-122. Be sure to update the default exploded view using the View Manager.

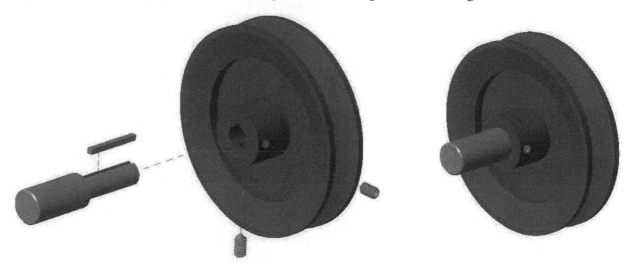

Figure 9-122 Exploded Assembly of the Step Shaft Components

9.3 Assemble the "**Adjustable_Shaft_Support**" from the pieces shown in Figures 9-123 through 9-128. Start with the base support (Figure 9-123) and use its coordinate system as the assembly's coordinate system. **Hint:** For the slanted setscrew in the fixed base align the centerlines, then use an edge constraint at the top of the hole instead of a surface constraint for the setscrew's placement. All dimensions are in inches.

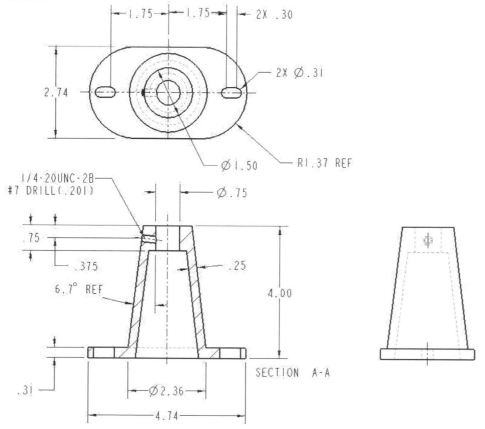

Figure 9-123   Base Support

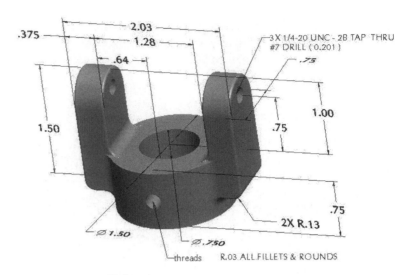

Figure 9-124   Vertical_Shaft

Figure 9-125   Yoke

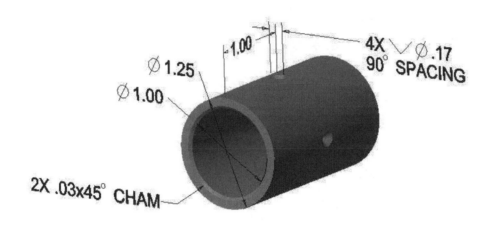

Figure 9-126   Bushing_Housing

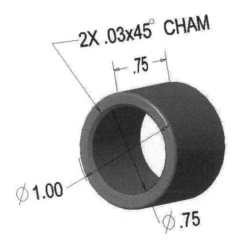

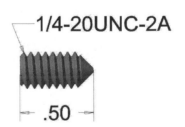

Figure 9-127   Bushing (2 required)

Figure 9-128 SetScrew-dogpt   (4 required)

9.4   Create an exploded assembly for the "**Adjustable_shaft_support**" assembly found in problem 9.3. Use the *Exploded View* icon to toggle between the two different views shown in Figure 9-129. Use a pin connection for the bushing housing (Figure 9-126) and limit its rotation to an angle of plus or minus 9 degrees (Figure 9-130). Be sure to update the default exploded view using the View Manager.

Figure 9-129   Exploded View

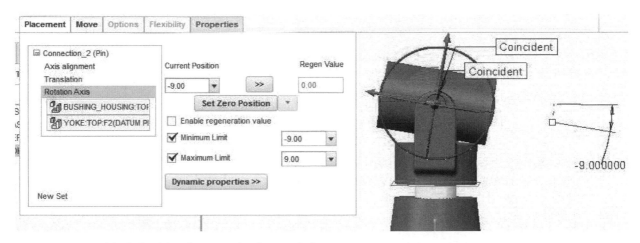

Figure 9-130   Limit Bushing Support Angle to ± 9 degrees

9.5 Create an Exploded assembly view of the **"take-up_frame"** that was put together in this chapter's exercise. Use the ***Exploded View*** icon to toggle between the two different views shown in Figure 9-131. Be sure to update the default exploded view using the View Manager.

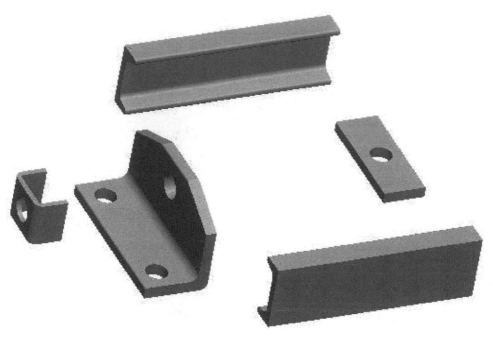

Figure 9-131 Exploded View of Take-up Frame

9.6 Create an Exploded assembly view of the **"take-up_piece_assembly"** that was put together in this chapter's exercise. Use the ***Exploded View*** icon to toggle between the two different views shown in Figure 9-132. Be sure to update the default exploded view using the View Manager. Be sure the ball bearing is not exploded, but rather only the setscrews are removed.

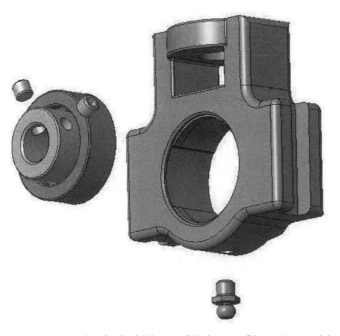

Figure 9-132 Exploded View of Take-up Piece Assembly

ASSEMBLY DRAWINGS

Objectives

▶ Create a template for making 1-view assembly drawings

▶ Create a template for making 2-view assembly drawings

▶ Adding a parts list to an assembly drawing

▶ Adding balloons to an assembly drawing

▶ Create assembly drawings

▶ Create exploded assembly drawings

▶ Create subassembly drawings

▶ Assembly Drawings Explored

A set of working drawings includes detailed drawings of each part and an assembly drawing showing how the parts fit together. An assembly drawing that shows only one part of a larger assembly is referred to as a subassembly drawing. An example of an assembly drawing is shown in Figure 10.1. An example of a subassembly drawing is shown in Figure 10.2.

The views shown in an assembly or subassembly drawing must show how the parts fit together and/or suggest how the device functions. Sometimes an exploded assembly view is used to show how the parts fit together. See Figure 10-3.

Assembly drawings can be made up of two orthographic views if this better shows how the parts fit together, as seen in Figure 10-4.

Since assembly drawings show how parts fit together, sectioning at least one view may make it easier to understand how the parts fit together. A full-section, half-section, or several removed-sections are the easiest way to show this. See Figure 10-5. When sectioning parts, be sure to section each part differently so the observer can distinguish between parts. Small sectioned parts should have their section lines close together. Relatively thin parts, such as gaskets or sheet metal, should be shaded solid or not at all. Often solid parts which fall in the cutting plane do not need to be sectioned. It is customary to show these parts

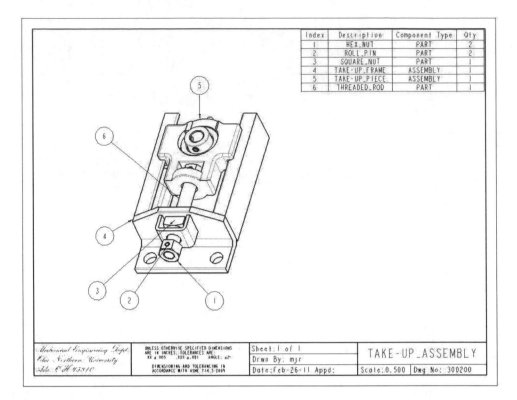

Index	Description	Component Type	Qty
1	HEX_NUT	PART	2
2	ROLL_PIN	PART	2
3	SQUARE_NUT	PART	1
4	TAKE-UP_FRAME	ASSEMBLY	1
5	TAKE-UP_PIECE	ASSEMBLY	1
6	THREADED_ROD	PART	1

TAKE-UP_ASSEMBLY

Sheet: 1 of 1   Drwn By: mjr   Date: Feb-26-11   Appd: Scale: 0.500   Dwg No: 300200

Figure 10-1   Assembly Drawing

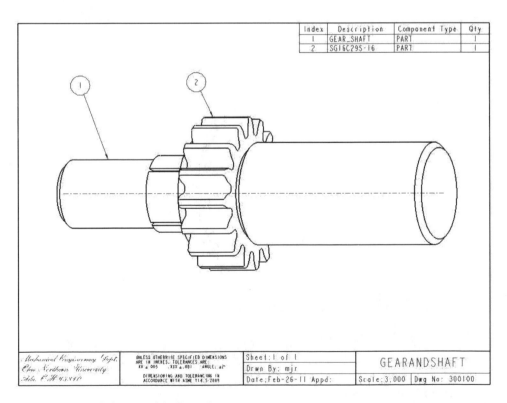

Index	Description	Component Type	Qty
1	GEAR_SHAFT	PART	1
2	SG16C29S-16	PART	1

GEARANDSHAFT

Sheet: 1 of 1   Drwn By: mjr   Date: Feb-26-11   Appd: Scale: 3.000   Dwg No: 300100

Figure 10-2   Subassembly Drawing

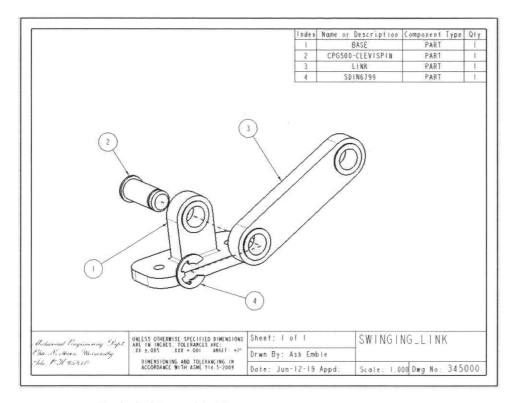

Index	Name or Description	Component Type	Qty
1	BASE	PART	1
2	CPG500-CLEVISPIN	PART	1
3	LINK	PART	1
4	SDIN6799	PART	1

Figure 10-3 Exploded Assembly View

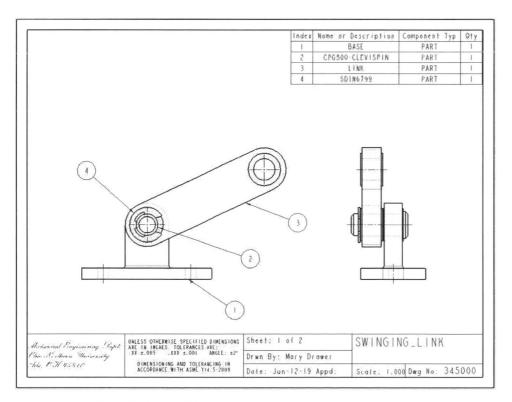

Index	Name or Description	Component Typ	Qty
1	BASE	PART	1
2	CPG500-CLEVISPIN	PART	1
3	LINK	PART	1
4	SDIN6799	PART	1

Figure 10-4 Two Orthographic Views

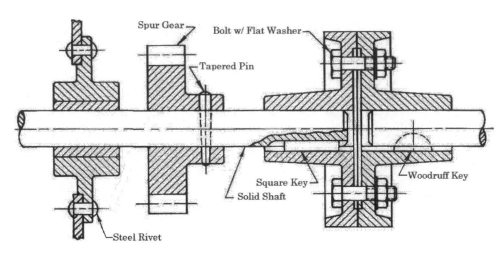

Figure 10-5   Assembly with Sectioned View

not sectioned or "in the round." These parts include screws, bolts, nuts, keys, pins, ball or roller bearings, gear teeth, and spokes. Hidden lines are often left out of an assembly drawing since the relationship between parts is the assembly drawing's main purpose. Hidden lines may tend to confuse the desired detail.

The purpose of a detailed drawing is to describe the size or shape of any part. Dimensions are shown on the detailed drawings; thus they are not included on the assembly drawing. A dimension can be shown on an assembly drawing if it indicates the amount of travel allowed by a given part or the overall height or width of the complete assembly. Keep this distinction in mind.

Parts are identified by placing a number inside a circle or balloon near the part with a leader line attaching it to the part. The circled numbers should be placed on the drawing in an orderly clockwise or counter-clockwise fashion. The balloon's leader lines should never cross one another. Leader lines close to each other should be approximately parallel.

A parts list, sometimes called a bill of materials, should include the part numbers, a description of the part, whether it is a subassembly or a part, and the required number of pieces. The parts list can be located in the upper right corner (Figure 10-6) or lower right corner of the drawing, or be located on a separate sheet.

No.	Name or Description	Component Type	Qty
1	HEX_NUT	PART	2
2	ROLL_PIN	PART	2
3	SQUARE_NUT	PART	1
4	TAKE-UP_FRAME	ASSEMBLY	1
5	TAKE-UP_PIECE	ASSEMBLY	1
6	THREADED_ROD	PART	1

Figure 10-6   Parts List

The border and title block for an assembly drawing should be the same as that for an engineering detailed drawing with one exception. In the title block for a detailed drawing, the part's material is listed. This does not apply for an assembly drawing; thus, this item should be replaced with the sheet number and total number of sheets for this assembly. See Figure 10-7.

Mechanical Engineering Dept. Ohio Northern University Ada, OH 45810	UNLESS OTHERWISE SPECIFIED DIMENSIONS ARE IN INCHES. TOLERANCES ARE: .XX ±.005 .XXX ±.001 ANGLE: ±2° DIMENSIONING AND TOLERANCING IN ACCORDANCE WITH ASME Y14.5-2009	Sheet: 1 of 2	SWINGING_LINK	
		Drwn By: Mary Drawer		
		Date: Jun-12-19 Appd:	Scale: 1.000	Dwg No: 345000

Figure 10-7   Title Block for Assembly Drawing

The procedure for creating engineering drawings was covered earlier. We will use these procedures in the creation of assembly and subassembly drawings.

Assembly Drawings Practice

Before we begin be sure the swinging linkage assembly, swinging_link.asm (Figure 10-8), the a_template_3dview.drw template (Figure 10-9), and the a-size_bordertitleblock.frm format file are in your working directory. If you did not create the exploded 3D view shown, go back to the assembly practice section and create it before continuing.

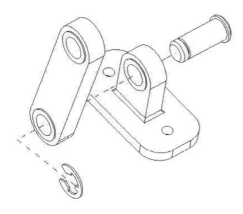

Figure 10-8   Exploded View of Swinging Link Assembly

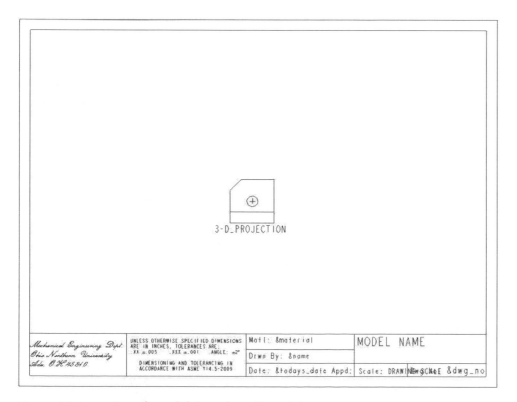

Figure 10-9   a_template_3dview.drw Template

Design Intent

Create a drawing template for an exploded assembly drawing with its parts list, border, and title block.

Step 1: Start Creo Parametric.

Step 2: Set your working directory.

Step 3: Open "a_template_3dview.drw."

Step 4: *Select* the Tools tab, then *select* the ***Template*** icon.

Step 5: **File>Save As>Save a copy**.

Step 6: Type "a_assembly_3dview" for the new name. Do not change the model name. *Pick* **OK.**

Step 7: **File>Close.**

Step 8: **File>Manage Session>Erase Not Displayed.** *Pick* **OK.**

Step 9: **File>Open.**

Step 10: *Select* "a_assembly_3dview.drw." *Pick* **Open.**

Step 11: *Select* the Tools tab, then *select* the ***Template*** icon.

Step 12: *Select* the Table tab.

Step 13: *Double-click* on the "Matl: &material" box in the title block.

Step 14: Replace the text with "Sheet: &sheet_number of &total_sheets" then use the LMB and pick anywhere in the graphics area.

The variable &sheet_number is a system parameter defined by Creo Parametric having the value of the current sheet number. The variable &total_sheets is defined as the total number of sheets present for this drawing.

Step 15: *Select* the Layout tab.

Step 16: *Pick* the one view in the graphics area, then *press* <Delete>.

Step 17: *Select* the ***Template View*** icon under the Layout tab.

Step 18: In the Template View Instructions window, name the view "ASSEMBLY" <Enter>. *Select* the **View States** option. To the right of Orientation type "3DVIEW" <Enter>. To the right of Explode type "Default" <Enter>. See Figure 10-10.

Step 19: For the **Model Display** option, *pick* **No hidden line.**

Step 20: For the **Tan Edge Display** option, *pick* **Dimmed.**

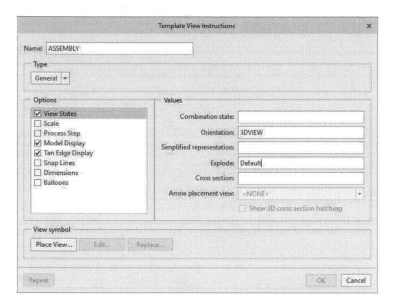

Figure 10-10   Set View State

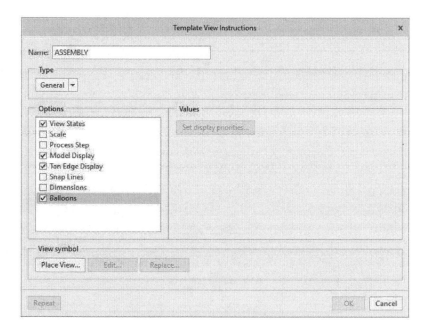

Figure 10-11   Check Balloons Option Box

Step 21: *Check* the **Balloons** option box. See Figure 10-11.

Step 22: *Pick* the **Place View** button, then *pick* a location on the left side of the graphics area above the tolerance title block information using one LMB click.

Step 23: *Pick* **OK** to close the Template View Instructions window.

Step 24: **File>Save.**

Design Intent

Create the parts list in the upper right corner of the drawing.

Step 25: *Select* the ⎢Table⎥ tab again.

Step 26: *Pick* the ***Table*** icon from the upper left corner of the screen, then highlight a 4 × 2 table. *Press* the LMB. Move the boxed area to the upper right corner of the border. Place it exactly on the upper and right border lines. The two lines will disappear when the table lines are directly over them. *Press* the LMB to place the table.

Table

Step 27: *Pick* the 4x2 table just placed, then *pick* the ***Properties*** icon from the pop-up menu. *Select* the second icon (***leftward and descending***) for the direction. For the column width (INCH), type "0.5" <Enter>. *Pick* **OK.** See Figure 10-12.

Step 28: *Pick* a 2nd column box in this table. *Pick* the arrow by the ***Select Table*** icon in the upper left corner of the ⎢Table⎥ tab, then *pick **Select Column***. With the cursor on a 2nd column box, press the LMB and a pop-up menu will appear. *Select* the ***Height and Width*** icon from this menu. Change the column width to 2.0 inches. *Press* <Enter>. *Pick* **OK.**

Step 29: *Pick* a 3rd column box with the LMB. With the cursor on the 3rd column of the table, press LMB again and a pop-up menu will appear. Select the ***Height and Width*** icon from this menu. Change the column width to 1.4 inches. Press <Enter>. *Pick* **OK.**

Step 30: ***Zoom In*** on the table. If the table lines do not lie on top of the border lines, select the table and move the table so that it lines up with the border lines in the upper right corner of the page. See Figure 10-13.

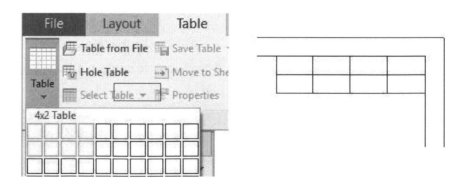

Figure 10-12 Pick Point for Table in Upper Right Corner of Border

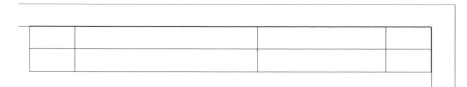

Figure 10-13 Create Parts List Table

Step 31: <u>*Double-clicking* on the table area allows you to enter text</u>. In the 1st (top) row of the table enter the following four descriptions in the four boxes: "Index," "Name or Description," "Component Type," and "Qty." (Do not press the <Enter> key.) See Figure 10-14.

Index	Name or Description	Component Type	Qty

Figure 10-14: Title Row of Parts List

Step 32: **>File>Save.**

Step 33: *Select* the ***Repeat Region…*** icon above the Data section of the Table tab. See Figure 10-15.

Step 34: In the Menu Manager window, *select* **Add,** then *select* **Simple.** *Pick* the 1st box in the second row. *Pick* the fourth box (under Qty) in the second row. <u>*Do not pick* the fourth box more than once</u> even if it looks like it wasn't selected. *S*croll down and *select* **Done** from the Menu Manager window. If this isn't done properly, the next step will not work.

Step 35: In the 2nd row, the 1st box of the table, *double-click* on the area to bring up the Report Symbol window. *Select* **rpt…,** then **Index.** Don't worry about the text not fitting inside the box area. Follow the same procedure for the rest of the second row.

Step 36: In the second box *select* **asm…, mbr…,** then **name.**

Step 37: In the third box *select* **asm…, mbr…,** then **type.**

Step 38: In the fourth box *select* **rpt…,** then **qty.**

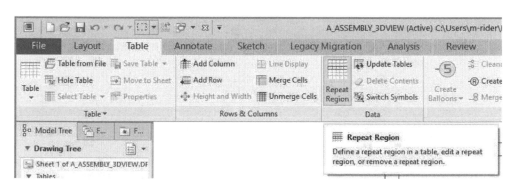

Figure 10-15 Repeat Region Tool

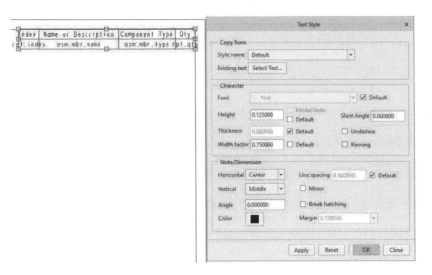

Figure 10-16: Text Style for Table

Step 39: Draw an imaginary box around the entire table to select it, then press down and hold the RMB until a pop-up menu appears. *Select* **Text Style** from this menu. Under Note/Dimensions set the horizontal setting to **Center** and the Vertical setting to **Middle.** *Pick* **Apply.** *Pick* **OK.** See Figure 10-16.

Step 40: *Select* the ***Repeat Region…*** icon under the Table tab. *Select* **Attributes** from the Menu Manager. Use the LMB to *pick* the second row of the table. *Select* **No Duplicates** from the menu manager window, then *select* **Done/Return.** Scroll down and *pick* **Done.**

Step 41: <u>With the Repeat Region still selected</u>, *select* the ***Create Balloons*** icon at the top of the screen. *Select* **Create Balloons—All** from the list.

Step 42: File>Save.

Step 43: File>Manage File>Delete Old Versions. *Pick Yes.*

Step 44: File>Close.

Now we are ready to try out our assembly drawing template. Note that our assembly needs a view called "**3DVIEW**" and this saved view needs an exploded view defined before we can use it with this assembly drawing template. Without these two features, our template will not create the desired results.

Design Intent

Create an exploded view assembly drawing of the swinging link assembly using the assembly drawing template just created. Be sure the saved orientation "**3DVIEW**" exists in your assembly.

Step 46: **File>New.**

Step 47: *Pick* the **Drawing** option in the New window. Enter its name as "swinging_link_assembly." *Pick* **OK.**

Step 48: With the **Use Template** button checked, Browse to your working directory and *select* the newly created assembly template, "a_assembly_3dview.drw."

Step 49: Since we erased all traces of the swinging linkage assembly from memory, *select* the **Browse** button by the default model, then *select* the "swinging_link.asm." *Pick* the **Open** button.

Step 50: In the New Drawing window, *pick* **OK.**

Step 51: For the parameter "name" type "your name"<Enter>.

Step 52: For the parameter "dwg_no" type "345000"<Enter>.

Step 53: Turn off axis, point, coordinate system, and plane display. *Select* the ***Repaint*** icon from the graphics toolbar if needed. See Figure 10-17.

Not bad for a starting point on an assembly drawing. Note that the scale in the title block shows "0.333," while the view could be drawn at full scale or "1.00" or "1:1." We will change this now.

Step 54: *Double-click* on the text "Scale: 0.333" in the lower left corner of your screen. In the window that appears, type "1.00" or "1/1" <Enter>. The exploded assembly drawing will change to full-size.

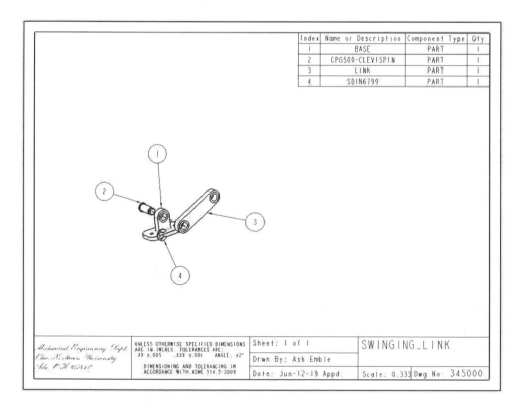

Index	Name or Description	Component Type	Qty
1	BASE	PART	1
2	CPG500-CLEVISPIN	PART	1
3	LINK	PART	1
4	SDIN6799	PART	1

UNLESS OTHERWISE SPECIFIED DIMENSIONS ARE IN INCHES. TOLERANCES ARE:
.XX ±.005   .XXX ±.001   ANGLE: ±2°
DIMENSIONING AND TOLERANCING IN ACCORDANCE WITH ASME Y14.5-2009

Sheet: 1 of 1

Drwn By: Ash Emble

Date: Jun-12-19 Appd:

SWINGING_LINK

Scale: 0.333 Dwg No: 345000

Mechanical Engineering Dept.
Ohio Northern University
Ada, OH 45810

Figure 10-17 Initial View of Assembly Drawing

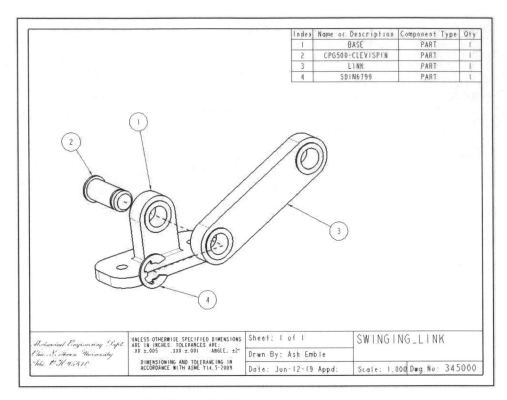

Figure 10-18   Exploded View at Full Scale

Step 55: *Select* the ⟨Layout⟩ tab. *Select* the view with the LMB, then press and hold down the RMB until a pop-up menu appears. *Select* **Lock View Movement** to unlock the view and move it toward the right so that the view is centered inside the drawing area (Figure 10-18).

We want to have the balloons going in a clockwise or counter-clockwise fashion around the assembly. If we move part 3's balloon below the link, the balloon numbers will go in a clockwise fashion. (Your balloons may be different.)

Step 56: *Select* balloon 4. *Press down and hold* the LMB while on top of the number 4. Drag the number away from the clevis pin 4. Move the arrowhead if necessary. See Figure 10-19.

Step 57: *Select* the balloon with the number 3 in it. *Press down and hold* the LMB, then drag the balloon above the link. *Release* the LMB.

Step 58: Press down and hold the RMB over the number 3 balloon until you get a pop-up menu. *Select* **Edit Attachment** from this menu. Using the LMB *pick* the top edge of the link. *Select* **Done/Return.**

Step 59: Move the other balloons around as needed. See Figure 10-19.

Step 60: **File>Save.** *Pick* **OK.**

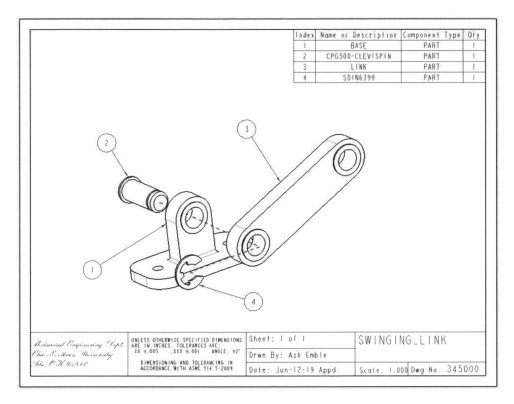

Index	Name or Description	Component Type	Qty
1	BASE	PART	1
2	CPG500-CLEVISPIN	PART	1
3	LINK	PART	1
4	SDIN6799	PART	1

	UNLESS OTHERWISE SPECIFIED DIMENSIONS ARE IN INCHES. TOLERANCES ARE: .XX ±.005 .XXX ±.001 ANGLE. ±2° DIMENSIONING AND TOLERANCING IN ACCORDANCE WITH ASME Y14.5-2009	Sheet: 1 of 1	SWINGING_LINK
Mechanical Engineering Dept. Ohio N. Northern University Ada, O.H. 45810		Drwn By: Ash Emble	
		Date: Jun-12-19 Appd:	Scale: 1.000 Dwg No: 345000

Figure 10-19   Final Assembly Drawing

Step 61: **File>Manage File>Delete Old Versions.** *Pick Yes.*

Step 62: **File>Print>Print.** *Pick* **OK** to print the assembly drawing to prove that you have completed this practice.

What if we wanted an unexploded assembly view instead of an exploded assembly view? The following three steps will provide this option.

Design Intent

Create an unexploded view assembly drawing of the swinging link assembly using the assembly drawing just created.

Step 63: *Select* the Layout tab.

Step 64: *Double-click* on the exploded view to bring up the Drawing View window. *Select* the **View States** category. *Uncheck* the **Explode components in view** box. *Pick* **Apply.** *Pick* **OK.** Move the view if necessary to center it in the drawing area.

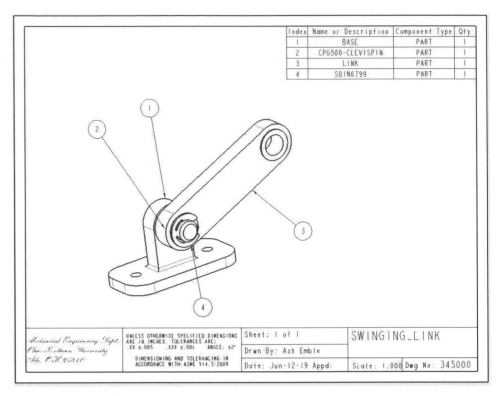

Figure 10-20   Assembly Drawing

Step 65: *Pick* the balloons one at a time, then move them and/or their attachments to clean up the drawing. *Select* **Done/Return** after locating each balloon. See Figure 10-20.

Step 66: **File>Print>Print.** *Pick* **OK** to print the assembly drawing to prove that you have completed this portion of the practice. (Save a copy if desired.)

Step 67: **File>Close.**

Step 68: Exit from Creo Parametric.

▶ Assembly Drawings Exercise

Before we begin, be sure the take-up frame assembly, "take-up_frame.asm" (Figure 10-21), the "a_template_2-view.drw" template (Figure 10-22), and the "a-size_bordertitleblock.frm" format file are in your working directory. If the take-up frame assembly doesn't have a FRONT (on the left in Figure 10-21) and RIGHT side view (on the right in Figure 10-21), go back to the assembly exercise section and create it using the View Manager before continuing. Don't worry if your two views are labeled differently than those shown in Figure 10-21; we can change the views that the assembly drawing uses later in this section.

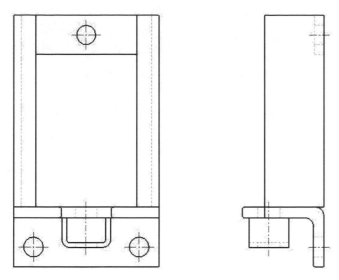

Figure 10-21   Front and Right Side View of
Take-up Frame Assembly

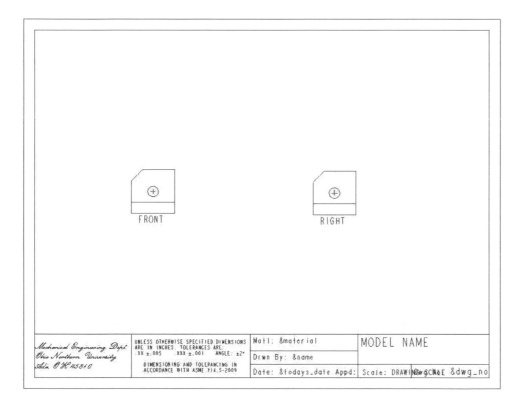

Figure 10-22   a_template_2-view.drw Template

Design Intent

Create a drawing template for a 2-view assembly drawing with its parts list, border, and title block.

Step 1: Start Creo Parametric.

Step 2: Set your working directory.

Step 3: Open "a_template_2-views.drw".

Step 4: *Select* the ⎾Tools⏋ tab, then *select* the **Template** icon.

Step 5: **File>Save As>Save a copy.**

Step 6: Type "a_assembly_2-views" for the new name. Do not change the model name. *Pick* **OK.**

Step 7: **File>Close.**

Step 8: **File>Manage Session>Erase Not Displayed.** *Pick* **OK.**

Step 9: **File>Open.**

Step 10: *Select* "a_assembly_2-views.drw". *Pick* **Open.**

Step 11: *Select* the ⎾Tools⏋ tab and *select* the **Template** icon.

Step 12: *Select* the ⎾Table⏋ tab.

Step 13: *Double-click* on the "Matl: &material" text in the title block to get a blinking cursor.

Step 14: Erase the text, then replace it with "Sheet: &sheet_number of &total_sheets". Use LMB and *pick* anywhere in the graphics area.

The variable &sheet_number is a system parameter defined by Creo Parametric having the value of the current sheet number. The variable &total_sheets is defined as the total number of sheets present for this drawing.

Step 15: *Select* the ⎾Layout⏋ tab.

Step 16: 🖱 *Pick* the front view in the graphics area, then use the RMB to bring up a pop-up menu. *Select* the **Properties** icon from this menu.

Step 17: In the Template View Instructions window, *select* the **View States** option. To the right of Orientation type "FRONT" <Enter> if it is not there.

Step 18: For the **Model Display** option, *pick* **Hidden line.**

Step 19: For the **Tan Edge Display** option, *pick* **Dimmed.**

Step 20: *Uncheck* **Snap Lines.**

Step 21: *Uncheck* **Dimensions.**

Step 22: *Check* the **Balloons** option box.

Step 23: *Pick* **OK** to close the Template View Instructions window.

Step 24: *Pick* the right side view in the graphics area, then use the RMB to bring up a pop-up menu. *Select* **Properties** from this menu.

Step 25: Leave the **View States** as is.

Step 26: For the **Model Display** option, *pick* **Hidden line.**

Step 27: For the **Tan Edge Display** option, *pick* **Dimmed.**

Step 28: *Uncheck* **Snap Lines.**

Step 29: *Uncheck* **Dimensions.**

Step 30: Leave **Balloons** *unchecked*.

Step 31: *Pick* **OK** to close the Template View Instructions window.

Step 32: File>Save.

Design Intent

Create the parts list in the upper right corner of the assembly drawing template.

Step 33: *Select* the Table tab again.

Step 34: *Pick* the *Table* icon from the upper left corner of the screen, then highlight a 4 × 2 table. *Press* the LMB. Move the boxed area to the upper right corner of the border. Place it exactly on the upper and right border lines. These two lines will disappear when the table lines are directly over them. *Press* the LMB to place the table.

Step 35: *Pick* the 4x2 table just placed, then *pick* the *Properties* icon from the pop-up menu. *Select* the second icon (***upper right to lower left***) for the direction. For the column width (INCH), type "0.5" <Enter>. *Pick* **OK.** See Figure 10-23.

Step 36: *Pick* a 2nd column box in this table. *Pick* the arrow by the *Select Table* icon in the upper left corner, then *pick* *Select Column*. With the

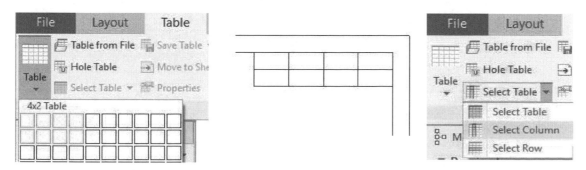

Figure 10-23 Pick Point for Table in Upper Right Corner of Border and Select Column

cursor on a 2nd column box, press the LMB and a pop-up menu will appear. Select the *Height and Width* icon from this menu. Change the column width to 2.0 inches. Press <Enter>. Pick OK.

Step 37: Pick a 3rd column box with the LMB. With the cursor on the 3rd column of the table, press LMB again and a pop-up menu will appear. Select the *Height and Width* icon from this menu. Change the column width to 1.4 inches. Press <Enter>. *Pick* **OK.**

Step 38: **Zoom In** on the table. If the table lines do not lie on top of the border lines, *select* the table and move the table so that it lines up with the border lines in the upper right corner of the page. See Figure 10-24.

Step 39: *Double-clicking* on the table area allows you to enter text. In the first (top) row of the table enter the following four descriptions: "Index," "Name or Description," "Component Type," and "Qty." (Do not press the <Enter> key.) See Figure 10-25.

Step 40: >**File**>**Save.**

Step 41: *Select* the *Repeat Region...* icon above the Data section of the Table tab. See Figure 10-26.

Figure 10-24 Create Parts List Table

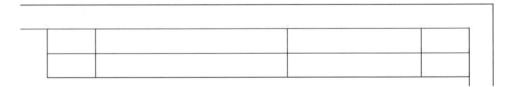

Figure 10-25 Title Row of Parts List

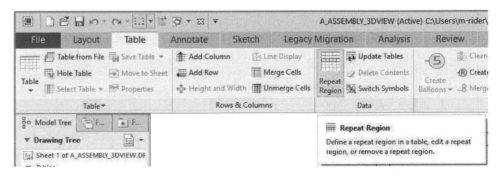

Figure 10-26 Repeat Region Tool

Step 42: In the Menu Manager window, *select* **Add,** then *select* **Simple.** *Pick* the first box in the second row. *Pick* the fourth box (under Qty) in the second row. <u>Do not pick the fourth box more than once</u> even if it looks like it wasn't selected. *Scroll* down and *select* **Done** from the Menu Manager window. If this isn't done properly, the next step will not work.

Step 43: In the second row, the first box of the table, *double-click* on the area to bring up the Report Symbol window. *Select* **rpt…,** then **Index.** Don't worry about the text not fitting inside the box area. Do the same for the rest of the second row.

Step 44: In the second box *select* **asm…, mbr…,** then **name.**

Step 45: In the third box *select* **asm…, mbr…,** then **type.**

Step 46: In the fourth box *select* **rpt…,** then **qty.**

Step 47: Draw an imaginary box around the entire table to *select* it, then press down and hold the RMB until a pop-up menu appears. *Select* **Text Style** from this menu. Under Note/Dimensions set the horizontal setting to **Center** and the Vertical setting to **Middle.** *Pick* **Apply.** *Pick* **OK.** See Figure 10-27.

Step 48: *Select* the ***Repeat Region…*** icon again. *Select* **Attributes** from the Menu Manager. Use the LMB to *pick* the second row of the table. *Select* **No Duplicates** from the menu manager window, then *select* **Done/Return.** Scroll down and *pick* **Done.**

Step 49: <u>With the Repeat Region still selected</u>, *select* the ***Create Balloons*** icon from the top of the screen. *Select* **Create Balloons—All** from the list. *Pick* the second row of the Parts List Table.

Step 50: **File>Save.**

Step 51: **File>Manage File>Delete Old Versions.** *Pick Yes.*

Step 52: **File>Close.**

Step 53: **File>Manage Session>Erase Not Displayed.** *Pick* **OK.**

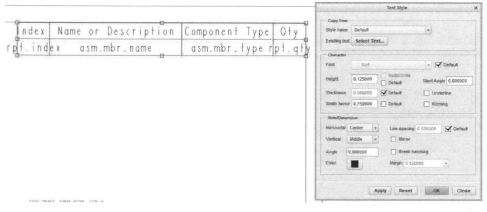

Figure 10-27 Text Style for Table

Now we are ready to try out our 2-view assembly drawing template. Note that our assembly needs a view called "**FRONT**" and a view called "**RIGHT**" before we can use it with this assembly drawing template. Without these two views, our template will not create the desired results.

Design Intent

Create a 2-view assembly drawing of the take-up frame assembly using the assembly drawing template just created.

Step 54: File>New.

Step 55: *Pick* the **Drawing** option in the New window. Enter its name as "take-up_frame_assembly". *Pick* **OK.**

Step 56: With the **Use Template** button checked, Browse to your working directory and *select* the newly created assembly template "a_assembly_2-views.drw". *Pick* **Open.**

Step 57: Since we erased all traces of the take-up frame assembly from memory, *select* the **Browse** button by the default model, then *select* the "take-up_frame.asm". *Pick* the **Open** button.

Step 58: In the New Drawing window, *pick* **OK.**

Step 59: For the parameter "name" type "your name"<Enter>.

Step 60: For the parameter "dwg_no" type "304100"<Enter>. See Figure 10-28.

If the left view is different, *select* it, *press* and hold the RMB to bring up a pop-up menu. *Select* **Properties.** Change the view to TOP, then *pick* **OK.** Answer **Yes** to "Do you want to modify the view orientation?" *Pick* **Yes.** See Figure 10-28.

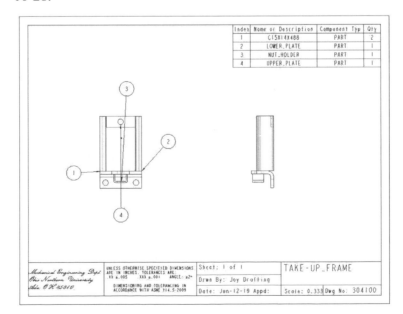

Figure 10-28 Initial View of Assembly Drawing

Not bad for a starting point on an assembly drawing. However, if you got the following error message (Figure 10-29), perform steps 61 through 75, then go back to step 54 and repeat. Go to step 76 if you didn't get an error message.

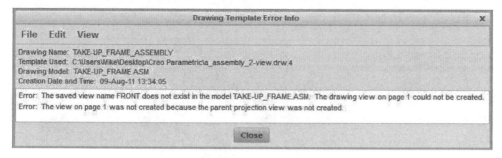

Figure 10-29   Drawing Template Error

Do steps 61 through 75 only if you got an error similar to Figure 10-29.

Step 61: *Pick* **Close** to close the error window.

Step 62: **File>Close.**

Step 63: **File>Manage Session>Erase Not Displayed.** *Pick* **OK.**

Step 64: **File>Open.**

Step 65: *Select* the "take-up_frame.asm". *Pick* the **Open** button.

Step 66: *Select* the *View Manager* in the graphics toolbar.

Step 67: *Select* the Orient tab in the View Manager window.

Step 68: *Pick* **New.** Type "Front"<Enter>.

Step 69: **Edit>Redefine** in the View Manager window. For reference 1 (Front), *pick* a TOP datum plane of the Lower_plate. For reference 2 (Right), *pick* a RIGHT datum plane of the Lower_plate. See Figure 10-30. (An alternate selection is for reference 1 (Front), *pick* a top surface of the take-up frame. For reference 2 (Top), *pick* the right side surface of the take-up frame.) *Pick* **OK.**

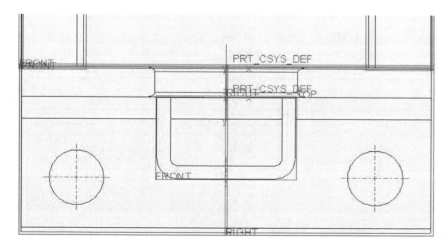

Figure 10-30   Defining Front View

Step 70: *Pick* **New.** Type "Right"<Enter>.

Step 71: **Edit>Redefine** in the View Manager window. For reference 1 (Front), *pick* a RIGHT datum plane of the Lower_plate. For reference 2 (Bottom), *pick* a FRONT datum plane of the Lower_plate. See Figure 10-31. *Pick* **OK.**

Step 72: *Pick* **Close** to close the View Manager window.

Step 73: **File>Save.**

Step 74: **File>Manage File>Delete Old Versions.** *Pick Yes.*

Step 75: **File>Close.** Go back to step 54.

Step 76: If your 2 views came in differently, such as in Figure 10-32, perform steps 77 through 83. If your drawing looks similar to Figure 10-28, go to step 84.

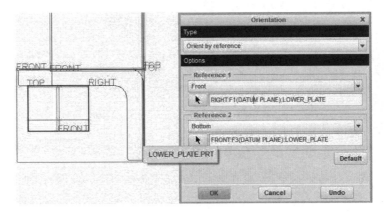

Figure 10-31   Defining Right View

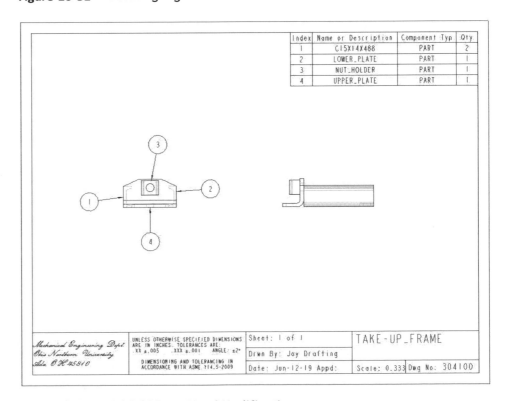

Figure 10-32   Initial Views Need Modification

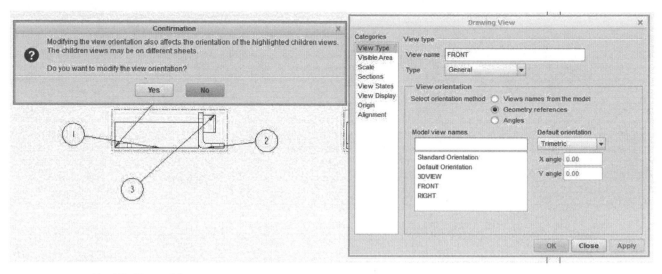

Figure 10-33   Modify Front View

Step 77: *Select* the Layout tab. *Pick* the front view so that it is outlined with a green dashed line. *Double-click* on the view to bring up the Drawing View window.

Step 78: Under **View orientation,** *select* **Geometry references.** When the confirmation window appears, *pick* **Yes.** See Figure 10-33.

Step 79: *Select* the **Default Orientation** button. See Figure 10-34.

Step 80: Set Reference 1 to **Front,** then using the LMB *pick* the top surface of the take-up frame.

Step 81: Set Reference 2 to **Right,** then *pick* the right side of the take-up frame. See Figure 10-35.

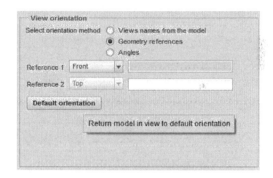

Figure 10-34   Default Orientation

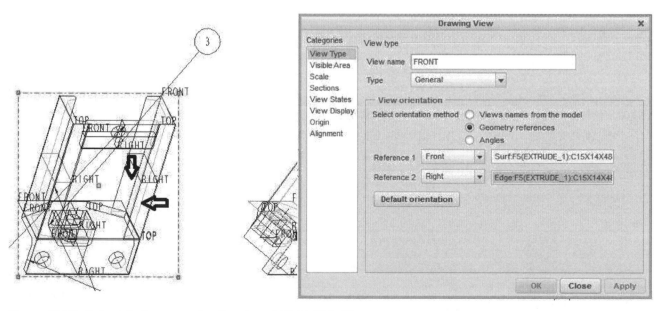

Figure 10-35   Select References for Front and Right Side Views

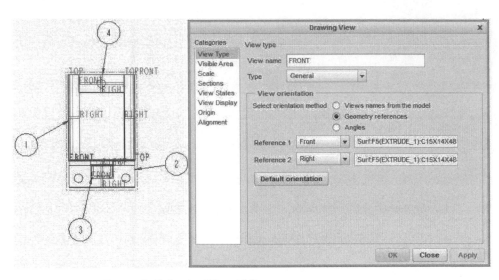

Figure 10-36   Properly Oriented Front View

Step 82: *Pick* the **Apply** button. If everything looks OK (Figure 10-36), *pick* the **Close** button. If not, *pick* the **Default orientation** button and repeat the above two steps.

Step 83: *Pick* anywhere on the drawing in an open area. Your drawing should look like Figure 10-28.

Step 84: *Select* the Table tab at the top of the screen. *Double-click* on the "Scale: 1:3" text in the lower left corner of the screen. Type "1/2" <Enter> to change the scale to half. The 2 views will increase in size. Turn OFF Axis, Point, Csys, and Plane Displays.

We want to have the balloons going in a clockwise or counter-clockwise fashion around the assembly. If we move part 2's balloon toward the lower left, the balloon numbers will go in a counter-clockwise fashion.

Step 85: *Select* the balloon with the number 2 in it. *Press down and hold* the LMB, then drag the balloon toward the lower left. *Release* the LMB.

Step 86: Press down and hold the **RMB** over the number 2 balloon until you get a pop-up menu. *Select* **Edit Attachment** from this menu. Using the LMB *pick* the left edge of the piece near the bottom hole. See Figure 10-37. *Select* **Done/Return** in the Menu Manager window.

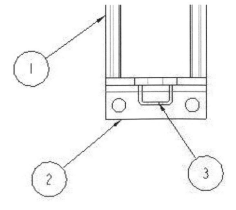

Figure 10-37   Modify Balloon 2's Location and Attachment

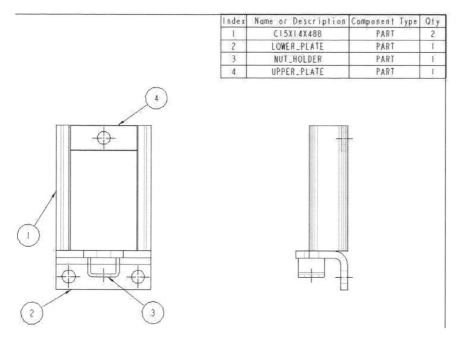

Index	Name or Description	Component Type	Qty
1	C15X14X488	PART	2
2	LOWER_PLATE	PART	1
3	NUT_HOLDER	PART	1
4	UPPER_PLATE	PART	1

Figure 10-38   Assembly Drawing with Balloons

Step 87: *Pick* the FRONT view so that a colored dashed line appears around it. Press down and hold the RMB until a pop-up menu appears. *Select* **Cleanup BOM Balloons** from the menu. *Pick* **Preview.** *Pick* **OK** in the Clean BOM Balloons window. Move balloons similar to Figure 10-38.

We need to add centerlines to our drawing.

Step 88: *Select* the ‾Annotate‾ tab.

Step 89: *Select* the ***Show Model Annotations*** icon at the top of the screen.

Step 90: *Select* the far right tab in the Show Model Annotations window.

Step 91: Using the LMB *pick* a hole in the front view, then pick the centerline that shows up. The box in the Show Model window will be checked.

Step 92: Use the LMB to *pick* each of the other centerlines in the front view. Use the RMB to search through the model until you locate the appropriate hole, then *select* it with the LMB.

Step 93: *Pick* the **Apply** button in the Show Model Annotations window.

Step 94: Using the LMB *pick* a hole in the right side view, then pick the centerline that shows up. See Figure 10-39.

Step 95: Using the LMB *pick* each of the other centerlines in the right side view.

Step 96: *Pick* the **Apply** button in the Show Model Annotations window.

Step 97: *Pick* the **Cancel** button in the Show Model Annotations window.

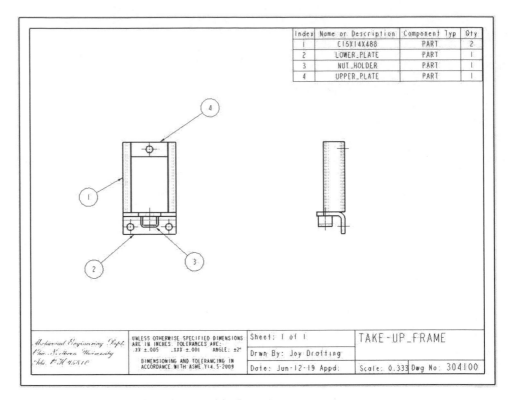

Figure 10-39   Completed Assembly Drawing

Step 98: **File>Save.**

Step 99: **File>Manage File>Delete Old Versions.** *Pick Yes.*

Step 100: **File>Print>Print** the assembly drawing to prove that you have completed this exercise. *Pick* **OK.** See Figure 10-39.

Step 101: **File>Close.**

Step 102: Exit from Creo Parametric.

▶ Review Questions

1. What is the purpose of an assembly drawing?

2. Why would you create different templates for assembly drawings and detailed drawings?

3. How do you add an existing part to an assembly?

4. What is a bill of materials? What information does it contain?

5. What feature of Creo Parametric is used to create the bill of materials so that it updates automatically?

6. Why would you add balloons to an assembly drawing?

7. Why would you create an exploded assembly view?

8. What is the difference between an assembly drawing and a subassembly drawing?

9. How many views are typically shown in a simple assembly drawing?

10. Name at least six of the ten ribbons at the top of the screen when making an assembly drawing. Are they different than the ribbons at the top when making a part drawing?

11. How do you add balloons to an assembly drawing?

Assembly Drawings Problems

10.1 Create an exploded assembly drawing of the adjustable shaft support shown in Figure 10-40. Drawing number 310100. Name this Creo assembly drawing 310100.

10.2 Create an unexploded 2-view assembly drawing of the adjustable shaft support shown in Figure 10-40. Drawing number 310101. Name this Creo assembly drawing 310101.

10.3 Create an unexploded assembly drawing of the adjustable shaft support shown in Figure 10-40. Drawing number 310102. Name this Creo assembly drawing 310102.

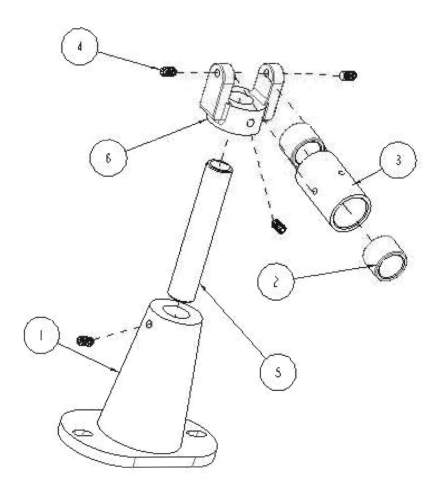

Figure 10-40 Adjustable Shaft Support (adjustable_shaft_support.asm)

10.4 Create an exploded subassembly drawing of the step shaft shown in Figure 10-41. Drawing number 410200. Name this Creo assembly drawing 410200.

10.5 Create an unexploded 2-view subassembly drawing of the step shaft shown in Figure 10-41. Drawing number 410201. Name this Creo assembly drawing 410201.

10.6 Create an unexploded subassembly drawing of the step shaft shown in Figure 10-41. Drawing number 410202. Name this Creo assembly drawing 410202.

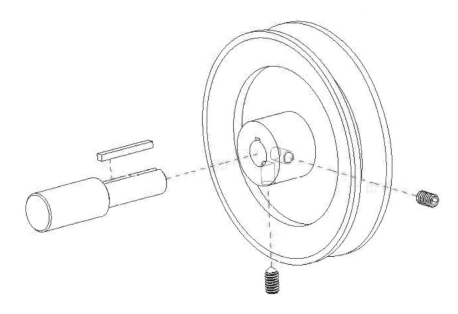

Figure 10-41 Step Shaft Subassembly (step_shaft_assembly.asm)

10.7 Create an exploded subassembly drawing of the take-up piece shown in Figure 10-42. Drawing number 410300. Name this Creo assembly drawing 410300.

10.8 Create an unexploded 2-view subassembly drawing of the take-up piece shown in Figure 10-42. Drawing number 410301. Name this Creo assembly drawing 410301.

10.9 Create an unexploded subassembly drawing of the take-up piece shown in Figure 10-42. Drawing number 410302. Name this Creo assembly drawing 410302.

Figure 10-42 Take-up Piece (take-up_ piece.asm)

10.10 Create an exploded assembly drawing of the take-up assembly shown in Figure 10-43. Drawing number 310400. Name this Creo assembly drawing 310400.

10.11 Create an unexploded 2-view assembly drawing of the take-up assembly shown in Figure 10-43. Drawing number 310401. Name this Creo assembly drawing 310401.

10.12 Create an unexploded assembly drawing of the take-up assembly shown in Figure 10-43. Drawing number 310402. Name this Creo assembly drawing 310402.

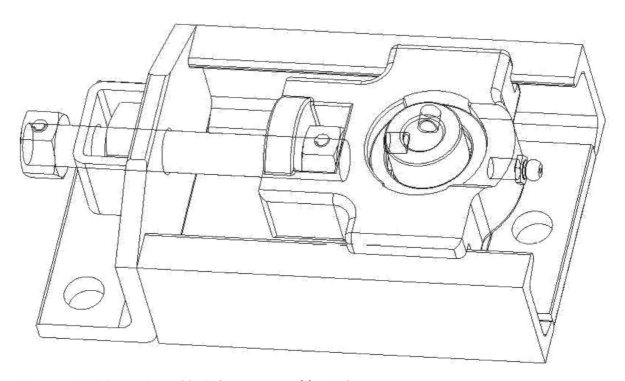

Figure 10-43 Take-up Assembly (take-up_ assembly.asm)

RELATIONS AND FAMILY TABLES

Objectives

▶ Become familiar with parameters
▶ Learn to use relationships

▶ Create similar-part family tables

Relations and Family Tables Explored

Parameters are variables that can be assigned to parts and assemblies which give additional information not found in the created dimensions. Parameters can drive dimensions through relations, or be driven by relations. A parameter can be used as a column in a family table such as THDPERINCH representing the number of threads per inch for a threaded fastener. Parameters can be real numbers, integers, strings of characters, or Yes/No answers.

Part relations provide a way to capture design intent in a model in addition to the methods used when creating sketches and features. The relations created exist as a mathematical function associated with the part and enforce these mathematical rules each time the part is regenerated. Relations automate the value of some dimensions each time other dimensions are changed.

Relations are written using either the symbolic form of the dimension (such as d1, d2, etc.) or by using customized dimension names (such as Diameter, Length, Width, etc.). The menu item at the bottom of the Model ribbon **Model Intent>Switch Dimensions** toggles between the dimensional names and the dimensional values. It is recommended that the dimensions used in relations be renamed for identification purposes.

There are several types of relations that can be used.

1. Part relations are written using any of the dimensions in the model. Part relations are calculated at the beginning of the regeneration cycle before the features are updated.
2. Feature relations are written in the context of a particular feature. They are calculated with the feature during its regeneration cycle.
3. Sketcher relations are calculated with the feature during its regeneration cycle just like feature relations. The difference is that sketcher relations are written within the sketcher using sketcher variables such as sd1, sd2, etc., instead of d1, d2, etc.

There are several operators that can be used when writing relations. The two basic types are assignment statements and conditional if statements. For assignment statements a variable is listed on the left followed by an equal sign and another already defined variable or a mathematical expression made using + - * / ^ ( ) [ ].

Variable = assignment statement
Width = Length/3
Outside_Dia = 2 * Inside_Dia
Depth = Width
Volume = pi*Dia^2*Length/4

The conditional if statement takes on the form of:

if logical_expression
do this if logical_expression is true
else
do this if logical_expression is not true
endif

The logical operators are:

>   Greater than
<   Less than
>= Greater than or equal to
<= Less than or equal to
== Equal to
!= Not equal to

Some of the built-in functions and parameters are listed in Table 11-1.

Table 11-1 Some Built-in Functions and Parameters				
abs	acos	asin	atan	atan2
bound	cable_len	ceil	cos	cosh
dead	eang	elen	exists	exp
floor	if	ln	log	massprop_param
material_param	max	min	mod	mp_mass
mp_surf_area	near	pow	sign	sin
sinh	sqrt	string_length	tan	tanh
false	no	pi	true	yes

In addition, there are options to help you verify the equations that you have written. They are

Verify—computes the relations to verify that they are valid.
Sort—sorts the relations according to their dependencies on previous lines of code.

The following example uses two parameters (DIAMETER & LENGTH) to determine the size of a solid cylinder, and one parameter (VOLUME) to store the volume of the solid cylinder. See Figure 11-1. Three relations are defined to connect the diameter and length of the cylinder to the actual dimensions, d1 and d0, and to calculate the volume of the solid cylinder. See Figure 11-2.

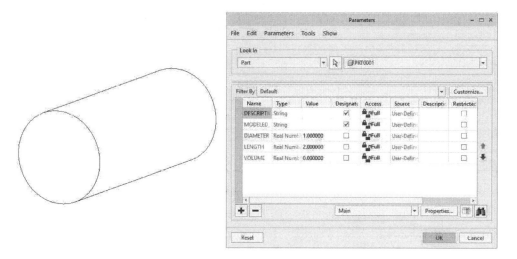

Figure 11-1 Cylinder Parameters Defined

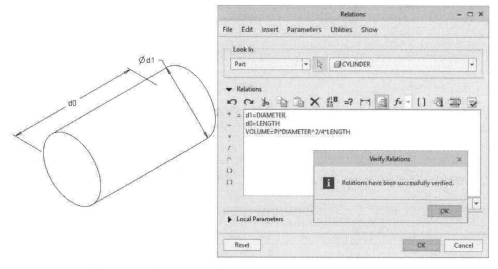

Figure 11-2 Cylinder Relations Defined

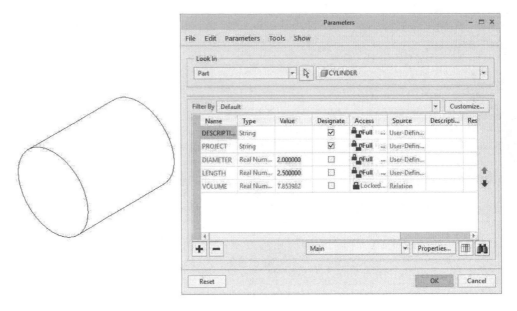

Figure 11-3   Modified Parameters

To change the size of the cylinder, simply change the parameter values DIAMETER and LENGTH, then *select* the ***regenerate*** icon from the top of the screen. The cylinder will change size accordingly and the volume will be updated. See Figure 11-3. With DIAMETER = 2 and LENGTH = 2.5, the VOLUME equals 7.853982.

A Family Table is a collection of parts that are similar but deviate slightly from each other. For example, roundhead machine screws come in various sizes but look similar except for their basic size (diameter) and length, and they perform the same function. Therefore, it is useful to think of them as a family of parts. Parts in a Family Table are known as table-driven parts.

When creating a family table, Creo Parametric doesn't create a *.prt file for each instance. The instances are virtual. When a table instance is opened, Creo Parametric actually opens the generic (original) model, then regenerates the table instance according to the information in the family table row. If the family of parts needs to be modified, the user must modify the generic model. The individual instances cannot be modified other than changing the defined values for the instance.

Family tables allow you to create and store large numbers of similar parts compactly within a single part file. They save time by standardizing part generation. They create variations in similar parts without using relations to change the model. Family tables can be included on a part drawing to show the members of the family.

Relations Practice

The base radius, R_1, can be calculated as follows:

$$Base\ Radius = R_1 = \frac{\left(\dfrac{3*height}{\tan(angle)}\right) + \sqrt{\dfrac{36*Volume}{\pi*height} - 3*\left(\dfrac{height}{\tan(angle)}\right)^2}}{6}$$

The cone frustum is best created using the revolve function. Let's begin by starting Creo Parametric and setting the working directory.

Step 1: Start Creo Parametric by *double-clicking* with the LMB on the **Creo Parametric** icon on the desktop, or from the Program list: Creo Parametric.

Step 2: Set your working directory, by *selecting* the **Select Working Directory** icon in the Home ribbon. Locate your working directory, and then *pick* **OK.**

Step 3: Create a new object by *picking* the **New** icon at the top of the screen, or by *selecting* **File>New.**

Step 4: *Select* **Part** from the window and name it "Cone_Frustum_Volume." The sub-type should be Solid. Make sure the default template is checked, and then *pick* **OK.**

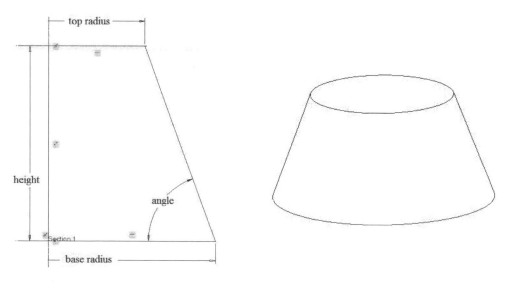

Figure 11-4 Frustum of a Right Cone

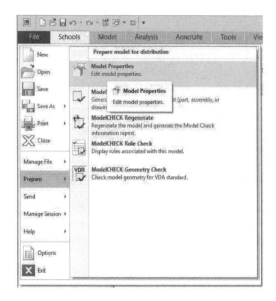

Figure 11-5   Model Properties

An XYZ coordinate system will appear in the middle of the screen along with FRONT, TOP, and RIGHT side planes. The words FRONT, TOP, and RIGHT will appear if the appropriate option is set. (**File>Option>Entity Display,** *check* **Show datum plane tags.**) We will create the frustum by sketching its 2D cross-section, then revolving it around an axis.

Step 5:   We need to set the units for our cone frustum to the IPS (Inch-Pound-Second) system of units. Use: **File>Prepare>Model Properties.** See Figure 11-5.

Step 6:   The following Model Properties window appears as shown in Figure 11-6. *Pick* the word **change** on the right end of the Units row. Note that the material property is not assigned to the part at this time.

Step 7:   The Units Manager window (Figure 11-7) appears. If the system of units, **IPS,** is not selected as shown below, *pick* it from the list, then *pick* the ⟶ Set... button.

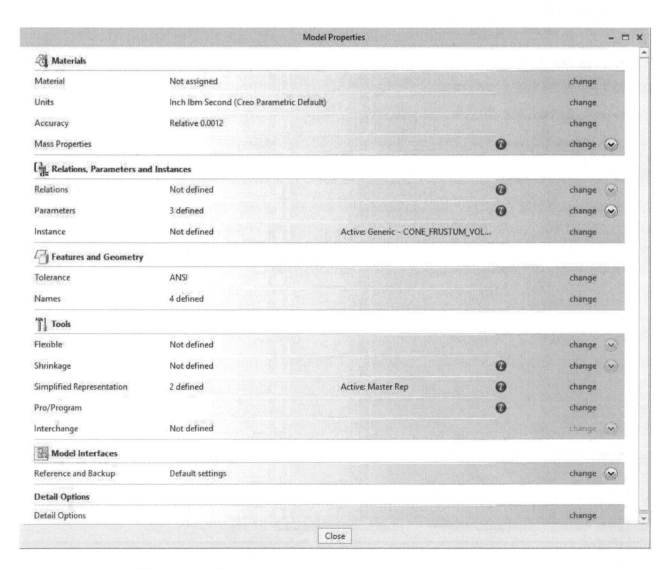

Figure 11-6   Model Properties Window

Step 8: After picking the [→ Set...] button, the following window appears. *Select* Convert dimensions (for example 1″ becomes 25.4mm), then *pick* **OK.** See Figure 11-8.

Step 9: *Pick* **Close** on the Units Manager window. *Pick* **Close** on the Model Properties window.

Let's create the cone frustum by picking the revolve tool, then sketching the cross-section.

Step 10: ◑ Revolve *Pick* the **revolve** tool from the top of the screen.

Step 11: Move the cursor onto the FRONT plane and *select* it by *pressing* the LMB.

Step 12: [icon] *Pick* the **Sketch View** icon if necessary to orient the sketch plane so that it is parallel with the display screen.

Step 13: [Centerline] Sketch a vertical geometry centerline through the origin.

Step 14: Sketch the cross-section of the cone frustum as shown in Figure 11-9, and then *pick* the **checkmark** to exit from sketcher. Use a reference dimension on the top.

Step 15: *Pick* the **green checkmark** to accept the revolved section. See Figure 11-10.

Step 16: *Select* [Tools] tab, then select **Parameters**.

Step 17: *Pick* the **plus sign,** then type "VOLUME" for the name. Tab twice, and then type "1000" for the value. Check the Designate box.

Step 18: *Pick* the **plus sign,** then type "CONE_ANGLE" for the name. Tab twice, and then type "60" for the value. Check the Designate box.

Step 19: *Pick* the **plus sign,** then type "CONE_HEIGHT" for the name. Tab twice, and then type "6" for the value. Check the Designate box.

Figure 11-7   Units Manager Window

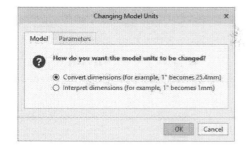

Figure 11-8   Changing Model Units

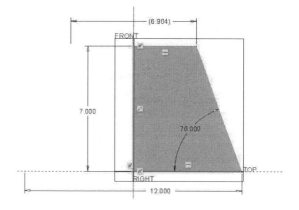

Figure 11-9   Cross-section of Cone Frustum Plate

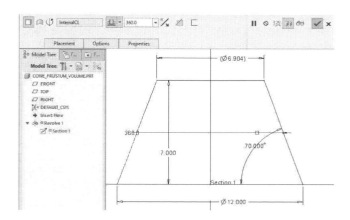

Figure 11-10   Revolved Section

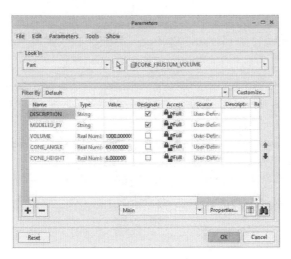

Figure 11-11   Define Parameters

Step 20: *Pick* **OK** to close the Parameters window. See Figure 11-11.

Step 21:   ⟨d1⟩ Right-click on **Revolve 1** in the model tree, then *select* **Edit** from the pop-up menu.

Step 22: Under Model tab, *select* **Switch Dimensions** under **Model Intent** at the far right end of the ribbon menu.

Step 23: Use LMB to pick a variable such as d5. In the upper left corner of the screen, change the name to reflect its function, such as "height," "angle," and "base_diameter." See Figure 11-12. The other variable names do not need to be modified.

Step 24: *Select* **Relations** under **Model Intent** at the far right end of the ribbon menu. Enter the comment (enclosed using /* … */) and the four lines of code shown below.

> height=CONE_HEIGHT
> angle=CONE_ANGLE
> /* Calculate required base diameter of cone frustum */
> htan=height/tan(angle)
> base_diameter = (3*htan+sqrt(36*VOLUME/(pi*height)-3*htan^2))/3

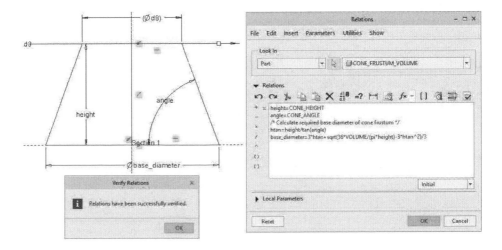

Figure 11-12   Rename Variables in Cross-section

Step 25: 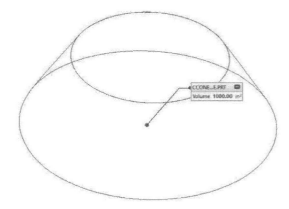 Disable Units check. Verify the relationships, then *pick* **OK.**

Step 26: *Pick* the regenerate icon in the upper left corner of the screen to make the cone frustum update to the parameters and relations you have set.

Step 27: *Select* the |Analysis| tab, then *pick* the **Volume tool** icon from the ribbon. Verify that the volume is correct. See Figure 11-13.

Step 28: Change the parameter, VOLUME, to a different value, then regenerate the part again. Verify the volume.

Figure 11-13   Regenerated Cone Frustum

Step 29: Change the variable, CONE_HEIGHT or CONE_ANGLE, to a different value, then regenerate the part. Verify the volume.

Step 30: *Select* the |Model| tab again.

Step 31: **File>Save.** *Pick* **OK.**

Step 32: **File>Manage File>Delete Old Versions.** *Pick* Yes.

Step 33: **File>Close.**

Step 34: Exit from Creo Parametric.

Relations and Pattern Practice

We will use relations and patterns to create a roller chain sprocket for any size chain between #40 and #160 and any number of teeth. The user will also define the diameter of the center hole.

Design Intent

For roller chain sizes #40 through #160 create a corresponding sprocket with a specified number of teeth. The sprocket will have a specified size diameter hole at its center along with two mounting holes. The pitch in inches is the chain number divided by 80. The thickness of the sprocket is 0.57 times the pitch. The pitch diameter is the pitch divided by the sine of 180 degrees divided by the number of teeth on the sprocket. The two ¼-inch mounting holes are located on a bolt circle equal to twice the shaft diameter minus ⅛ inch.

Step 1: Start Creo Parametric by *double-clicking* with the LMB on the **Creo Parametric** icon on the desktop, or from the Program list: Creo Parametric.

Step 2: Set your working directory by *selecting* the **Select Working Directory** icon in the Home ribbon. Locate your working directory, and then *pick* **OK.**

Step 3: Create a new object by *picking* the *New* icon at the top of the screen, or by *selecting* **File>New.**

Step 4: *Select* **Part** from the window and name it "Sprocket_Design." The subtype should be Solid. Make sure the default template is checked, and then *pick* **OK.**

An XYZ coordinate system will appear in the middle of the screen along with FRONT, TOP, and RIGHT datum planes. The words FRONT, TOP, and RIGHT will appear if the appropriate option is set. (**File>Option>Entity Display,** *check* **Show datum plane tags.**) We will create the sprocket blank by sketching a circle equal to the outside diameter of the sprocket, then extruding it to the appropriate thickness.

Step 5: Set the units for the sprocket to the IPS (Inch-Pound-Second) system of units. Use **File>Prepare>Model Properties.**

Step 6: *Pick* the word **change** on the right end of the Units row in the Model Properties window. When the Units Manager window appears, *select* **IPS** from the list, then *pick* the ➔ Set... button.

Step 7: *Select* Convert dimensions (for example 1″ becomes 25.4mm), then *pick* **OK.**

Step 8: *Pick* **Close** on the Units Manager window.

Step 9: Change material from not assigned to "Standard-Materials_Granta-Design", "Ferrous-metals", **Steel_high_carbon**. *Pick* **OK.** *Pick* **Close** on the Model Properties window.

Step 10: *Select* the Tools tab, then *select* **Parameters**. In the Value column type "sprocket" for the Description and "Your Initials" for Modeled by.

Step 11: *Pick* the **plus sign** in the lower left corner to add a new entry. Under Name, type "TEETH," under Type *select* **Integer,** under Value type "18," then *check* the **Designate** box. See Figure 11-14.

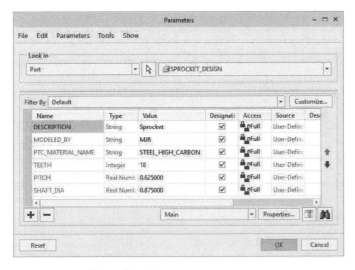

Figure 11-14 Sprocket Parameters

Step 12: *Pick* the ***plus sign*** again to add a new entry. Under Name, type "PITCH", under Type *select* Real, under Value type "0.625", then *check* the **Designate** box.

Step 13: *Pick* the ***plus sign*** again to add a new entry. Under Name, type "SHAFT_DIA", under Type *select* Real, under Value type "0.875", then *check* the **Designate** box.

Step 14: *Pick* **OK** to close the Parameters window.

Step 15: Under the |Tools| tab, *select* ***Relations***. In the Relations window type the following code and then *select* the ***Execute/Verify Relations*** icon. See Figure 11-15. Correct any errors, then *pick* **OK.**

R1=0.272*PITCH+0.036
R2=2.07*PITCH+0.255
THICKNESS=0.57*PITCH
DIAROLLER=0.628*PITCH+0.012
PITCHDIA=PITCH/SIN(180/TEETH)
OD=PITCHDIA+0.625*PITCH-0.25*DIAROLLER
BOLTCIRCLE=2*SHAFT_DIA-0.125

Step 16: *Pick* the |Model| tab. *Pick* the ***Extrude*** tool.

Step 17: Move the cursor onto the FRONT plane and *select* it by *pressing* the LMB.

Step 18: 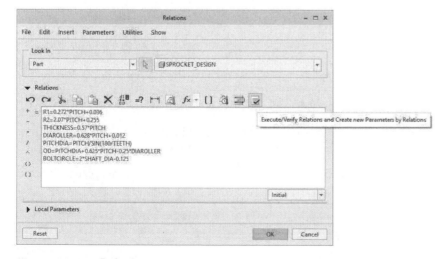 *Pick* the ***Sketch View*** icon if necessary to orient the sketch plane so that it is parallel with the display screen.

Step 19: Sketch a circle centered at the origin. *Double-click* on the dimensional value, then type "OD" <Enter>. When the question, "Do you want to add this relation – sd0 = OD?" appears, *pick* **Yes.** The value should change to 3.889 inches. *Pick* the ***green checkmark*** to exit from sketcher and keep the sketch. In the thickness box at the top of the screen, type "THICKNESS" <Enter>. *Pick* **Yes** when asked if you want to add THICKNESS as a feature relation. *Pick* the ***green checkmark*** to accept the extrusion.

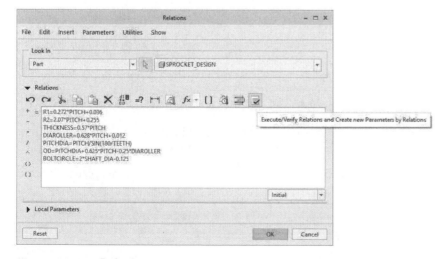

Figure 11-15 Relations

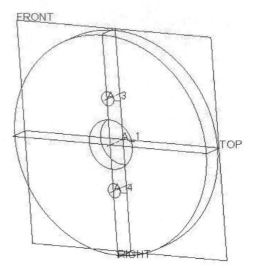

Step 20: *Select* the **Hole** tool, then add a hole at the origin. For its diameter, type "SHAFT_DIA" <Enter> and add it as a relation. Make it a through-hole. Accept the feature.

Step 21: *Select* the **hole** tool again, then add a hole directly above the center hole. For its diameter, type "0.25". Pick the |Placement| tab. Select the Offset References area with the LMB. Pick the TOP datum plane, hold down <Ctrl> and pick the RIGHT datum plane. For the TOP datum plane, set the Offset to "BOLTCIRCLE/2" and add the relation. For the RIGHT datum plane, select Align. Make it a through-hole. Accept the feature.

Figure 11-16   Sprocket Blank with Holes

Step 22: With Hole 2 highlighted, *select* the **Pattern** tool. In the first box, *select* **Axis.** Rotate the view a bit so you can *pick* the A_1 axis at the center of the sprocket blank. Set the number of patterns to "2" and the incremental angle to "180" degrees. Accept the pattern by *picking* the **green checkmark.** See Figure 11-16. Regenerate the model, <Ctrl>G.

Step 23: **File>Save.** *Pick* **OK.**

Step 24: *Pick* the **Extrude** tool. Move the cursor onto the front surface of the sprocket (not FRONT datum plane) and *select* it by pressing the LMB.

Step 25: ⬚ *Pick* the **Sketch View** icon if necessary to orient the sketch plane so that it is parallel with the display screen.

Step 26: Sketch a circle centered at the origin. *Double-click* on the dimensional value, then type "PITCHDIA" <Enter>. When the question, "Do you want to add this relation – sd0 = PITCHDIA?" appears, *pick* **Yes.** The value should change to 3.599 inches. *Select* the circle with the LMB. When a pop-up menu appears, *select* the Construction icon ⬚ from this menu to make this circle a construction circle. It will change to a dotted circle.

Step 27: **Zoom in** on the upper area of the sprocket blank so we can add a circle where the construction circle crosses the RIGHT datum plane.

Step 28: Use the **Dimension** tool to dimension the circle as a radius instead of a diameter. For its value, type "R1" and add the relation created.

Step 29: Use the **Line** tool to add two lines, each offset from the RIGHT datum plane by 62.85 degrees so the total enclosed angle is 125.7 degrees. Trim the lines at the edge of the small circle. See Figure 11-17.

Step 30: *Select* the ***References*** tool, then *pick* the outside diameter of the sprocket blank. *Pick* **Close.**

Step 31: *Select* the ***three-point arc*** tool. Draw two arcs as shown in Figure 11-17 so that each arc is tangent to the small circle or perpendicular to the previously drawn lines.

Step 32: *Double-click* on one of the arc radii values, then type "R2" <Enter> and accept the feature relation. The value should change to 1.549 inches.

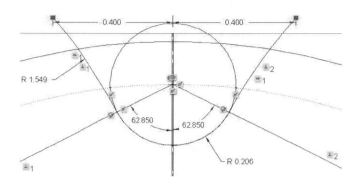

Figure 11-17 Initial Tooth Profile Sketch

Step 33: Use the ***Equal*** constraint tool to make the two arcs equal size.

Step 34: Use the ***Delete Segment*** tool to remove the upper portion of the small circle.

Step 35: Change the two short angled lines to Construction lines.

Step 36: *Select* the ***Project*** tool. *Pick* the outside diameter of the sprocket blank.

Step 37: Use the ***Delete Segment*** tool to remove the lower portion of the outside sprocket circle. The final sketch should appear as shown in Figure 11-18.

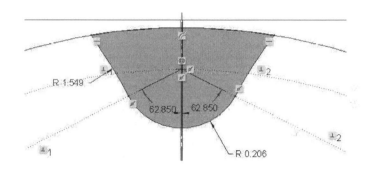

Figure 11-18 Completed Tooth Profile Sketch

Step 38: *Select* the green checkmark to exit from sketcher and keep the sketch.

Step 39: Select the ***Remove Material*** icon in the Extrude toolbar. *Select **Through All Surfaces*** for the thickness.

Step 40: Orient the part so you can see if the arrows point into the part. If they don't, then pick the Arrow icon to reverse its direction. See Figure 11-19.

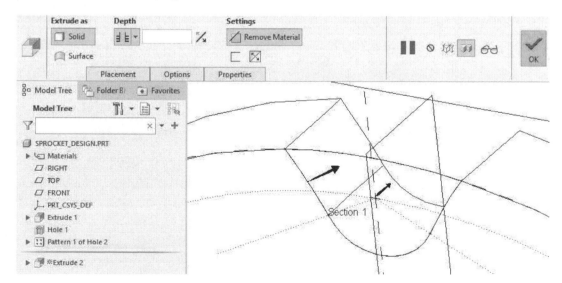

Figure 11-19 Extruded Section

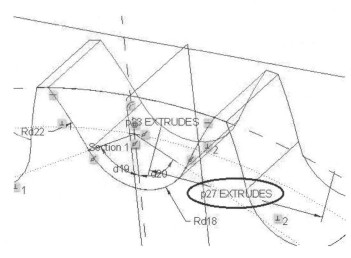

Figure 11-20   Variable Name for Number of Extrusions

Step 41: *Pick* the **green checkmark** to accept the extrusion cut.

Step 42: *Highlight* **Extrude 2** in the Model tree. *Select* the **Pattern** tool.

Step 43: *Pick* **Axis** in the first box of the Pattern ribbon. Rotate the sprocket blank a bit so you can *pick* the A_1 axis located at the origin. Set the number of patterns to "18" and the incremental angle to "360/TEETH" <Enter> degrees. *Pick* **Yes** to accept the feature relation. Accept the pattern by *picking* the **green checkmark.**

Step 44: *Pick* the arrow in front of Pattern 2 of Extrude 2 in the Model tree so that it shows the 18 extrusion instances. *Select* the first instance with the LMB. When a pop-up menu appears, *pick* the **Edit Dimensions** icon from this menu.

Step 45: *Select* the ⎍Tools⎍ tab. *Select* **Switch Dimensions** to view the variable names for the sprocket parameters. Note the name of the variable that represents the number of extrusions. In my case, it is called "p27". See Figure 11-20.

Step 46: *Select* the **Relations** icon. Add a new line to the relations, "p27=TEETH," then *pick* **OK** to close the window.

Step 47: ⬚ Fit the part to the screen. See Figure 11-21.

Step 48: **File>Save.**

Step 49: **File>Manage File>Delete Old Versions.** *Pick Yes.*

Step 50: **File>Print>Print** to prove that you have completed this practice.

Step 51: *Select* the **Parameters** icon. Change the number of teeth from 18 to 24. For #60 chain, change the PITCH to 0.750 inches (pitch = 60/80; the pitch in inches is the chain number divided by 80). Change the SHAFTDIA from 0.875 to 1.00 inches. *Pick* **OK** to close the window.

Step 52: *Pick* the **Regenerate** icon in the upper left corner of the screen or type <Ctrl>-G. See Figure 11-22.

Step 53: Repeat steps 50 and 51 with other values to verify your model works correctly.

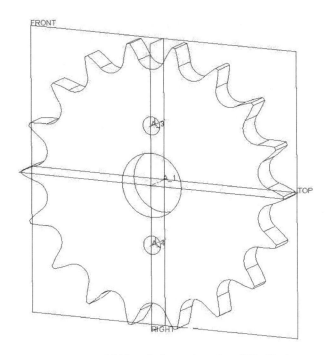

Figure 11-21 18-Tooth Sprocket for #50 Chain

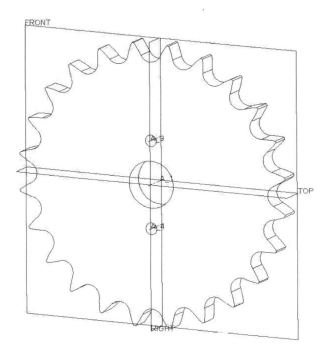

Figure 11-22 24-Tooth Sprocket for #60 Chain

Relations and Family Tables Practice

There are many mechanical parts used in industry that are scaled copies of each other. For example, a normal flat washer's outside diameter, hole diameter, and thickness follow a set pattern. See Table 11-2.

Table 11-2 Normal Flat Washer Dimensions				
Basic Size (inches)	Inside Diameter (inches)	Outside Diameter (inches)	Outside Diameter2 (inches)	Washer Thickness (inches)
0.138	0.156	0.375		0.049
0.164	0.188	0.438		0.049
0.19	0.219	0.500		0.049
0.188	0.250	0.562		0.049
0.216	0.250	0.562		0.065
0.25	0.281	0.625		0.065
0.3125	0.344	0.688		0.065
0.375	0.406	0.812		0.065
0.4375	0.469	0.922		0.065

Table 11-2 Continued

Basic Size (inches)	Inside Diameter (inches)	Outside Diameter (inches)	Outside Diameter2 (inches)	Washer Thickness (inches)
0.5	0.531	1.062		0.095
0.5625	0.594	1.156		0.095
0.625	0.656	1.312		0.095
0.75	0.812	1.469		0.134
0.875	0.938	1.750		0.134
1	1.062	2.000		0.134
1.125	1.250	2.250		0.134
1.25	1.375	2.500		0.165
1.375	1.500	2.750		0.165
1.5	1.625	3.000		0.165
1.625	1.750		3.750	0.180
1.75	1.875		4.000	0.180
1.875	2.000		4.250	0.180
2	2.125		4.500	0.180
2.25	2.375		4.750	0.220
2.5	2.625		5.000	0.238
2.75	2.875		5.250	0.259
3	3.125		5.500	0.284

For basic sizes less than or equal to 5/8 inch, the hole diameter in the normal flat washers is approximately the basic size +0.031 inches. For basic sizes greater than 5/8 inch and less than or equal to 1 inch, the hole diameter in the normal flat washer is approximately the basic size +0.062 inches. For basic sizes greater than 1 inch and less than or equal to 3 inches, the hole diameter in the normal flat washer is the basic size +0.125 inches. If we plot this on a graph (Figure 11-23), it shows that the washer's hole diameter is approximately:

$$\text{Hole\_diameter} = 1.0457 * \text{BASIC\_SIZE} + 0.0267 \text{ inches}$$

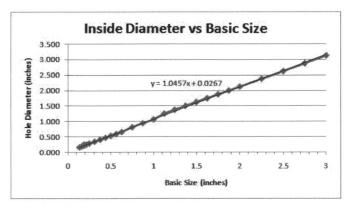

Figure 11-23 Inside Diameter versus Basic Size

If we plot the outside diameter versus the basic size on a graph (Figure 11-24), it shows that the washer's outside diameter is approximately:

Outside_ diameter = 1.8905*BASIC_SIZE+ 0.1243 inches (for BASIC_SIZE <1.6 inches)

Outside_ diameter = 1.2237*BASIC_SIZE + 1.9098 inches (for 1.6 inches < BASIC_SIZE < 3.1 inches)

When we plot the washer thickness versus basic size, we notice that it is more discrete than linear, as shown in Figure 11-25. A set of conditional "if statements" will work better for the washer thickness. Let's begin.

Step 1: Start Creo Parametric.

Step 2: Set the Working directory.

Step 3: **File>New.**

Step 4: *Select* **Part.** Name it "flat_washer."
Check **Use default template.** *Pick* **OK.**

Step 5: **File>Prepare>Model Properties.**

Step 6: Change the material from not assigned to "Standard-Materials_Granta-Design", "Ferrous-metals", **Steel low carbon.**

Step 7: Change the units to **Inch-Pound-Seconds (IPS).**

Step 8: Close the Model Properties window.

Step 9: *Select* the **Extrude** tool, then the FRONT datum plane. *Select* the **Sketch View** icon from the toolbar to make the FRONT datum plane parallel with the screen. Draw the flat washer's inside and outside diameters, as shown in Figure 11-26.

Step 10: *Pick* **OK** to exit Sketcher.

Step 11: Set the washer thickness to 0.134 inches. *Pick the green checkmark.*

Step 12: *Select* **EXTRUDE1** in the model tree, then rename it **BODY.**

Step 13: *Select* the washer model. When a pop-up menu appears, *select* **Edit Dimension** icon from this menu.

Step 14: At the bottom of the Model ribbon *select* **Model Intent>Switch Dimensions.**

Step 15: Change the washer's parameter names to "ID," "OD," and "thickness" and change the variable's accuracy to 3-decimal places if it isn't already set. For example, *pick* the 2.00 dimension, then name it "OD"

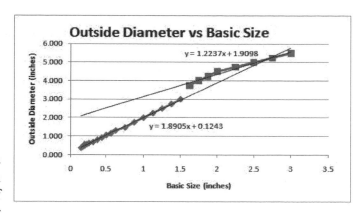

Figure 11-24   Outside Diameter versus Basic Size

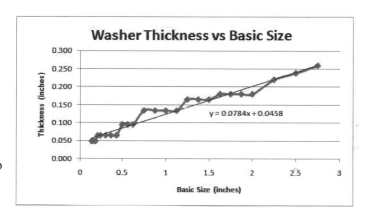

Figure 11-25   Washer Thickness versus Basic Size

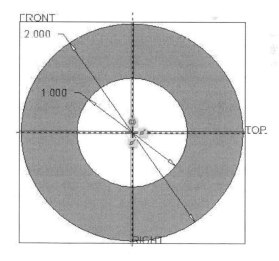

Figure 11-26   Sketch Flat Washer

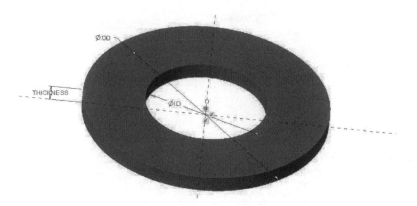

Figure 11-27   Washer with Parameter Names

by typing the new name in the box in the upper left corner of the window. Change the number of decimal places to "3" by selecting 0.123 from the pull-down menu. See Figure 11-27.

Step 16: At the bottom of the Model ribbon *select* **Model Intent>Parameters.**

Step 17: *Pick* the ***plus sign*** to add a new parameter. (Note that PTC_MATERIAL_ NAME has a string value of "Steel_low_ carbon" since we assigned this property to the part.)

Step 18: Type "BASIC_SIZE." Type "1.0" for its real value. See Figure 11-28. *Check* the Designate box. *Pick* **OK.**

Step 19: At the bottom of the Model ribbon *select* **Model Intent>Relations** which brings up the relations window.

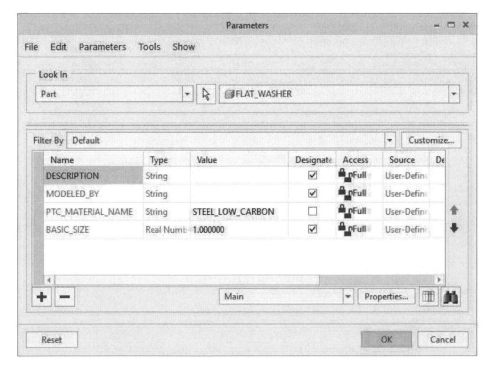

Figure 11-28   Add BASIC_SIZE Parameter

Step 20: In the blank area type the following. (Comments can be added to the code by preceding them with a slash and a star (/*). Indenting the lines can help you see the different sets of if statements. See Figure 11-29.)

/* Set Inside Diameter
ID=1.0457*BASIC_SIZE+0.0267

/* Set Outside Diameter
if BASIC_SIZE < 1.6
  OD=1.8905*BASIC_SIZE+ 0.1243
else
  OD= 1.2237*BASIC_SIZE + 1.9098
endif

/* Set washer thickness
if BASIC_SIZE < 0.2
  thickness = 0.049
else
  if BASIC_SIZE < 0.5
  thickness = 0.065
  else
   if BASIC_SIZE < 0.75
   thickness = 0.095
   else
    if BASIC_SIZE < 1.25
    thickness = 0.134
    else
     if BASIC_SIZE < 1.6
     thickness = 0.165
     else
      if BASIC_SIZE < 2.25
      thickness = 0.180
      else
       if BASIC_SIZE < 2.5
       thickness = 0.220
       else
        thickness = 0.092*BASIC_SIZE+0.0073
       endif
      endif
     endif
    endif
   endif
  endif
endif

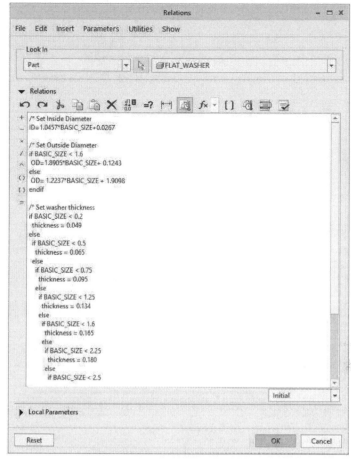

Figure 11-29 Relations for Flat Washer

Step 21: *Pick* the ***checkmark*** icon in the upper right corner of the relations window to verify your code. Fix any errors that appear.

Step 22: *Pick* **OK** to close the relations window.

Step 23: ⬜ *Pick* the **regenerate** icon at the top of the screen. The washer should change size slightly.

Step 24: Highlight the BODY of the flat washer in the model tree. When a pop-up menu appears, *select* the **Edit Dimensions** icon. (If the parameter names appear instead of their values, at the bottom of the Model ribbon *select* **Model Intent>Switch Dimensions** to change them to their values.)

Step 25: Compare the flat washer's size to Table 11-2. The values are off a bit, but close. OD is 2.015 compared to 2.000 inches. ID is 1.072 compared to 1.062 inches. Thickness is .134 compared to .134 inches.

Step 26: At the bottom of the Model ribbon *select* **Model Intent>Parameters** which brings up the parameters window. Change the BASIC_SIZE to 3.0 inches. *Pick* **OK.** *Pick* the **Regenerate** icon again. Highlight the washer body, then *select* **Edit Dimensions** from the pop-up menu. Compare the washer's values with Table 11-2. Once again the values are quite close. OD is 5.581 compared to 5.500 inches. ID is 3.164 compared to 3.125 inches. Thickness is .283 compared to .284 inches.

Step 27: Repeat the above step checking several more basic sizes. What do you think? Can we use this part model for standard flat washers? (Maybe. Maybe not.)

Step 28: **File>Save As>Save a Copy...** Type "flat_washer_formula" as the new name. Do not change the model name. *Pick* **OK.**

If we need to place standard (nominal) flat washers in our prototype model, then this part will do fine. If we are doing precision work where size and weight are critical, then this model will not work. What if we want the exact values that are found in Table 11-2? How can we do this? How about using a family table?

Step 29: At the bottom of the Model ribbon *select* **Model Intent>Relations** which brings up the relations window. Highlight all the code in the open window, then press <Delete> or *pick* the **Scissors** icon. Make sure all code is gone, then *pick* **OK** to close the window.

Step 30: At the bottom of the Model ribbon *select* **Model Intent>Parameters** which brings up the parameters window. Change the BASIC_SIZE to 0.5 inches. *Pick* **OK.** *Pick* the **Regenerate** icon at the top of the screen. Why didn't the flat washer change size? (We deleted the relationship between the variable, BASIC_SIZE, and the dimensions of the flat washer.)

Step 31: At the bottom of the Model ribbon *select* **Model Intent>Parameters** which brings up the parameters window. Highlight BASIC_SIZE, then *pick* the **Minus Sign** to delete the parameter. *Pick* **OK.**

Step 32: **File>Save.** *Pick* **OK.**

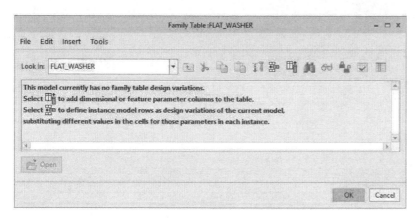

Figure 11-30 Family Table for Flat_Washer

Now let's create a family table for the flat washers.

Step 33: At the bottom of the Model ribbon *select* **Model Intent>Family Table.**

Step 34: *Select **Add/delete table columns*** icon to get the following Family Items window to appear. See Figure 11-30.

Step 35: *Pick* the BODY in the model tree or pick the washer on the graphics screen to get the dimensions to show up. Your values may be different.

Step 36: With Add Item set to **Dimension,** *pick* the inside diameter of the washer, the outside diameter of the washer, and the washer thickness dimensions. They will turn color. See Figure 11-31. *Pick* **OK.**

Step 37: *Select* the ***Insert New Instance icon at the selected row*** icon to add a new row to the family table.

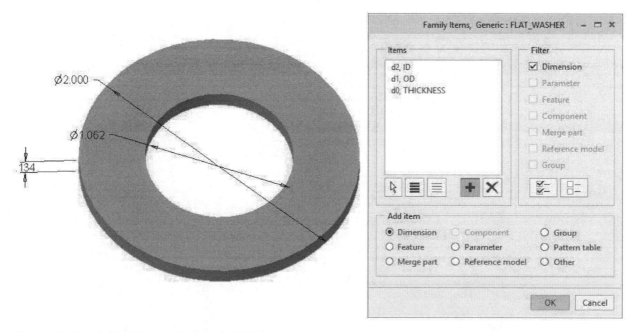

Figure 11-31 Add Columns to Family Table

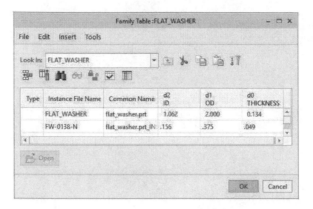

Figure 11-32   Enter First Row of Family Table

Step 38: Type "FW-0138-N" for the Instance Name; <Tab> (leave Common Name); <Tab> "0.156" for the ID; <Tab> "0.375" for the OD, and <Tab> "0.049" for the thickness. See Figure 11-32.

Step 39: Press the <Enter> key to start a new row.

Step 40: Type "FW-0164-N" for the Instance Name; <Tab>; <Tab> "0.188" for the ID; <Tab> "0.438" for the OD, and <Tab> "0.049" for the thickness.

Step 41: Continue to add rows until your family table looks like Figure 11-33. Refer to Appendix J, page 509, for flat washer details.

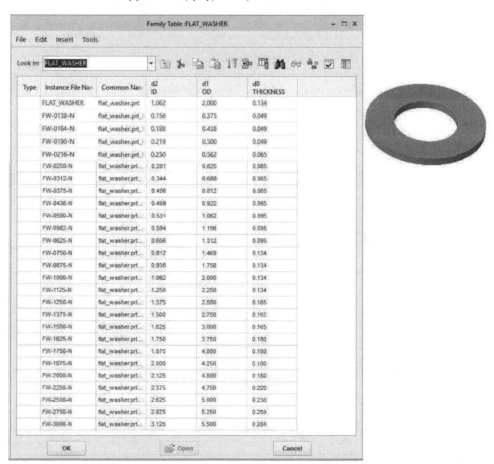

Figure 11-33   Family Table with Values

Step 42: 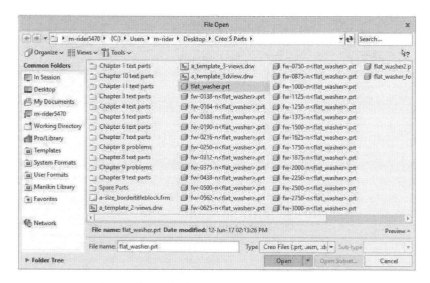 *Pick* the *Verify Instances* of Family icon.

Step 43: In the Family Tree window, *pick* **VERIFY.**

Step 44: If all is successful, *pick* **CLOSE** to close this window. (If not, repeat the previous steps to correct the problem.)

Step 45: *Pick* **OK** to close the Family Tree: FLAT_WASHER window.

Step 46: **File>Save.**

Step 47: **File>Manage File>Delete Old Versions.** *Pick Yes.*

Step 48: **File>Close.**

Step 49: **File>Manage Session>Erase Not Displayed…** *Pick* **OK.**

Step 50: **File>Open.**

Step 51: *Select* the "flat washer.prt" (Figure 11-34). *Pick* **Open.**

Note that each instance of the normal flat washers appears as a part if you wanted to select them individually. However, if you look in your working directory, the family table members do not appear. They are inside the part file called "flat_washer.prt". They are shown in the **File Open** window for convenience.

Step 52: When the Select Instance window appears, *select* a normal flat washer with a basic size of ¾ inch (FW-0750-N). See Figure 11-35.

Step 53: *Pick* **Open.**

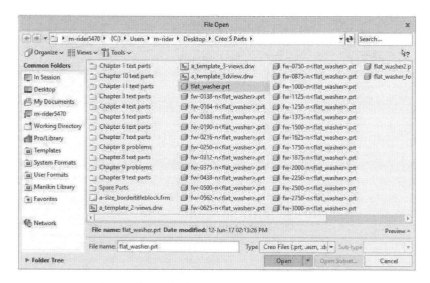

Figure 11-34 Open Flat_Washer Part File

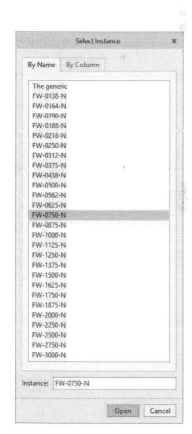

Figure 11-35 Select Instance

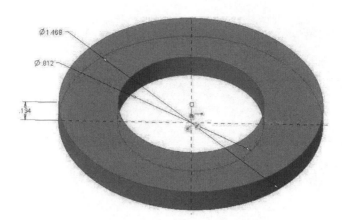

Figure 11-36   ³/₄-inch Flat Washer

Step 54: In the graphics window, *pick* the flat washer. When a pop-up menu appears, *pick* the **Edit Dimensions** icon. Verify that the dimensions shown are the correct sizes for the ³/₄-inch flat washer. See Figure 11-36.

Step 55: **File>Close.**

Step 56: Try several other flat washer sizes to verify that each flat washer is created at the proper size.

Note that each flat washer is the correct size as compared to the formula method done earlier. Creating a family table was more work; however, we have total control over the part's dimensions when we use it.

End of Relations and Family Tables Practice

Relations and Family Tables Exercise

A round head machine screw (Figure 11-37) is a very common part when assembling jigs, fixtures, dies, and small household appliances. Table 11-3 shows the dimensions for a series of round head machine screws. After curve fitting the round head screw's dimensions, we note that they conform closely to a set of linear equations. We will use these equations to generate all round head machine screws with a basic size between 0.060 inches and 0.750 inches in diameter. Standard lengths are specified as:

Length: ¼″ to 1″, increments of ⅛″
Length: 1″ to 4″, increments of ¼″
Length: 4″ to 6″, increments of ½″

Basic_Size = standard diameters from 0.060" (#0) to 0.750″
Rotate_360 = 360 degrees   /* Not necessary */
Curvature rho (elliptical) = Hd_Dia
slot_depth = 2/3 * Hd_Height
chamfer = 1 / ThdPerInch (Threads per inch)

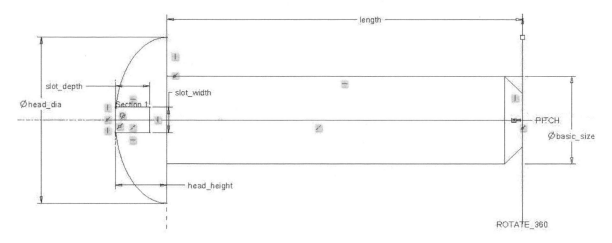

Figure 11-37 Round Head Machine Screw Variables

Table 11-3	Round Head Machine Screw Dimensions			
Standard Size	**Basic_Size (inches)**	**Head Diameter (inches)**	**Head Height (inches)**	**Slot Width (inches)**
#0	0.060	0.113	0.053	0.023
#1	0.073	0.138	0.061	0.026
#2	0.086	0.162	0.069	0.031
#3	0.099	0.187	0.078	0.035
#4	0.112	0.211	0.086	0.039
#5	0.125	0.236	0.095	0.043
#6	0.138	0.260	0.103	0.048
#8	0.164	0.309	0.12	0.054
#10	0.190	0.359	0.137	0.06
#12	0.216	0.408	0.153	0.067
¼	0.250	0.472	0.175	0.075
5/16	0.313	0.590	0.216	0.084
⅜	0.375	0.708	0.256	0.094
7/16	0.438	0.750	0.328	0.094
½	0.500	0.813	0.355	0.106
9/16	0.563	0.938	0.41	0.118
⅝	0.625	1.000	0.438	0.133
¾	0.750	1.250	0.547	0.149

The governing curve fit equations for round head machine screw are:

For Basic Size Less than 0.40 inches
  Hd_Dia = 1.8889 * Basic_size - 0.0003 (inches)
  Hd_height = 0.6456 * Basic_size - 0.014 (inches)
For Basic Size greater than 0.40 inches and less than 0.80 inches
  Hd_Dia = 1.6074 * Basic_size +0.026 (inches)
  Hd_height = 0.7053 * Basic_size + 0.0101 (inches)
For Basic Size Less than 0.15 inches
  slot_width = 0.3214 * Basic_size + 0.0032 (inches)
For Basic Size greater than 0.15 inches and less than 0.40 inches
  slot_width = 0.1874 * Basic_size + 0.0253 (inches)
For Basic Size greater than 0.40 inches and less than 0.80 inches
  slot_width = 0.1795 * Basic_size + 0.0168 (inches)

Let's begin.

Step 1: Start Creo Parametric.

Step 2: Set the Working directory.

Step 3: **File>New.**

Step 4: *Select* **Part.** Name it "roundhead_machine_screw". *Check* **Use default template.** *Pick* **OK.**

Step 5: **File>Prepare>Model Properties.**

Step 6: Change the material from not assigned to "Standard-Materials_Granta-Design", "Ferrous-metals", **Steel_low_carbon.**

Step 7: Change the units to **Inch-Pound-Seconds (IPS).**

Step 8: Close the Model Properties window.

Step 9: *Select* the **Revolve** tool. *Pick* the FRONT datum plane.

Step 10: *Pick* the **Sketch View** icon if necessary to orient the sketch plane so that it is parallel with the display screen.

Step 11: Draw a horizontal geometry centerline on top of the TOP datum plane.

Step 12: Using the **Rectangle** or **Line** tool, draw the body of the machine screw in the first quadrant as shown in Figure 11-38.

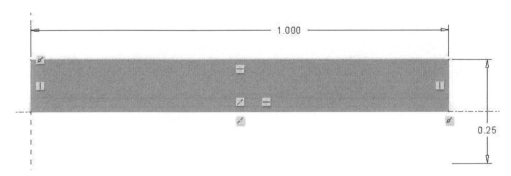

Figure 11-38 Body of Machine Screw

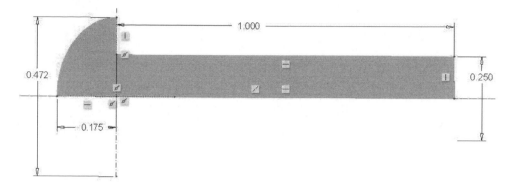

Figure 11-39 Add Head to Roundhead Machine Screw

Step 13: Add the head, as shown in Figure 11-39. Use an elliptical arc (or a rectangle with an elliptical fillet) for the curved section of the head. Delete the two separate horizontal lines on top of the centerline, and then replace them with one continuous horizontal line from the top of the head to the bottom of the screw. (If you don't make this one horizontal line, you will not be able to revolve this section.) Set the values as shown in the figure.

Step 14: Verify that you have a closed section, then exit from Sketcher.

If you cannot create the revolved machine screw, go back and make sure there is only one solid line on top of the centerline.

Step 15: *Pick* the **green checkmark** to build the revolved section. See Figure 11-40.

Create an extruded cut (slot) in the top of the rounded head for a Robert's screwdriver.

Step 16: *Select* the **Extrude** tool. *Pick* the FRONT datum plane.

Step 17: Draw a horizontal construction centerline through the TOP datum plane.

Figure 11-40 Revolve Machine Screw

Step 18: *Select* **References** in the Sketch ribbon. *Pick* the rounded head of the screw in the graphics area. *Pick* **Close.** Draw a vertical construction centerline tangent to the top of the roundhead screw.

Step 19: Draw a rectangle, as shown in Figure 11-41. Width equals 0.117 inches. Height equals 0.075 inches and symmetric about the horizontal centerline.

Step 20: Exit Sketcher.

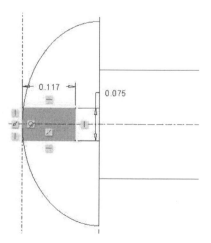

Figure 11-41 Sketch Slot in Head

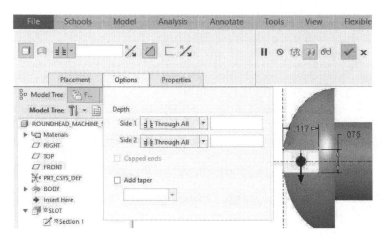

Figure 11-42   Remove Material in Both Directions   **Figure 11-43**   Add Chamfer to Screw

Step 21: *Select* the ***Remove Material*** icon. *Select* the Options tab. *Select*
Through All for Side 1 and Side 2. *Select* the ***glasses*** icon to see the
resulting cut. See Figure 11-42. If all looks correct, *pick* the ***green
checkmark.***

Step 22: *Select* the ***chamfer*** tool from the Model ribbon at the top of the screen.
Pick the right end of the screw body and add a chamfer to the end of the
screw. See Figure 11-43. *Pick* the ***green checkmark*** to accept the
chamfer.

Step 23: Rename Revolve1 in the model tree as BODY. Rename Extrude1 in the
model tree as SLOT. Rename Chamfer1 as THD_CHAMFER.

Now we need to change the variable names for the machine screw to something more meaningful.

Step 24: *Select* **BODY** in the model tree. When a pop-up menu appears, *select*
the **Edit Dimensions** icon.

Step 25: At the bottom of the Model ribbon *select* **Model Intent>Switch
Dimensions.**

Step 26: *Select* a dimension using LMB. In the variable name area in the upper
left corner of the screen, change the variable's name such as "d29"
becomes "head_dia". Your initial variable may be different.

Step 27: Change all variable names according to Figure 11-44.

Step 28: *Select* SLOT in the model tree. When a pop-up menu appears, *select*
the **Edit Dimensions** icon again.

Step 29: Change the two variable names according to Figure 11-45.

Step 30: *Select* **THD_CHAMFER** in the model tree. When a pop-up menu
appears, *select* the **Edit Dimensions** icon again.

Step 31: Change the chamfer variable name to PITCH. See Figure 11-46.

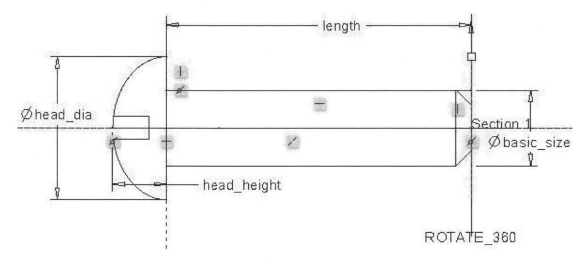

Figure 11-44 Screw Body Variable Names

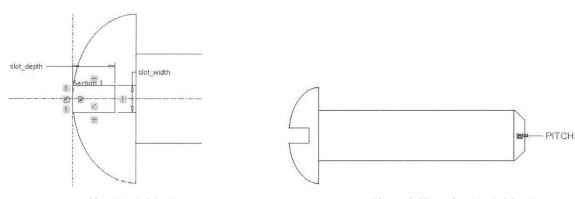

Figure 11-45 Slot Variable Names **Figure 11-46** Thread Chamfer Variable Name

Now it is time to add a parameter and the relationships between the variables.

Step 32: At the bottom of the Model ribbon *select* **Model Intent>Parameters.**

Step 33: *Select* the ***plus sign.*** For name, type "THDSPERINCH." Change the type to integer. Type a value of "28" for its value. *Check* the designate box. *Pick* **OK.**

Step 34: At the bottom of the Model ribbon *select* **Model Intent>Relations.** Type the following code. See Figure 11-47.

```
/* Determine head size and slot width
if BASIC_SIZE < 0.40
 head_dia=1.8889*BASIC_SIZE-0.0003
head_height=0.6456*BASIC_SIZE-0.014
slot_width=0.1874*BASIC_SIZE+0.0253
else head_dia=1.6074*BASIC_SIZE+0.026
head_height=0.7053*BASIC_SIZE+0.0101
slot_width=0.1795*BASIC_SIZE+0.0168
endif
```

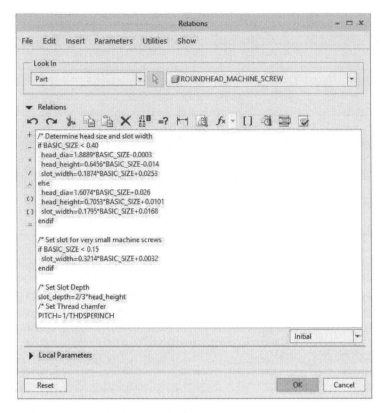

Figure 11-47   Relations Code

```
/* Set slot for very small machine screws
if BASIC_SIZE < 0.15
 slot_width=0.3214*BASIC_SIZE+0.0032
endif
/* Set Slot Depth
 slot_depth=2/3*head_height
 /* Set Thread chamfer
PITCH=1/THDSPERINCH
```

Step 35: Execute/Verify the lines of code. Correct any errors.

Step 36: *Pick* **OK** to close the relations window.

Step 37: *Select* the ***regenerate*** icon to update the roundhead machine screw according to its parameters and relations. See Figure 11-48. The basic size is equal to .25 inches. The threads per inch is set to 28. The length is set to 1.00 inches.

Figure 11-48 $^1/_4$-28UNF x 1.00

Step 38: At the bottom of the Model ribbon *select* **Model Intent>Parameters.** Change the THDSPERINCH value from 28 to 10 <Enter>. *Pick* **OK.**

Step 39: *Select* BODY in the model tree. When a pop-up menu appears, *select* the **Edit Dimensions** icon. *Pick* the "basic_size" dimension.

Step 40: Change the BASIC_SIZE from 0.25 inches to 0.75 inches. Change the LENGTH from 1.00 inch to 2.00 inches.

Step 41: If the screw doesn't regenerate automatically, *select* the *regenerate* icon to update the roundhead machine screw according to its parameter and relations. See Figure 11-49. Isn't this easier than drawing this size roundhead screw from scratch?

Step 42: Try several other combinations, such as: Basic_size – Threads per Inch (Thread series UNC or UNF) × Length

0.060-80 UNF × 0.25
0.190-24 UNC × 0.75
0.500-13 UNC × 1.50

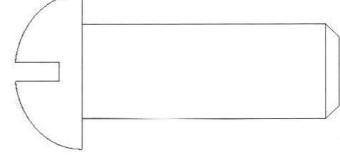

Step 43: Change the values to: "0.250-20 UNC × 1.00."

Figure 11-49   ³⁄₄-10 UNC x 2.00

Step 44: **File>Save.**

Step 45: **File>Manage File>Delete Old Versions.** *Pick Yes.*

Now let's make a family table for the roundhead machine screw just created.

Step 46: At the bottom of the Model ribbon *select* **Model Intent>Family Table.**

Step 47: *Select* ***Add/delete table columns*** icon to get the following Family Items window to appear. See Figure 11-50.

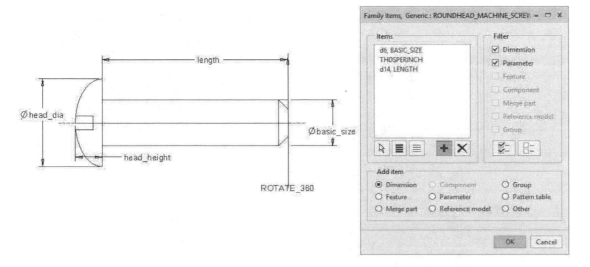

Figure 11-50   Add Columns to Family Table

Step 48: *Pick* the **BODY** in the model tree or on the graphics screen to get the dimensions to show up.

Step 49: With Add Item set to **Dimension,** *pick* the BASIC_SIZE (ø.25) of the roundhead machine screw.

Step 50: With Add Item set to **Parameter,** *select* THDSPERINCH from the parameter list. *Pick* the **Insert Selected** button. *Pick* the **Close** button.

Step 51: With Add Item set to **Dimension** again, *pick* the LENGTH (1.00) of the roundhead machine screw. *Pick* **OK** to close the Family Items window. (Your dxx variables may be different.)
NOTE: To change the order of two columns, *pick* the 1st column, hold down <Ctrl> and *pick* the 2nd column. Under the Edit option in the Family Table window, *pick Swap Two Column*.

Step 52: ⊞ *Select* the **Insert New Instance icon at the selected row** icon to add a new row to the family table.

Step 53: Type "RH0250-20UNCx1000" for the Instance Name; <Tab>; <Tab> "0.250" for the BASIC_SIZE; <Tab> "20" for the THDPERINCH, and <Tab> "1.00" for the LENGTH. See Figure 11-51.

Step 54: Press the <Enter> key to start a new row.

Step 55: Type "RH0250-20UNCx1250" for the Instance Name; <Tab>; <Tab> "0.250" for the BASIC_SIZE; <Tab> "20" for the THDPERINCH, and <Tab> "1.25" for the LENGTH. Press the <Enter> key to start a new row.

Step 56: Type "RH0250-20UNCx1500" for the Instance Name; <Tab>; <Tab> "0.250" for the BASIC_SIZE; <Tab> "20" for the THDPERINCH, and <Tab> "1.50" for the LENGTH.

Step 57: ☑ *Select* the **Verify instances of the family** icon to check to see if the data entered will correctly create the desired part. *Pick* the VERIFY button. Correct any errors.

Step 58: *Pick* **CLOSE** to close the Family Tree window.

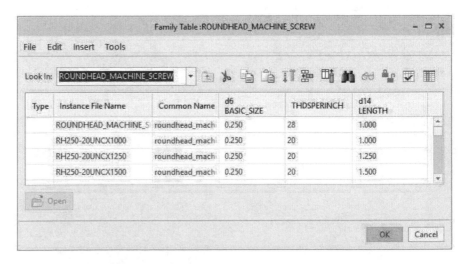

Type	Instance File Name	Common Name	d6 BASIC_SIZE	THDSPERINCH	d14 LENGTH	
	ROUNDHEAD_MACHINE_S	roundhead_machi	0.250	28	1.000	
	RH250-20UNCX1000	roundhead_machi	0.250	20	1.000	
	RH250-20UNCX1250	roundhead_machi	0.250	20	1.250	
	RH250-20UNCX1500	roundhead_machi	0.250	20	1.500	

Figure 11-51   Enter the First Three Rows of Family Table

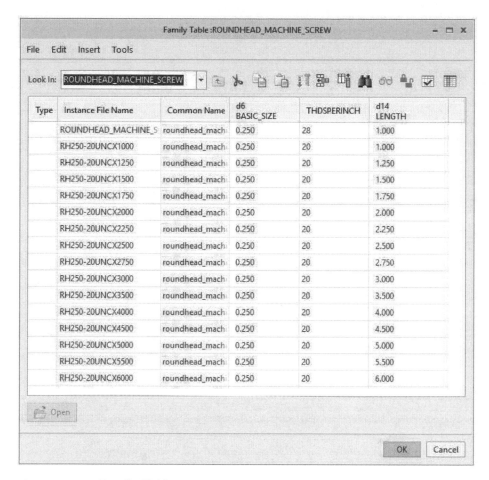

Figure 11-52   Family Table

Step 59: *Pick* **OK** to close the Family Table window.

Step 60: File>Save.

Step 61: At the bottom of the Model ribbon *select* **Model Intent>Family Table.**

Step 62: Continue to add rows until your family table looks like Figure 11-52. To edit with Excel, *pick* the Excel icon, add/edit the rows, then close Excel without saving the file. *Pick* **Yes** when asked, "Do you want to update the changes you made?"

Step 63: *Select* the ***Verify instances of the family*** icon to check to see if the data entered will correctly create the desired part. *Pick* the VERIFY button. Correct any errors.

Step 64: *Pick* **CLOSE** to close the Family Tree window.

Step 65: *Pick* **OK** to close the Family Table window.

Step 66: File>Save.

Step 67: File>Manage File>Delete Old Versions. *Pick Yes.*

We will stop at this point. More table entries can be added as needed and as time permits.

#2 machine screws 0.086-56 UNC × (0.250,0.375,0.500, 0.750)

#3 machine screw 0.099-48 UNC × (0.250,0.312,0.375,0.437,0.500,0.562,0.625,0.750,0.875,1.000)

#4 machine screw 0.112-40 UNC × (0.250,0.312,0.375,0.437,0.500,0.562,0.625,0.750,0.875,1.000,1.250,1.500)

#5 machine screw 0.125-40 UNC × (0.250,0.312,0.375,0.437,0.500,0.562,0.625,0.750,0.875,1.000,1.250,1.500,2.000)

#6 machine screw 0.138-32 UNC × (0.250,0.312,0.375,0.437,0.500,0.562,0.625,0.750,0.875,1.000,1.250,1.500,1.750,2.000,2.250, 2.500,2.750,3.000,3.500,4.000)

#8 machine screw 0.164-32 UNC × (0.250,0.312,0.375,0.437,0.500,0.562,0.625,0.750,0.875,1.000,1.250,1.500,1.750,2.000,2.250, 2.500,2.750,3.000,3.500,4.000,5.000,6.000)

#10 machine screw 0.190-24 UNC × (0.250,0.312,0.375,0.437,0.500,0.562,0.625,0.750,0.875,1.000,1.250,1.500,1.750,2.000,2.250, 2.500,2.750,3.000,3.500,4.000,4.500,5.000,5.500,6.000)

#12 machine screw 0.216-24 UNC × (0.250,0.312,0.375,0.437,0.500,0.562,0.625,0.750,0.875,1.000,1.250,1.500,1.750,2.000,2.250, 2.500,2.750,3.000,3.500,4.000,4.500,5.000,5.500,6.000)

¼″ machine screw 0.250-20 UNC × (0.250,0.312,0.375,0.437,0.500,0.562,0.625,0.750,0.875,1.000,1.250,1.500,1.750,2.000,2.250, 2.500,2.750,3.000,3.500,4.000,4.500,5.000,5.500,6.000)

5/16" machine screw 0.3125-18 UNC × (0.375,0.500,0.625,0.750,0.875,1.000,1.250,1.500,1.750,2.000,2.250,2.500,2.750,3.000,3.500, 4.000,4.500,5.000,5.500,6.000)

3/8″ machine screw 0.375-16 UNC × (0.500,0.625,0.750,0.875,1.000,1.250,1.500,1.750,2.000,2.250,2.500,2.750,3.000,3.500,4.000, 4.500,5.000,5.500,6.000)

½″ machine screw 0.500-13 UNC × (1.000,1.250,1.500,1.750,2.000,2.250,2.500,2.750,3.000,3.500,4.000,5.000,6.000)

There is also a series of UNF (Unified National Fine) machine screws. These can be purchased at any hardware store or *www.mscdirect.com.* If you add any additional screws to the family table, perform the following.

Step 68: *Select* the ***Verify instances of the family*** icon to check to see if the data entered will correctly create the desired part. *Pick* the **Verify** button. Correct any errors if present.

Step 69: *Pick* **CLOSE** to close the Family Tree window.

Step 70: *Pick* **OK** to close the Family Table window.

Step 71: File>Save.

Step 72: File>Manage File>Delete Old Versions. *Pick Yes.*

Step 73: File>Close.

Step 74: File>Manage Session>Erase Not Displayed. *Pick* **OK.**

End of Relations and Family Table Exercise.

▶ Review Questions

1. What is a family table?

2. Why would one use a family table?

3. What are the steps necessary when creating a family table?

4. What command do you use to verify that the family table entries are valid?

5. How do you change the value of a dimension in a family table member without affecting the generic part?

6. What is a parameter?

7. What is a relation?

8. What is the purpose of a relation?

9. What is the difference between a sketcher relation and a part relation?

10. Can relations use decision-making tools when calculating values?

11. Name at least 5 built-in parameters.

12. Name at least 5 built-in functions.

13. Name an advantage and a disadvantage of relations compared to family tables.

14. How do you see the name of the parameters in part creation?

15. How do you rename the parameters in part creation? Why would you want to do this?

16. What must be done if you change the value of a part parameter and the part does not change size?

Relations and Family Tables Problems

11.1   Design a commercial grade "**mixing_bowl**" (Figure 11-53) for the European market using 1.00-millimeter thick stainless steel sheet material. The center of the radial portion of the metric bowl is located at the origin. The bottom of the bowl is located "bottom_locate" below the origin. The designer wants to specify the desired volume in milliliters and the tallness of the bowl in millimeters. (Use values assigned by your instructor.) The base diameter must be ½ of the top diameter for stability. Use the **Volume** tool under the |Analysis| tab to verify that the volume is correct before using the shell feature to make the bowl hollow. Be sure to add the shell to the outside of the volume so the bowl holding capacity does not change. (You cannot check the volume after performing the shell operation.) All dimensions are in millimeters.

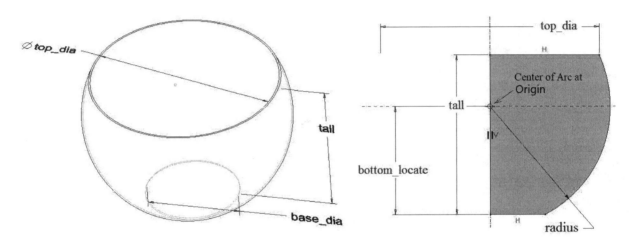

Figure 11-53—Metric Bowl

Necessary relations:

    tall=TALLNESS
    top_dia=sqrt((VOLUME*6000/(pi*TALLNESS)-TALLNESS^2)/0.9375)
    base_dia=top_dia/2
    b=(0.75*top_dia^2+4*TALLNESS^2)
    base_height=(-b+sqrt(b^2+4*(TALLNESS*top_dia)^2))/(TALLNESS*8)
    radius=(base_dia^2+4*base_height^2)/(8*base_height)
    bottom_locate=Radius-base_height

A solution: With Volume = 1000 ml and tallness = 88 mm, the top diameter calculates to 122.02 mm, bottom locate variable calculates to 59.86 mm, and the radius calculates to 67.19 mm.

Note: 1 milliliter = 1 cm³ = 1000 mm³

11.2 Use relations to design the "**extender_plate**" shown in Figure 11-54. The user specifies the height, width, and depth only. The four corner holes' diameters and the four corner radii are equal to the depth. The center hole's diameter is twice the depth. Initially, let width = 8 inches, height = 6 inches, and depth = 1 inch. Use other values specified by your instructor. All dimensions are in inches.

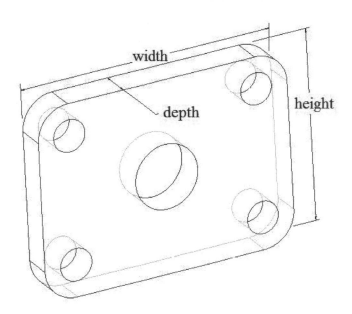

Figure 11-54—Extender Plate

11.3 Create a "**flathead_machine_screw**" (Figure 11-55) with the following parameters and relations.

Basic_size - Threads per inch = ¼-28, ¼-20, 5/16-24, 5/16-18, 3/8-24, 3/8-16, ½-20, ½-13, 9/16-18, 9/16-12, 5/8-18, 5/8-11, ¾-16, ¾-10.

Length = 1.00″, 1.25″, 1.50″, 1.75″, 2.00″, 2.25″, 2.50″, 2.75″, 3.00″

Hd_Dia = 1.8374 * Basic_size + 0.0196″

Slot_depth = (Hd_Dia − Basic_size)/3.5

Slot_width = 0.15 * Basic_size + 0.035″

Angle = 82 degrees

Chamfer size = 1/(Number of threads per inch)

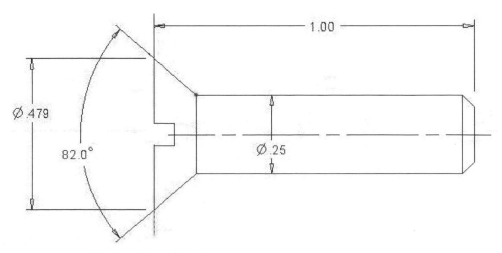

Figure 11-55—Flathead Machine Screw

11.4 Create a Family Table for the **"flathead_machine_screw"** (Figure 11-55) varying the basic size and the number of threads per inch. Be sure to give a unique name to each instance in the flathead screw table such as 250-28x1.00, 250-28x125, etc. All dimensions are in inches.

11.5 Create a generic **"woodruff_key"** (Figure 11-56) from its parameter KEY_SIZE where the width of the key is represented by the left two digits and the diameter of the key is represented by the right two digits. For example, woodruff key number 204 (or 0204) is 02 times 1/32th inches wide and 04 times 1/8th inches in diameter. The diameter is cut off approximately 0.06 inches below the center of the circle. All dimensions are in inches.

diameter = mod(KEY_SIZE,100)/8
width = (KEY_SIZE-8*diameter)/3200

KEY_SIZE = 202, 204, 304, 305, 404, 405, 406, 505, 506, 507, 606, 607, 608, 609, 807, 808, 809, 810, 811, 812, 1008, 1009, 1010, 1011, 1012, 1210, 1211, and 1212.

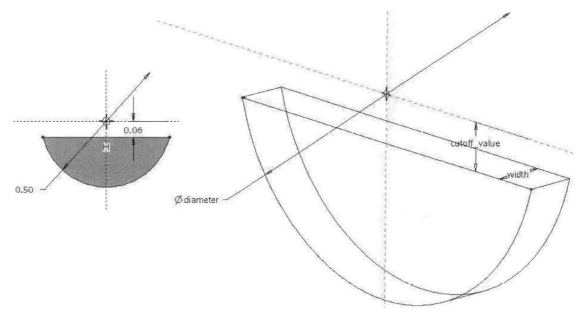

Figure 11-56—Woodruff Key

11.6 Create a Family Table for the **"woodruff_key"** (Figure 11-56) varying the KEY_SIZE shown in the previous problem. Be sure to give a unique name to each instance in the table such as 202, 204, etc.

11.7 Look up the correct value for the diameter cutoff value (approximately 0.06 inches) for each of the woodruff key sizes specified. Create a Family Table for the **"woodruff_key"** (Figure 11-56) varying the KEY_SIZE and the "cutoff_value." Be sure to give a unique name to each instance in the table such as 202, 204, etc.

11.8 Create a generic part of your choice.

11.9 Create a Family Table using your generic part.

TOLERANCING AND GD&T

Objectives

▶ Learn about size tolerancing
▶ Learn about Geometric Dimensioning and Tolerancing (GD&T)

▶ Learn how to add size tolerances to features of size on engineering drawings
▶ Learn how to add GD&T feature control frames for position, orientation and form on engineering drawings

---

▶ Tolerancing and GD&T Explored

Tolerance is the amount of variation permitted in the size of a part or in the location of a point, axis or surface. For example, a size dimension given as 1.375 inches ± .003 inches means that the size may be any value between 1.378 inches and 1.372 inches. The tolerance, or total amount of variation, is .006 inches. It becomes the job of the designer to specify the allowable error that may be acceptable for a given feature and still permit the part to function satisfactorily. Since greater accuracy costs more, it is better to specify a generous tolerance.

Without interchangeable part production, the manufacturing industry could not exist, and without effective size and location control by the design engineer, interchangeable manufacturing would not happen. For example, a car manufacturer subcontracts the production of many of the car's parts to other companies. These parts plus replacement parts must be similar so that any one of them will fit properly in the car. It might be thought that if the dimensions are given on the blueprint, such as 4½, 3.125, etc., all the parts will be exactly alike and will fit properly. Unfortunately, it is impossible to make any part to an exact size or locate a feature of size exactly. The part can be made very close to the specified dimensions, but such accuracy is expensive. Luckily, exact sizes are not required, only varying degrees of accuracy according to the functional requirements. The maker of children's building blocks would soon go out of business if they insisted on making the blocks with high accuracy. No one would be willing to pay such a high price for building blocks. What is needed is a means of specifying dimensions with whatever accuracy is required. The answer to this situation is the specification of a tolerance on each dimensional value.

In order to control the dimensions of two parts so that any two mating parts will be interchangeable, it is necessary to assign tolerances to the dimensions, as shown in Figure 12-1. The diameter of the hole must be machined to a diameter between 1.250 inches and 1.252 inches. These two values become the acceptable dimension limits and the difference between them, .002 inches, becomes the tolerance. The shaft must be produced between the limits of 1.248 inches and 1.246 inches in diameter. The shaft tolerance is the difference between these two values or .002 inches.

The maximum and minimum shaft diameters are shown in Figure 12.1(a). The loosest fit or maximum clearance occurs when the smallest shaft is mated with the largest hole, as shown in Figure 12-1(b). The tightest fit or minimum clearance occurs when the largest shaft is mated with the smallest hole, as shown in Figure 12-1(c). The allowance is the difference between these two dimensions on the tightest fit or .002 inches. Any shaft will fit any hole interchangeably.

Before continuing, let's define some terms.

▶ Nominal Size—the value that is used for general identification of a size.

▶ Basic Size—the value from which limits of size are derived by applying the allowance and tolerances. It is the theoretical value from which the size limits are calculated.

▶ Limits—the maximum and minimum values indicated by a toleranced dimension. Allowance—the minimum clearance or maximum interference between two mating parts. The allowance represents the tightest permissible fit and is simply the smallest hole minus the largest shaft. For clearance fits, this difference will be positive. For interference fits, the allowance will be negative.

There are three general types of fits between parts:

1. Clearance fit—an internal member fits into an external member leaving an air space or clearance between the parts. In Figure 12-1(b) the largest shaft is 1.248 inches and the smallest hole is 1.250 inches, which gives the smallest acceptable air space between the parts as .002 inches. This space is the allowance; in a clearance fit the allowance is positive.

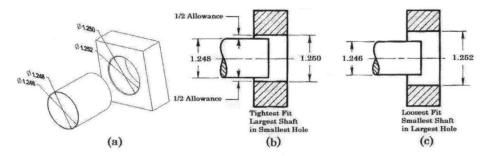

Figure 12-1 Limit Dimensions on a Shaft and Hole

2. Interference fit—the internal member is larger than the external member such that there is always interference. If the smallest shaft is 1.506 inches, and the largest hole is 1.502 inches, there would be an interference of .004 inches.

3. Transition fit—the fit might be either a clearance fit or an interference fit. If the smallest shaft was 1.503 inches and the largest hole was 1.506 inches, there would be .003 inches to spare (positive allowance). However, if the largest shaft, 1.509 inches, was forced into the smallest hole, 1.500 inches, an interference (negative allowance) of .009 inches would occur.

If allowances and tolerances are properly specified and feature of size are located properly, mating parts will be interchangeable. However, for close fits, it may be necessary to specify very small tolerances and allowances, thus the cost would increase. To avoid this expense, selective assembly is used. In selective assembly, all parts are inspected and classified into several classes according to their actual sizes, so that "small" shafts can be matched with "small" holes, "medium" shafts with "medium" holes, and "large" shafts with "large" holes. In this way, satisfactory fits may be obtained with less expense.

Basic Hole System

Standard broaches, reamers, and other standard tools are used to make holes. Standard plug gages are used to check their size. Shafts can be machined to any desired diameter. Therefore, toleranced dimensions are commonly calculated based on the basic hole system. In this system, the minimum hole is taken as the basic size and a plus or minus allowance is assigned to the shaft. Then the hole and shaft tolerances are applied to these parts.

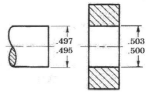

Figure 12-2 Basic Hole Fit with Clearance

The minimum size of the hole in Figure 12-2 is .500 inches and is taken as the basic size. An allowance of .003 inches is decided upon and subtracted from the basic hole size, giving the maximum shaft diameter of .497 inches. Tolerances of .003 inches for the hole and .002 inches for the shaft are applied to obtain the maximum hole of .503 inches and the minimum shaft of .495 inches. Thus, the minimum clearance (allowance) between the parts becomes .500 inches − .497 inches = .003 inches (smallest hole minus largest shaft), and the maximum clearance is .503 inches − .495 inches = .008 inches (largest hole minus smallest shaft).

For an interference fit, the maximum shaft size is calculated by adding the desired maximum interference to the basic size. In Figure 12-3, the basic size is 1.2500 inches. The maximum interference was set to .0024 inches, which is added to the basic size making the largest shaft 1.2524 inches.

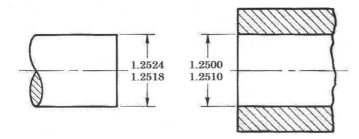

Figure 12-3 Basic Hole Fit with Interference

For common-fractional dimensions, the operator is not expected to work closer than he can be expected to measure with a steel ruler. It is customary to indicate an overall general tolerance for all common-fraction size dimensions by means of a note in the title block, such as ALL FRACTIONAL DIMENSIONS ±1/64″ UNLESS OTHERWISE SPECIFIED. General angular tolerances may be given as ANGULAR TOLERANCE ±2°. Without GD&T control feature frames, tolerances on decimal size dimensions may be given in a general note, such as:

> SIZE TOLERANCES ARE:
> .X ± .05
> .XX ± .005
> .XXX ± .001

Every dimension on an engineering drawing must have a tolerance, either direct or implied by a general tolerance note. Two methods of expressing tolerances on size dimensions follow:

1. Limit Dimensioning—the preferred method, specifies the maximum and minimum limits of size and location. Generally, the maximum material condition limit is placed above the minimum material condition limit. In note form, the maximum material condition limit is given first, thus for a hole: Ø.500- Ø.502. For a shaft, the note would read as Ø.504- Ø.503. (Note that many CAD systems always show the largest value first.)

2. Plus and Minus Dimensioning—the basic size is followed by a plus and minus expression of the tolerance resulting in either a unilateral (one nonzero value) or bilateral (two nonzero values) tolerance. If two unequal tolerance numbers are given, one plus and one minus, the plus value is placed above the minus. One of the tolerances may be zero. If a single tolerance value is given, it is preceded by the plus-or-minus symbol. This method can be used when the plus and minus values are equal. The unilateral system of tolerances allows variations in only one direction from the specified size. This method is advantageous when a critical size is approached as material is removed during manufacturing, as in the case of close-fitting holes and shafts. The bilateral system of tolerances allows variations in both directions from the specified size. *This method was typically used prior to GD&T to locate features of size; however, it does not account for orientation or form errors.*

When using either method above, it is important to be aware of the effect of one tolerance on another. When the location of a surface is affected by more than one tolerance, the tolerances are accumulative. On the left in Figure 12-4, if dimension C is omitted, surface Y would be controlled by both dimensions A and B, and there could be a total variation of .010 inches instead of the variation of .005 inches permitted by the shown dimension C. If the part is made using the minimum tolerances of A, B, and C, the total variation in the length of the part from 3.750 inches would be .015 inches; the part could be as short as 3.735 inches. However, the tolerance on the overall dimension D is only .005 inches, thus allowing the part to be only as short as 3.745 inches. Note that the part is over-dimensioned as shown. In some cases, for functional reasons, it may be desired to hold all three small dimensions A, B, and C close without regard to the overall length. In such

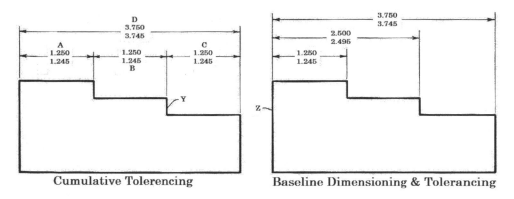

Figure 12-4 Accumulated Tolerances and Their Affects

a case, the overall dimension should be marked as reference (REF). It may be desirable to hold the two small dimensions, A and B, and the overall dimension close without regard to dimension C. In this case, dimension C should be omitted or marked as reference.

It is better to dimension each surface so that it is affected by only one dimension and tolerance. This is done by referring all dimensions from a single datum surface, like Z, as shown on the right in Figure 12-4.

As has been stated prior, size tolerances should be as large as possible and still permit the proper functioning of the part. This allows for less expensive tools, lower labor and inspection costs, and reduced material scrap, thus reducing the overall cost of the part. A chart of some shop processes with expected tolerances is shown in Figure 12-5.

Shop Process	Tolerance Range for Various Shop Processes								
Drilling									
Milling									
Turning, Boring									
Planing, Shaping									
Reaming									
Broaching									
Grinding, Boring									
Lapping, Honing									
From	**Up to**	Tolerances (inches)							
0.000	0.600	0.00015	0.0002	0.0003	0.0005	0.0008	0.0012	0.002	0.003	0.005
0.600	1.000	0.00015	0.00025	0.0004	0.0006	0.0010	0.0015	0.0025	0.004	0.006
1.000	1.500	0.0002	0.0003	0.0005	0.0008	0.0012	0.002	0.003	0.005	0.008
1.500	2.800	0.00025	0.0004	0.0006	0.0010	0.0015	0.0025	0.004	0.006	0.010
2.800	4.500	0.0003	0.0005	0.0008	0.0012	0.002	0.003	0.005	0.008	0.012
4.500	7.800	0.0004	0.0006	0.0010	0.0015	0.0025	0.004	0.006	0.010	0.015
7.800	13.600	0.0005	0.0008	0.0012	0.002	0.003	0.005	0.008	0.012	0.020
13.600	21.000	0.0006	0.0010	0.0015	0.0025	0.004	0.006	0.010	0.015	0.025

Figure 12-5 Various Shop Processes Tolerances

Coordinate Tolerancing Issues (Use GD&T instead)

Coordinate tolerancing can be applied to the location of a hole from a specified reference plane. It is important to determine the location of the reference planes or datum planes before dimensioning the part. Reference planes must be machined surfaces, the centerline of a machined hole, or the center plane of a feature of size. In Figure 12-6 the left and bottom edges of the part are defined as two of the part's datum planes. The center of the hole is located relative to these two datum planes. The center of the hole can be anywhere inside the square region of .010 inches by .010 inches. The positional tolerance of the hole is the length of the diagonal of this square region or .014 inches. If the maximum size of the bolt to go into this hole and a corresponding hole in a similar part is .500 inches, the hole diameter should be at least .514 inches to avoid interference. In equation form, H = S + T, where:

T = maximum positional tolerance of the hole's center = $\sqrt{.010^2 + .010^2} = .014$
S = maximum diameter of the bolt or shaft
H = minimum diameter of the hole so there is no interference
H = .500 + .014 = .514 inches

The same calculations would hold true if you were locating a second clearance hole relative to the first, as shown in Figure 12-7. The above equation can also be used to determine the allowable tolerance on the position of the second hole's center if the maximum diameter of the bolt and the minimum diameter of the hole are known, the holes are perfectly round, the holes are perpendicular to the surface, and the mating surfaces are perfectly flat. (*WOW, I see an issue.*)

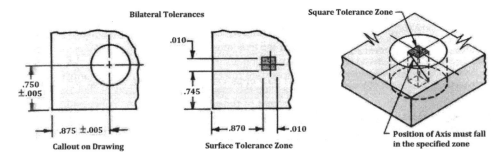

Figure 12-6   Coordinate Tolerancing of a Hole

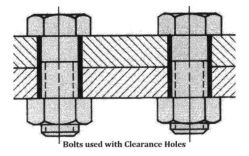

Figure 12-7   Coordinate Tolerancing for a Second Bolted Hole

When one of the mating parts is to be screwed to a second part using a threaded hole (Figure 12-8) using a cap screw or stud, the method is called fixed-fastener. When the threaded fastener is the same basic size as the clearance hole the positional tolerance for the two holes' centers becomes:

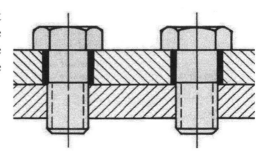

$T = (H - S)/2$. The hole's minimum diameter must be at least:
$H = S + T_H + T_S$

where:

Figure 12-8 Coordinate Tolerancing of a Second Threaded Hole

T_S = shaft's positional center tolerance (threaded hole's center)
T_H = clearance hole's positional center tolerance
S = maximum diameter of bolt or shaft
H = minimum diameter of hole so there is no interference if perfectly formed

A lid is to be screwed onto a box using ½-13UNC-2A × 1.00 long cap screws as shown in Figure 12-8. The distance between the two threaded holes is 2.000 ± .003 inches. The clearance holes are located 2.000 ± .004 inches apart. The maximum size for the cap screw is .500 inches. Determine the minimum diameter for the clearance holes so that there is no interference if the holes are perfectly round and perpendicular to the mating surfaces. (Also, we are ignoring the misalignment of the holes into or out of the page *so this may not assemble*.)

$H = S + T_H + T_S$
$H = .500 + .008 + .006$
$H = .514$ inches in diameter (minimum)

Geometric Tolerances

This chapter is not meant to be a comprehensive coverage of Geometric Dimensioning & Tolerancing (GD&T), but rather a simple introduction to it and how to add GD&T to your engineering drawings as needed. You might ask, "Why GD&T?" Geometric dimensioning and tolerancing is recognized around the world as the only effective way to define a part geometry.

Geometric tolerances are used to control location, orientation, and form. "Tolerance of form" specifies how far surfaces are permitted to vary from the perfect geometry implied by the engineering drawings. Theoretically, planes, cylinders, cones, etc. are perfect forms, but since it is impossible to produce a perfect form, it is necessary to specify the amount of variation permitted. Geometric tolerances define such conditions as flatness, parallelism, perpendicularity, angularity, cylindricity, and straightness. When geometric tolerances are not indicated on the drawing, the actual part is understood to be acceptable if it is within the dimensional limits shown, regardless of variations in form.

In the case of fabricated bars, sheets, and tubing established industry standards prescribe acceptable conditions of straightness, flatness, etc., and these standards are understood to hold if geometric tolerances are not shown on the drawing.

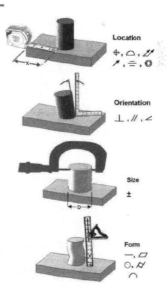

Figure 12-9 Major Geometric Features

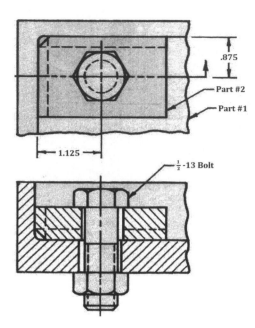

Figure 12-10   Block Bolted into Corner of Piece

Let's look at a simple example using a positional GD&T. A steel block is to be bolted into the corner of a steel box using a ½-13UNC-2A bolt. See Figure 12-10. The clearance hole for the bolt has a minimum diameter of ½ inch +1/64- inch or .516 inches. Determine the positional tolerances for the clearance hole.

Rearranging the positional tolerance equation and solving for the positional tolerance becomes:

$$T = H - S$$
$$T = .516 - .500$$
$$T = .016 \text{ inches (diameter)}$$

Basic dimensions represent exact theoretical values. This is the basis from which permissible variations are allowed. They are enclosed in a rectangular box and shown without tolerances. The corresponding control frame shows the allowable tolerance. Since we are locating the hole using GD&T, the .875-inch and 1.125-inch dimensions in Figure 12-10 will become basic dimensions in Figure 12-11. Also, these basic dimensions need to be given relative to datum planes A, B, and C. In Figure 12-11 the left and top edges of the block are shown as datum planes B and C. Datum plane A is located where the block would sit on the box and is specified as the primary datum.

The hole is dimensioned using limit dimensions with the maximum material condition (MMC) dimension given first. The hole can be as small as .516 inches but must be less than .520 inches. (Many CAD systems list the larger value first.) The control block under this dimension is the GD&T. The first area of this control block represents the type of geometric tolerance; in this case, it is a positional tolerance. The second area gives the allowable tolerance for the position with an additional qualifier. The center of the hole and must be located within a .016-inch diameter of its exact center location at maximum material condition. The next three areas represent the order in which the block is placed on the datum surfaces during its quality inspection. The block is set on datum A, then pushed left toward datum B, and finally pushed back toward datum C.

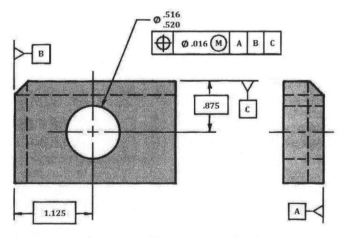

Figure 12-11 Geometric Dimensions and Tolerances for Clearance Hole

Feature	Tolerance Type	Characteristic	Symbol
Individual Features	Form	Straightness	▬
		Flatness	⟋⟋
		Circularity (Roundness)	◯
		Cylindricity	⌭
Individual or Related Features	Profile	Profile of a Line	⌒
		Profile of a Surface	⌓
Related Features	Orientation	Angularity	∠
		Perpendicularity	⊥
		Parallelism	∥
	Location	Position	⊕
		Concentricity	◎
		Symmetry	≡
	Runout	Circular Runout	↗
		Total Runout	↗↗

Figure 12-12 Geometric Characteristic Symbols

Feature-control symbols should be used in place of notes to specify positional and form tolerances. The symbols for the geometric characteristic in which the tolerances apply are given in Figure 12-12.

Before we begin an exercise, we need to review a few features of geometric tolerancing. Straightness controls how close to a straight line an edge or surface must be to be acceptable. See Figure 12-13. For a cylinder, it controls the straightness of the surface, not the straightness of its axis center. The straightness tolerance must be less than the size tolerance.

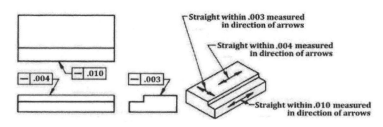

Figure 12-13 GD&T Straightness

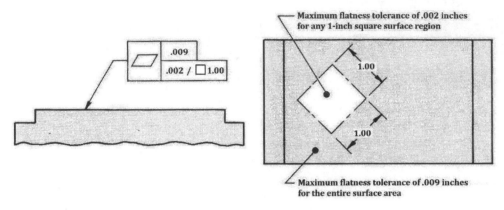

Figure 12-14   GD&T Flatness and Flatness per Unit Area

Flatness controls how close to a flat plane the surface must be to be acceptable. The flatness tolerance must be less than the size tolerance. Flatness per unit area can also be specified when it is necessary to control the flatness in a given area closer than the overall flatness requirement. See Figure 12-14. The overall flatness of the surface must be within .009 inches; however, the flatness of any given square inch must be within .002 inches.

"Features of size" reference shafts, holes, and parallel surfaces. With features of size, the control frame is associated with the size dimension. For example, if we wanted to control the straightness of a shaft's axis center, we would use the straightness symbol as before and include a diameter symbol with the specified form tolerance. We may also add a modifier such as MMC (maximum material condition), LMC (least material condition), or RFS (regardless of feature size). RFS is the default modifier if none is shown. See Figure 12-15.

Geometric tolerances for straightness, flatness, circularity, and cylindricity do not need to be located relative to any datum planes; however, geometric tolerances for orientation and position need to be referenced from specified datum planes. These datum planes must be specified on the drawing. The primary datum is a flat surface against which the part is first placed. The part is moved so that it touches the secondary datum next. Finally, the part is moved perpendicular to the previous motion until it contacts the tertiary datum. Note that a hole may be used as a datum. If a single datum is established using two other datums, then both datums are referenced in the same area of the control frame with a dash between them. See Figure 12-16.

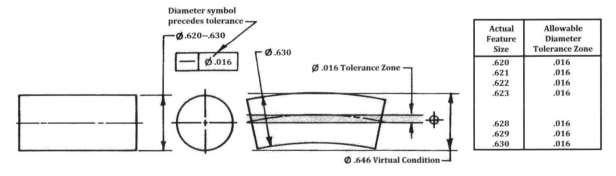

Figure 12-15   Straightness of a Shaft's Centerline Regardless of Feature Size

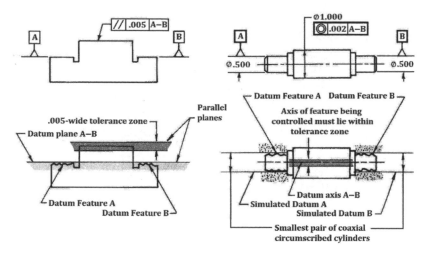

Figure 12-16 Dual Datums Referenced for Surface and Cylinder

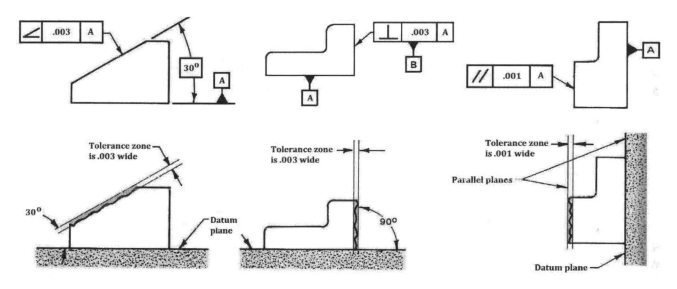

Figure 12-17 Orientation Tolerances

Orientation refers to the angular relationship between two lines or surfaces. Orientation tolerances control angularity, perpendicularity, and parallelism. A tolerance of form or orientation should be specified when the size and location tolerances do not provide proper control of the feature. See Figure 12-17. A perpendicularity tolerance should be used on the secondary and tertiary datum planes to indicate how close to perpendicular they need to be.

Positional tolerances are used on groups or patterns of holes at MMC. See Figure 12-18. This method meets the functional requirements for the feature in most cases. Positional tolerancing refers to the tolerance zone away from the exact location of the centerline for the hole or shaft. At MMC the

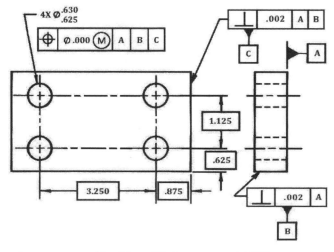

Figure 12-18 Positional Tolerances at MMC

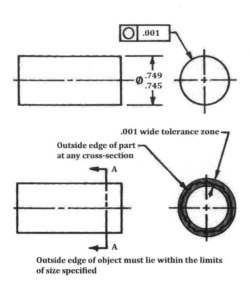

Figure 12-19 a Circular Tolerance for Round Part

tolerance cannot exceed the value specified; however, the positional tolerance allowed is dependent upon the actual size of the considered feature. If the holes are made at MMC, the smallest acceptable (.625 inches), then the hole centers must be dead on location. If a hole is .005 inches larger at .630 inches, then the center location for this hole may be off by .0025 inches or half of the hole size tolerance.

Circularity tolerance is measured radial and specifies the distance between two circles that a particular circular cross-section must stay in for it to be acceptable. It is shown without modifiers; thus it is assumed to be regardless of feature size (RFS). See Figure 12-19.

That's enough review for now. For a more detailed look at GD&T see "Geometric Dimensioning and Tolerancing," based on ANSI/ASME Y14.5M-1994, by David Madsen or "The GD&T Hierarchy, Y14.5M-2009," by Don Day. Let's get started.

We will create the part, and then make an engineering drawing for the part with some appropriate geometric tolerances.

▶ Tolerancing and GD&T Practice

Before beginning this section, be sure "base.prt" from the assembly practice session (Figure 12-20), "a-size_bordertitleblock.frm," and "a_template_3-views.drw" are in your working directory.

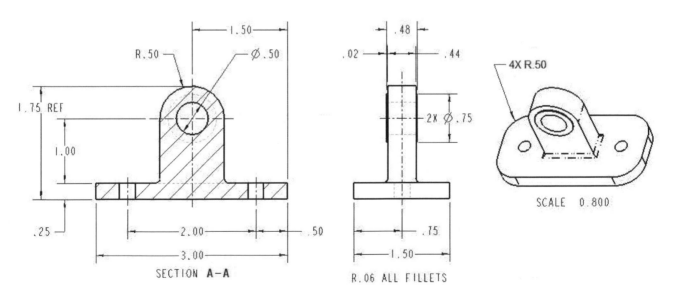

Figure 12-20   Base.prt

Design Intent

Create an engineering drawing of "base.prt" with its border and title block. Use GD&T to make sure the ½-inch hole is positioned within a diameter of .001 inches relative to datums A, B, and C.

Step 1: Start Creo Parametric.

Step 2: Set your working directory.

Step 3: **File>Open.**

Step 4: *Select* "a-size_bordertitleblock.frm" from the working directory. *Pick* **OK.**

Step 5: *Double-click* on the text in the title block area so that it can be edited.

Step 6: Modify the text in the title block area according to Figure 12-21.

Step 7: **File>Save.**

Step 8: **File>Manage File>Delete Old Versions.** *Pick* the *Yes.*

Step 9: **File>Close.**

Step 10: **File>New.**

Step 11: *Select* **Drawing** from the window and name it "base." Make sure the default template is checked, and then *pick* **OK.**

Step 12: Make sure "base.prt" is the default model. If it is not, *pick* the **Browse...** button, then *select* the "base.prt."

Step 13: *Pick* the **Browse...** button inside the Template area. *Pick* **Working Directory** from the left side of the open window. *Select* "a_template_3-views.drw" from the directory. *Pick* the **Open** button. *Pick* the **OK** button in the New Drawing window.

Step 14: At the top of the window, enter "1020 CD STEEL" for the material, then *pick* the *green checkmark.* Enter your name, then press the <Enter> key. (This is the same as *picking* the *green checkmark.*) Enter the drawing number, "123456" <Enter>.

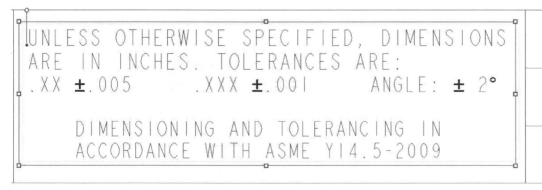

Figure 12-21 Modify A-Size Format Sheet

Step 15:   Turn off *Axis display, Point display, Csys display,* and *Plane display*.

Step 16:   [icon] *Pick* the **Repaint** icon to update the display <u>if necessary</u>.

Step 17: File>Prepare>Drawing Properties.

Step 18:   In the Drawing Properties window, *select* **change** to the right of Detail Options.

Step 19:   In the Option box, enter "tol_display." Set its value to "yes." *Select* the **Add/Change** button.

Step 20:   In the Option box, enter "gtol_datums." Set its value to "std_asme." *Select* the **Add/Change** button. *Pick* **Close** to close the drawing options window. *Pick* **Close** to close the Model Properties window.

Step 21:   In the lower right corner of the screen, *select* "Annotation" from the list. Draw an imaginary box around all three views, thus highlighting all the dimensions. *Press* the <Delete> key to remove all the dimensions.

Step 22:   In the lower right corner of the screen, change the option back to "General." *Pick* the FRONT view with the LMB, then *press and hold* the RMB until a pop-up menu appears. *Select* "Lock View Movement" to unlock the views. Rearrange the view similar to Figure 12-22.

Step 23:   *Select* the Annotate tab. Select **Show Model Annotations**, then add the necessary centerlines to all three views. Be sure to connect the centerlines of the two smaller holes in the top view.

Step 24:   [icon] *Pick* the **Geometric Tolerance** icon. Move the cursor to the bottom line of the FRONT view so it is highlighted, then *press* the LMB. Move the cursor down and to the left, then press the MMB to place the control frame. Change the **Geometric Characteristic** to Flatness. Set the value to 0.005 inches, then *pick* anywhere on the screen to finalize the control frame.

Step 25:   *Select* the **Datum Feature Symbol** icon. Move the cursor over the flatness control frame and *press* the LMB. Move the cursor down, then *press* the MMB to place datum tag, A.

Step 26:   *Select* the **Show Model Annotations** icon, then *pick* the small right hole in the TOP view with the LMB. *Select* the diameter dimension that shows up. Under the Dimension tab, *pick* the **Tolerance** icon, then *select* Limits. Change the upper limit to .258 and the lower limit to .253. In the **Precision box**, set the number of decimal places to three. Next, *pick* the **Display** icon. *Check* the **ISO Tolerance Display Style** box. *Pick* anywhere in the graphics area to finalize the settings.

Step 27:   *Pick* the **Regenerate** icon at the top of the screen.

Step 28:   [icon] *Select* the **Geometric Tolerance** icon again. *Pick* the .258-.253 dimension with the LMB. Change the **Geometric Characteristic** to perpendicularity. Change the value to 0.004 and add a diameter symbol in front of this value. Add datum A as the primary datum. See Figure 12-23.

Step 29:   *Select* the **Datum Feature Symbol** icon. *Pick* the Perpendicularity control frame with the LMB. Press the MMB to place datum tag B.

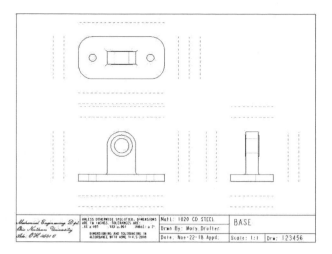

Figure 12-22 Initial Layout of Views

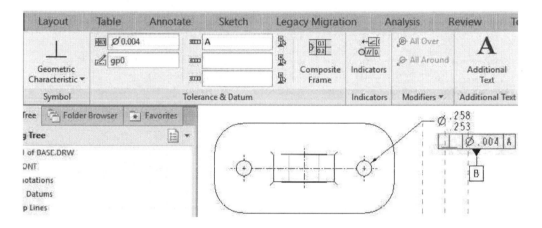

Figure 12-23 Perpendicularity Control Frame and Datum Tag B

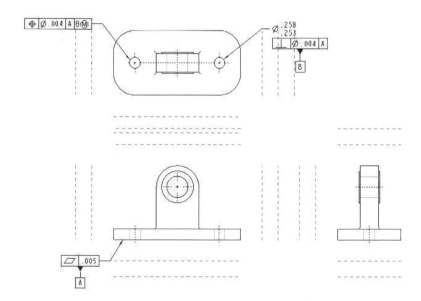

Figure 12-24 Datums A, B, and C Defined

Step 30: *Select* the ***Geometric Tolerance*** icon again. *Pick* the small left hole in the TOP view with the LMB. Change the ***Geometric Characteristic*** to Position. Change the value to 0.004 and add a diameter symbol in front of this value. Add datum A as the primary datum, and datum B at MMC as the secondary datum. See Figure 12-24.

Step 31: *Select* the ***Datum Feature Symbol*** icon. *Pick* the Position control frame with the LMB. Press the MMB to place datum tag C.

Step 32: *Select* the Show Model Annotations icon, then add the fillet radius dimension to the drawing. At the bottom of the Annotate ribbon *select* **Format>Switch Dimensions.** Note the variable for the small fillet radius on the base. My variable showed up as d22. See Figure 12-25.

Step 33: *Select* the ***Unattached Note*** icon at the top of the screen. *Pick* a point on the drawing below the RIGHT view. Type "R&d22 ALL FILLETS", then *pick* any area on the drawing sheet. (Use the correct variable name for the fillet radius.) *Delete* the R.06 dimension from the drawing. *Select* **Format>Switch Dimensions** again.

Step 34: Add needed dimensions, then clean up the views as shown in Figure 12-26. Since the right hole in the TOP view uses a geometric position tolerance, the 2.00 dimension needs to be a Basic dimension with 3-decimal places.

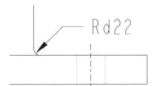

Figure 12-25   Small Fillet Radius Variable

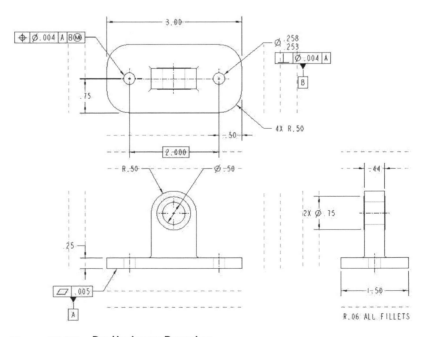

Figure 12-26   Preliminary Drawing

Step 35: *Move* the cursor on top of the dimension for the .50-inch diameter hole in the FRONT view, then *press* the LMB. *Pick* the ***Tolerance*** icon, then *select* Limits from the list. *Set* the Precision to three decimal places. *Enter* 0.506 for the upper limit and 0.500 for the lower limit. *Pick* the Display icon, then *check* the **ISO Tolerance Display Style** box. *Pick* anywhere in the graphics area.

Step 36: *Select* the ***Geometric tolerance*** icon in the **Annotate** ribbon. *Move* the cursor on top of this limit dimension for in the FRONT view, then *press* the LMB. The geometric tolerance box should attach itself to this dimension.

Step 37: In the Geometric Tolerance ribbon, *select* the ***Position*** symbol. See Figure 12-27.

Step 38: *Set* the tolerance value to 0.003, then add a ***diameter*** symbol in front and a ***Maximum Material Condition*** symbol after the value. See Figures 12-28 and 12-30.

Step 39: *Set* the Primary Datum to "A". Set the Secondary datum to "B" and the Tertiary datum to "C". See Figure 12-29.

Step 40: *Pick* the ***Regenerate*** icon at the top of the screen to update the model.

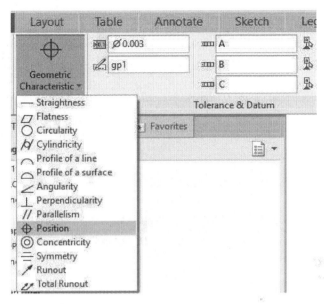

Figure 12-27   Select Position Symbol

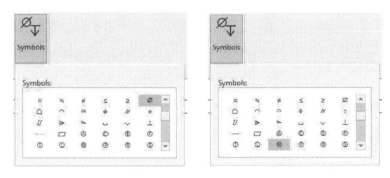

Figure 12-28   Add Diameter and MMC symbols

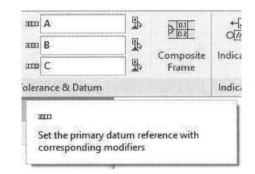

Figure 12-29   Select Primary Datum

Figure 12-30   Completed Geometric Tolerance Option

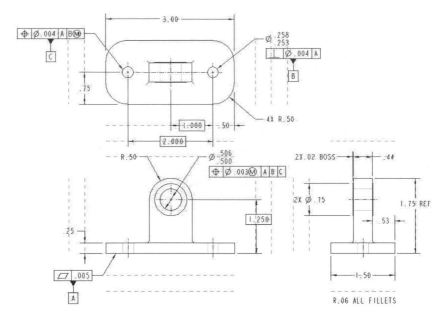

Figure 12-31   GD&T Added to Drawing

Step 41: *Select* the Dimension tool, then add dimensions as shown above. Convert the 1.00-inch and 1.25-inch dimensions to basic dimensions with 3-decimal places. Be sure to add the .53-inch dimension that locates the upper section and the .02 BOSS on each side of the upper section. See Figure 12-31.

Step 42: Add a 2X in front of the .258 over .253 dimension in the TOP view.

Now let's add a 3D view of the part in the upper right corner of the drawing.

Step 43: *Select* the ⎢Layout⎢ tab at the top of the screen. *Select* the **General View** icon. *Select* **DEFAULT ALL** in the Select Combined State window. *Pick* **OK.**

Step 44: *Pick* a point in the upper right corner of the drawing area. Set the scale to 0.750 under Custom scale. *Pick* **Apply.** Set **No Hidden** as the Display style under the View Display category. *Pick* **Apply.** *Pick* **OK.**

Step 45: Unlock the FRONT view, then move the views as shown in Figure 12-32 if necessary. Lock the views again.

Step 46: **File>Save.**

Step 47: **File>Manage File>Delete Old Versions.** *Pick Yes.*

Step 48: **File>Print>Print.** Pick the Preview icon. Your drawing should look similar to Figure 12-32. *Pick Print,* then follow your environment instructions. *Pick Close Print Setup.*

Step 49: **File>Close.**

Step 50: With base.prt visible, **File>Manage File>Delete Old Versions.** *Pick Yes* to delete any old versions of the 3D model.

Step 51: **File>Manage Session>Erase Not Displayed.** *Pick* **OK.**

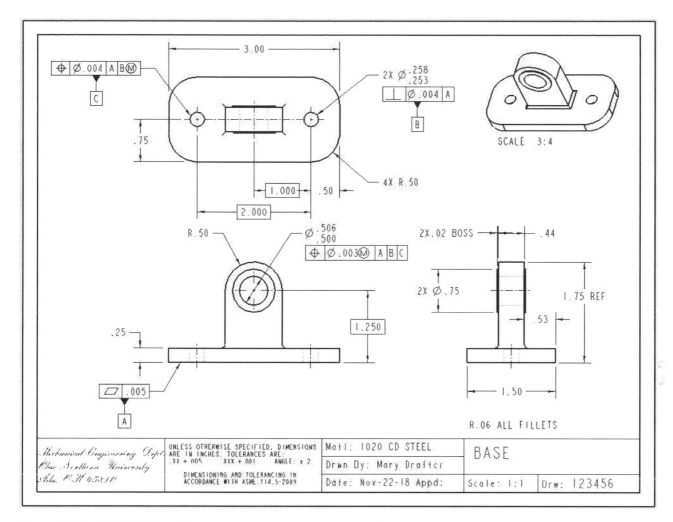

Figure 12-32 Finished Drawing

Design Intent

Create a 12-inch by 9-inch by 1.00-inch steel cover plate. Add a ⅞-inch diameter clearance hole in each of the four corners 2.00 inches from each edge. Place 2-inch rounds in each of its four corners. Round the upper lip of the cover plate using a 0.12-inch radius.

Step 1: **File>New.**

Step 2: With the type set to **Part** and **Solid,** name the new part "cover_plate.prt." *Pick* **OK.**

Step 3: **File>Prepare>Model Properties.** When the Model Properties window appears, *pick* the word **change** on the right end of the Material row. Set the material to **"Standard-Materials_Granta-Design", "Ferrous-metals", Steel_low_carbon.** *Pick* **OK.** Change the units to **IPS.** *Pick* **Close** to close the model properties window.

Step 4: Create an extrusion in the fourth quadrant on the FRONT datum plane that is a 12-inch by 9-inch rectangle, 1.00-inch thick.

Step 5: Place a ⅞-inch diameter hole through the cover plate in the upper left corner 2.00 inches from each edge, then pattern it into the other three corners using a pattern distance of 5 inches and 8 inches.

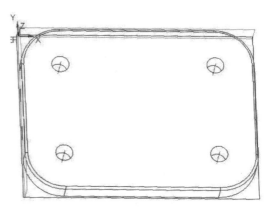

Figure 12-33   Cover Plate

Step 6: Round the four corners using a 2.00-inch radius (outside of sketcher). Remember to hold down <Ctrl> while *picking* the other three corners. Round the top edge using a 0.12-inch radius. See Figure 12-33.

Step 7: **File>Save.** *Pick* **OK.**

Step 8: **File>Close.**

Design Intent

Create a two-view engineering drawing of the cover plate with the following specifications:

▶ Bottom of the cover plate flat within .004 inches. Call this datum A.

▶ The left and back faces of the cover plate are datums B and C, respectively.

▶ Four ⅞-inch diameter clearance holes have an upper and lower limit of 0.880 inches and 0.875 inches. Locate these four holes using a geometric positioning tolerance as large as possible.

▶ The four studs that go through these four clearance holes have a maximum diameter of 0.813 inches and will have the same positional tolerance as the clearance holes.

Step 9: **File>New.**

Step 10: *Select* **Drawing** from the window and name it "cover_plate." Make sure the default template is checked, and then *pick* **OK.**

Step 11: Make sure "cover_plate.prt" is the default model. If it is not, *pick* the **Browse...** button, then *select* the "cover_plate.prt."

Step 12: *Pick* the **Browse...** button inside the Template area. *Pick* **Working Directory** from the left side of the open window. *Select* "a_template_2-views.drw" from the directory. *Pick* the **Open** button. *Pick* the **OK** button in the New Drawing window.

Step 13: At the top of the window, enter "1018 HR STEEL" for the material, then *pick* the **green checkmark.** Enter your name, then press the <Enter> key. (This is the same as *picking* the **green checkmark.**) Enter the drawing number, "212120."

Step 14: Turn off **Axis display, Point display,** and **Csys display.** Make sure **Plane display** is turned on.

Step 15: **File>Prepare>Drawing Properties.** *Select* **change** to the right of Detail Options.

Step 16: In the Option box, enter "tol_display." Set its value to "yes." *Select* the **Add/Change** button.

Step 17: In the Option box, enter "gtol_datums." Set its value to "std_asme." *Select* the **Add/Change** button. *Pick* the **Close** button to close the Options window. *Pick* **Close** to close the Drawing Properties window.

Step 18: *Select* the Annotate tab.

Step 19: *Double-click* on the "FRONT" datum plane in the RIGHT side view to bring up the Datum window. Change the Name from FRONT to "A." *Select* the rightmost datum symbol type. *Pick* **OK.**

Step 20: Do the same for the RIGHT and TOP datum planes in the FRONT view. Change RIGHT to "B." Change TOP to "C."

Step 21: Erase duplicate datum symbols so that datums B and C are shown in the front view only, and datum A is shown in the right side view. (If any of the datums are not visible, *select* the Datums item in the Drawing Tree area for the appropriate view, either FRONT or RIGHT, then *select* **Model A**, **B**, or **C.** Press down the RMB, then *select* **Unerase** to make the datum visible again.)

Step 22: In the lower right corner of the screen change the filter to Annotation. Draw an imaginary box around both views, thus selecting all dimensions in these two views, then change the accuracy of the dimensions to 2-decimal places. (0.123) goes to (0.12) in Dimension ribbon if needed.

Step 23: *Select* the 5.00-inch dimension. Hold down <Ctrl> and *select* the 8.00-inch dimension and both 2.00-inch dimensions related to the four small holes. In the Dimension ribbon, *select* the **Tolerance** icon, then *pick* **Basic.** There should be a box around these four dimensions to indicate they are basic dimensions. The variation for any basic dimension is controlled by the geometric tolerance found in the related control frame.

Step 24: *Select* the 0.88-inch dimension with LMB and when a pop-up menu appears, *pick* the **Flip arrows** icon. Also, change its accuracy back to 3-decimal places. *Pick* the **Dimension Text** icon in the ribbon, then add "4X " in front of the diameter symbol. Pick anywhere in the graphics window. Pick the **Show Model Annotations** icon, then add centerlines to all the holes in the two views.

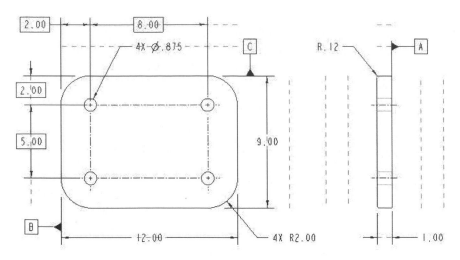

Figure 12-34   Cover Plate with Datums and Basic Dimensions

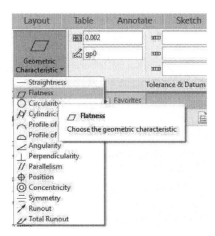

Figure 12-35   Flatness Selected

Step 25: Modify the dimensioning scheme similar to the one shown in Figure 12-34. Clean up extension lines so there is a gap between them and the part feature. Be sure to show all appropriate centerlines and the connections between them. <u>Be sure the Datum Feature Symbols do not line up with any dimension because that means something different.</u>

Now let's add the necessary geometric tolerancing control frames.

Step 26: Select the **Geometric Tolerancing** icon from the top of the screen.

Step 27: *Move* the cursor to the RIGHT side view and *pick* the rightmost vertical line, then back up and away and *press* the MMB to place the geometric tolerance box.

Step 28: *Select* the **Flatness** symbol (parallelogram) in the Geometric Tolerance ribbon. See Figure 12-35.

Step 29: *Set* the **Tolerance Value** to "0.004". See Figures 12-36 and 12-

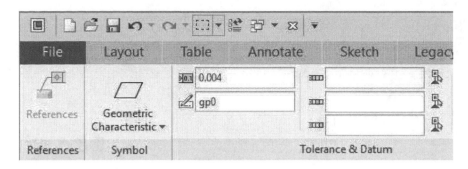

Figure 12-36 Tolerance Value Set

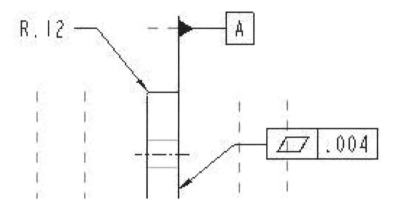

Figure 12-37 Flatness Specified on Part

Step 30: *Select* the **Geometric Tolerancing** icon from the top of the screen again.

Step 31: Move the cursor on top of the "4X ϕ. 875" and press the LMB to connect the geometric tolerance box to the dimension.

Step 32: *Select* the **Position tolerance** symbol. See Figure 12-38.

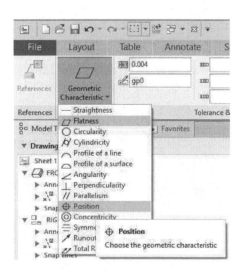

Figure 12-38 Position Characteristic Selected

How big of positional tolerance can we use and still slide the cover plate over the four threaded studs? Let's calculate it now. Let's assume the positional tolerance for the hole is the same as for the threaded cap screw or stud. (See the Tolerance and GD&T Explored section for details.)

$$T = (H - S)/2$$
$$T = (0.875 - 0.813)/2$$
$$T = 0.031 \text{ inches}$$

Step 33: Change the tolerance value to 0.031 inches. See Figure 12-39. Also, change the four basic dimensions to 3-decimal places to match the accuracy of the geometric tolerance.

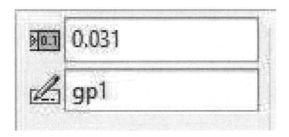

Figure 12-39 Set Tolerance Value

Step 34: Add the diameter symbol in front of the 0.031 value. See Figure 12-40. Add the Material Condition, **MMC** after the 0.031 value.

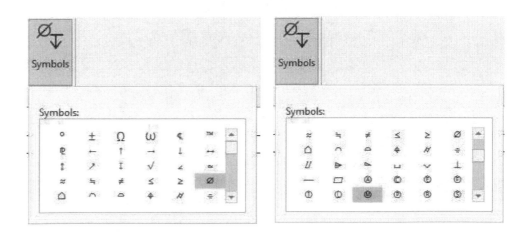

Figure 12-40 Add Diameter Symbol Before and MMC Symbol After

Step 35: *Set* the **Primary** datum to "A" and the **Secondary** datum to "C". Set the **Tertiary** datum to "B." See Figure 12-41.

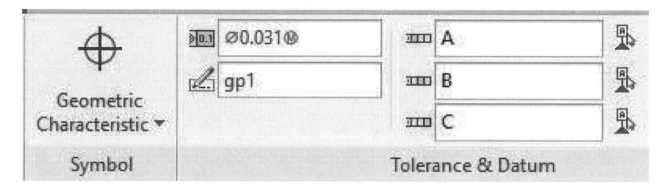

Figure 12-41   Datum References Set

Step 36: *Pick* the .875-inch diameter dimension on the drawing. In the Dimension ribbon, change the tolerance mode from Nominal to **Limits.** Type "0.880" <Enter> for the upper limit. Type "0.875" <Enter> for the lower limit. *Pick* the **Display** icon. *Check* the **ISO Tolerance Display Style** box. *Pick* anywhere in the graphics window.

Step 37: Move the 2.00-inch and 8.00-inch basic dimension to the bottom of the FRONT view as shown in Figure 12-42.

The drawing is not complete without specifying how close to perpendicular datums B and C must be relative to datum A. We will do this now.

Step 38: Add a perpendicularity control frame to the left side of the FRONT view. Set its value to 0.004 inches and have it reference datum tag A. See Figure 12-42.

Step 39: Add a perpendicularity control frame to the top of the FRONT view. Set its value to 0.004 inches and have it reference datum tag A as the primary datum and datum B as the secondary datum. See Figure 12-42.

Step 40: File>Save. *Pick* **OK.**

Step 41: File>Print>Print. *Pick* the **Preview** icon. Your drawing should look similar to Figure 12-42. *Pick* **Print,** then follow your environment instructions. *Pick* **Close Print Setup.**

Step 42: File>Manage File>Delete Old Versions. *Pick* **Yes.**

Step 43: File>Close.

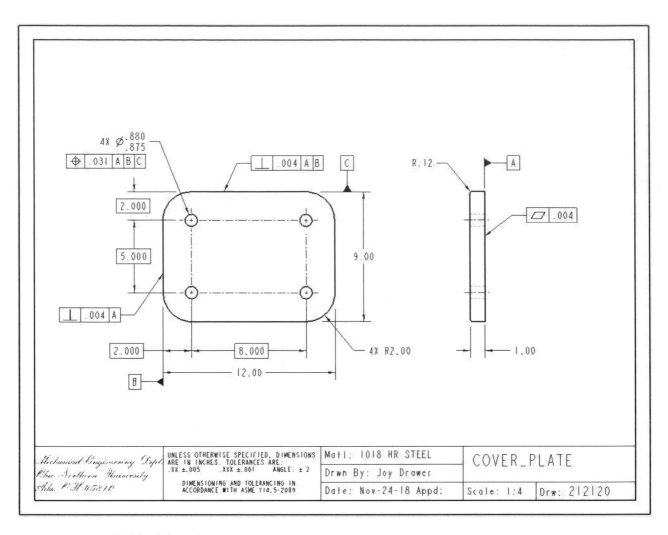

Figure 12-42   Finished Drawing

Step 44: File>Open "Cover_plate.prt" *Pick* **Open.**

Step 45: File>Manage File>Delete Old Versions. *Pick Yes.*

Step 46: File>Close.

Step 47: File>Manage Session>Erase Not Displayed. *Pick* **OK.**

Step 48: Exit from Creo Parametric.

End of Tolerancing and GD&T Practice

Note that we created the datum feature symbols using a different technique (the existing FRONT, RIGHT, and TOP planes) and we specified their relationship to the primary datum using separate feature control frames.

▶ ## Tolerancing and GD&T Exercise

Design Intent

Create a .75-inch thick round cover plate with six equally spaced ½-inch holes on a bolt circle of 6.00 inches. The six holes, having an upper limit of 0.506 inches and a lower limit of 0.502 inches, must be countersunk to accommodate ½-13UNC-2A flathead screws. The cover plate's outer diameter must have an upper limit of 8.000 inches and a lower limit of 7.995 inches. Round the top edge of the cover plate using a .06-inch radius. Datum A is the bottom surface of the cover plate and must be flat within 0.002 inches. Datum B is the outside diameter of the cover plate and must be within 0.002 inches of perpendicular relative to the bottom surface. Datum C is one of the holes in the pattern. The hole pattern must be located within a diameter of 0.002 inches at MMC relative to datum A and B. See Figure 12-43. Show the cover plate round in the FRONT view on the drawing.

Step 1: Start Creo Parametric.

Step 2: Set the Working directory.

Step 3: **File>New.**

Step 4: With the type set to **Part** and **Solid,** name the new part "round_cover", then *pick* **OK.**

Step 5: **File>Prepare>Model Properties.** Set the material to "Standard-Materials_Granta-Design", "Ferrous_metals", **Steel_low_carbon** and the units to **IPS.** *Close* the model properties window.

Step 6: Create an extrusion above the FRONT datum plane centered about the origin that is 8 inches in diameter and .75 inches thick. *Pick* the ***Extrude*** icon, then *pick* the FRONT datum plane to enter sketcher directly.

Step 7: Add a .06-inch round along the top edge.

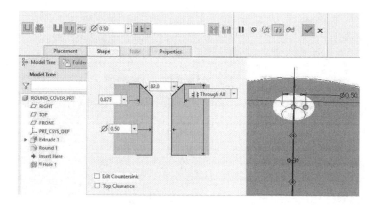

Figure 12-43 Counter Sink Hole

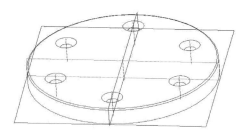

Figure 12-44 Round Cover

Step 8:   Add a 0.500-inch diameter countersunk hole (O.D.= 0.875) using a 6.00-inch <u>Diameter</u> placement. Pick the |Placement| tab, then set the type to Diameter. Specify the center axis of the cover plate with a 0.0° offset from the RIGHT datum plane. See Figure 12-43.

Step 9:   Pattern the hole axially six times equally spaced at 60° intervals about the center axis. See Figure 12-44.

Step 10: File>Save. *Pick* **OK.**

Step 11: File>Close.

Now we will make the engineering drawing for this part.

Step 12: File>New.

Step 13: *Select* **Drawing** from the window and name it "round_cover." Make sure the default template is checked, and then *pick* **OK.**

Step 14: Make sure "round_cover.prt" is the default model. If it is not, *pick* the **Browse…** button, then *select* the "round_cover.prt."

Step 15: *Pick* the **Browse…** button inside the Template area. *Pick* **Working Directory** from the left side of the open window. *Select* "a_template_2-views.drw" from the directory. *Pick* the **Open** button. *Pick* the **OK** button in the New Drawing window.

Step 16: At the top of the window, type "1018 HR STEEL" for the material, then *pick* the **green checkmark.** Enter your name, then press the <Enter> key. (This is the same as *picking* the **green checkmark.**) For the drawing number, type "200120" <Enter>.

Step 17: Turn off *Axis display, Point display, Csys display,* and *Plane display*.

Step 18: File>Prepare>Drawing Properties. *Select* **change** to the right of Detail Options.

Step 19: In the Option box, enter "tol_display." Set its value to "yes." *Select* the **Add/Change** button.

Step 20: In the Option box, enter "gtol_datums." Set its value to "std_asme." *Select* the **Add/Change** button. *Pick* the **Close** button to close the Options window. *Pick* **Close** to close the Drawing Properties window.

Step 21: *Select* the |Annotate| tab.

Step 22: *Double-click* on the scale value in the lower left corner of the screen, then change the value to ".500" or "1/2" <Enter>.

Step 23: *Select* the |Layout| tab. Unlock the views, then reposition them if necessary. When you are finished, lock the views again, then *select* the |Annotate| tab.

Step 24: *Pick* the **Geometric Tolerance** icon, then *pick* the right edge of the RIGHT view with the LMB. Move up and to the right, then *press* the MMB to place the feature control frame. Change the **Geometric Characteristic** to flatness with a value of 0.002 inches.

Step 25: [Datum Feature Symbol] *Select* the **Datum Feature Symbol** icon, then *pick* the flatness control frame you just placed. Move the cursor down, then *press* the MMB to place the datum feature symbol. It will be datum "A."

Step 26: *Pick* the 8-inch dimension in the front view to bring up the Dimension ribbon. Be sure the number of decimal places is "3." *Pick* the Tolerance icon, then *select* **Limits** from the list. Set the upper limit to "8.000" inches. Set the lower limit to "7.995" inches. Pick the Display icon, then check the **ISO Tolerance Display Style** box.

Step 27: *Pick* the **Geometric Tolerance** icon, then *pick* the 8.000-7.995 dimension. Change the **Geometric Characteristic** to perpendicularity with a value of 0.002 inches. Make the primary datum to "A", then press <Enter>.

Step 28: *Select* the **Datum Feature Symbol** icon, then *pick* the perpendicularity control frame you just placed. Move the cursor down, then *press* the MMB to place the datum feature symbol. It should be datum "B."

Step 29: Flip the arrows on the 1/2-inch and 3/4-inch dimensions.

Step 30: At the bottom of the Annotate ribbon *select* **Format>Switch Dimensions.** Repaint the screen. Move the dimensions similar to Figure 12-45. Note the variable names for the dimensions.

    a. ½-inch diameter hole = d15 → _____

    b. Countersink diameter = d16 → _____

    c. 82° Countersink angle = d17 → _____

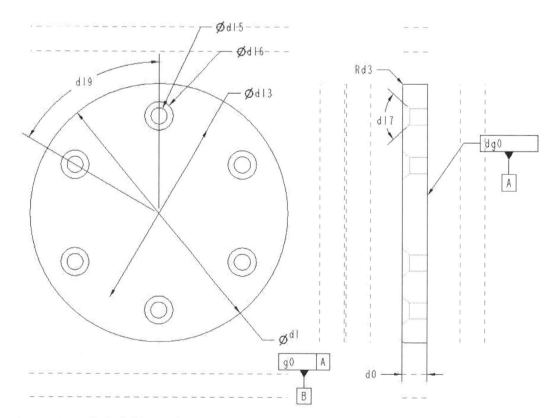

Figure 12-45 Switch Dimensions

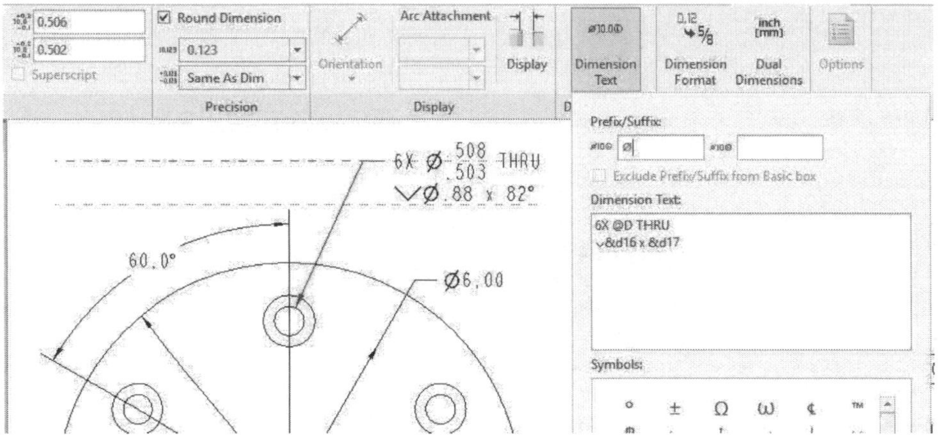

Figure 12-46   Countersunk Hole Dimensioned

Step 31: Change the angle dimension to zero-decimal places (82°), the 7/8-inch to 2 decimal places, and the ½-diameter to 3 decimal places, then delete the 82-degree and the 7/8-inch diameter dimensions (d17, d16).

Step 32: *Pick* the 1/2-inch diameter dimension (d15) to bring up the Dimension ribbon. *Select* the ***Dimension Text*** icon. *Type* "6X @D THRU" <Enter> " ∨&d16 x &d17." At the bottom of the Annotate ribbon *select* **Format>Switch Dimensions** again. *Pick* the ***Tolerance*** icon, then *select* Limits from the list. Set the upper limit to 0.508 and the lower limit to 0.503 inches. *Pick* the ***Display*** icon, then *check* the ***ISO Tolerance Display Style*** box. *Pick* the ***Regenerate*** icon at the top of the screen. See Figure 12-46.

Step 33: **File>Prepare>Drawing Properties.** *Select* **change** to the right of Detail Options. In the Option box, type "radial_pattern_axis_circle." Change the value to **yes.** *Pick* the **Add/Change** button. *Pick* **Close** twice.

Step 34: *Select* the FRONT view. *Pick* the ***Show Model Annotations*** icon at the top of the screen. *Pick* the far right tab in the Show Model Annotations window. *Select* all the centerlines. *Pick* **Apply.**

Step 35: *Select* the RIGHT side view. *Pick* the five visible centerlines in the right side view. *Pick* **Apply.** *Pick* **Cancel** to close the window. See Figure 12-47.

Step 36: **File>Save.** *Pick* **OK.**

Step 37:  *Select* the geometric tolerance icon at the top of the screen. (See practice session if the following instructions are not clear.)

Step 38:  *Pick* the 1/2-inch diameter dimension with the LMB. *Set* the Geometric Characteristics to ***Position***.

Step 39: *Set* the ***Primary datum*** as **A.** *Set* the ***Secondary datum*** as B.

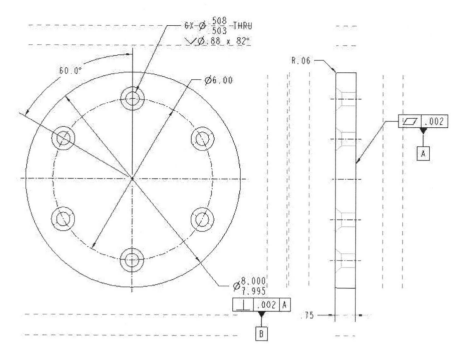

Figure 12-47　Centerlines Added

Step 40: *Set* the **Tolerance Value** to a diameter of "0.003". *Select* the **Symbols** icon. Place the *diameter* symbol before and the Maximum Material Condition (*circled M*) after. See Figure 12-48.

Step 41: *Pick* 6.00-inch diameter dimension, then set it as 3-decimal places if it isn't already. *Select* the **Tolerance** icon, then pick **Basic** from the list. A box will appear around 6.000. See Figure 12-48.

Step 42: *Pick* anywhere in the graphics window.

Step 43: *Pick* the geometric tolerance icon again.

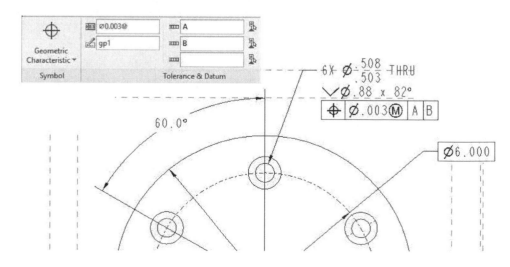

Figure 12-48　First Positional Tolerance Added and a Basic Dimension

Step 44:   ⊥   *Pick* the 1/2-inch diameter dimension with the LMB again. *Set* the Geometric Characteristics to ***Perpendicularity***.

Step 45:   *Set* the ***Primary datum*** as **A.** Leave the ***Secondary*** and ***Tertiary datums*** blank.

Step 46:   *Set* the **Tolerance Value** to a diameter of "0.001". *Select* the ***Symbol*** icon. *Place* a diameter symbol before the 0.001 value. See Figure 12-49.

Step 47:   *Select* the ***Datum Feature Symbol*** icon, then pick the ½ inch hole dimension. Move the cursor down, then press the MMB to place the Datum feature symbol, "C." See Figure 12-50.

Step 48:   *Pick* the 60-degree dimension, then set it as 0-decimal places if it isn't already. *Pick* the ***Dimension Text*** icon, then add "6X " as its prefix. *Select* the ***Tolerance*** icon, then pick **Basic** from the list. A box will appear around **6X 60°**. See Figure 12-51.

Step 49:   *Pick* anywhere in the graphics window.

Step 50:   **File>Save.** (In case you mess up the next section, you can start from here and try again.)

Step 51:   **File>Close.**
File>Manage Session>Erase Not Displayed. *Pick* **OK.**

Step 52:   **File>Open**. *Select* "round_cover.drw." *Pick* **Open**. *Select* the Annotate tab.

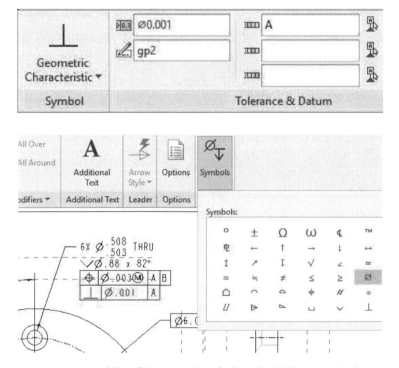

Figure 12-49 Adding Diameter Symbol to the Tolerance Value

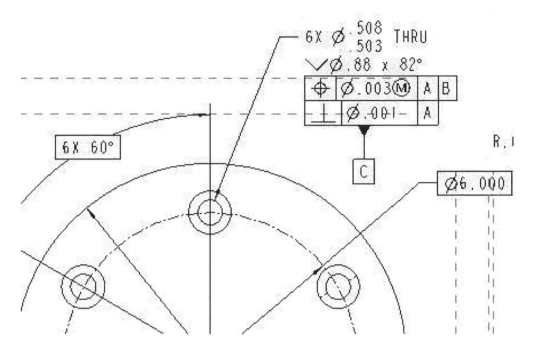

Figure 12-50 Adding the "C" Datum Feature Symbol

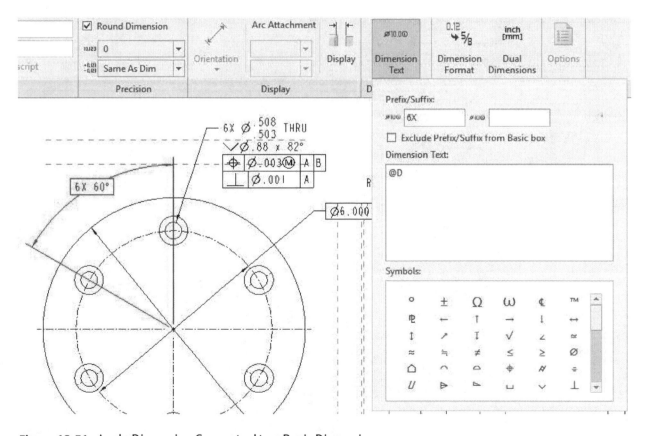

Figure 12-51 Angle Dimension Converted to a Basic Dimension

Step 53: Rearrange the views as necessary. See Figure 12-52.

Step 54: File>Save.

Step 55: File>Manage File>Delete Old Versions. *Pick **Yes.***

Step 56: File>Print>Print. *Pick* the ***Preview*** icon. Your drawing should look similar to Figure 12-52. *Pick **Print,*** then follow your environment instructions. *Pick **Close Print Setup.***

Step 57: File>Close.

Step 58: *Reopen* "round_cover.prt", to get rid of older versions. **File>Manage File>Delete Old Versions.** *Pick* **Yes.**

Step 59: File>Close.

Step 60: File>Manage Session>Erase Not Displayed. *Pick* **OK.**

Step 61: Exit from Creo Parametric.

End of Tolerancing and GD&T Exercise

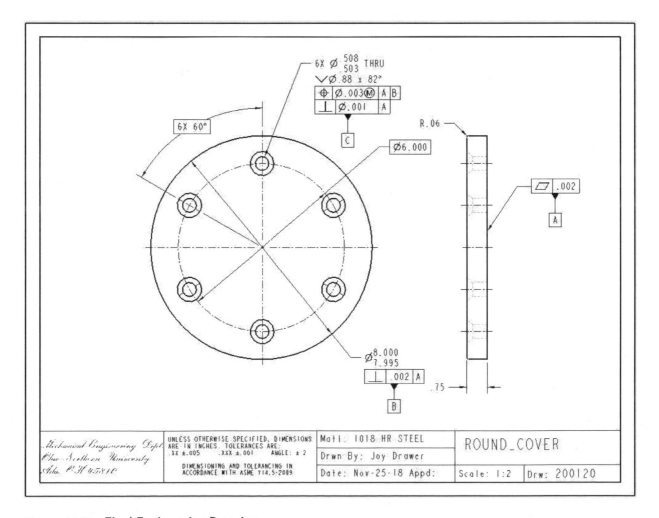

Figure 12-52 Final Engineering Drawing

▶ Review Questions

1. What is a tolerance?

2. Why when would you use geometric dimensioning and tolerancing compared to a standard tolerancing system?

3. What are the three basic types of fits?

4. Why must every dimension on a part drawing have a tolerance?

5. What is an implied tolerance?

6. What is the Basic Hole System of tolerancing?

7. What is meant by "tolerance of form"?

8. What is a basic dimension?

9. What is MMC? RFS? LMC?

10. What is a primary datum? What is a secondary datum?

11. What drawing options do you set to get tolerances to show up on an engineering drawing?

12. Is it possible to rename the default datum planes FRONT, RIGHT, and TOP?

13. What does a rectangular box around a dimension indicate?

14. How many datums are required for a positional geometric dimension and tolerance?

15. How many datums are required for a flatness geometric dimension and tolerance?

16. Which dimensional limit is at LMC? Hole: 1.4625 to 1.4621 inches. Shaft: 1.4620 to 1.4615 inches.

17. What Geometric Tolerances can appear without a datum reference?

18. What Geometric Tolerances must have at least one datum reference?

Tolerancing and GD&T Problems

12.1   Draw the "**table_block**" made from 1030 HR steel shown (Figure 12-53). Create a 3-view engineering drawing at ½-scale, drawing number 121001. The bottom of the table block needs to be flat within 0.003 inches and will become datum A. The front surface of the table block is datum B and it needs to be perpendicular to the bottom within 0.004 inches. The right side in the figure below is datum C and it must be perpendicular to datums A and B within 0.004 inches. All dimensions are in inches.

The 1.00-inch hole goes all the way through the block and must be located within a diameter of 0.003 inches relative to its true position. Convert the 3.25-inch and 1.25-inch dimensions to basic dimensions. The hole's center must be perpendicular to datum A within a diameter of 0.001 inches at MMC. The hole has an H7 fit, which means it has an upper limit of 1.0008 and a lower limit of 1.0000 inches.

Be sure to convert dimensions 3.25 and 1.25 to 3-place decimal basic dimensions to match the accuracy of the geometric tolerance used to locate the 1-inch diameter hole.

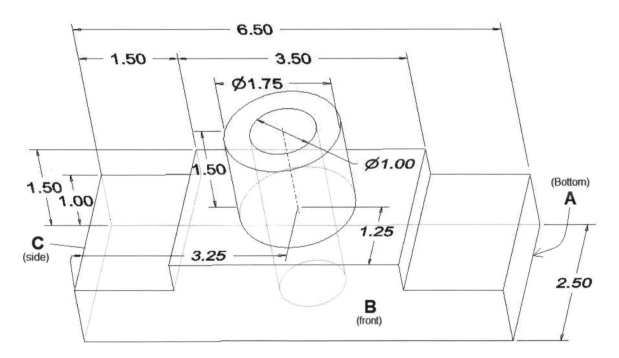

Figure 12-53 Problem 1—Table Block

12.2 Create the "**Support_Bracket**" made from 1030 HR steel, then create a 2-view engineering drawing at ½-scale, drawing number 122002. Datums A, B, and C are shown in Figure 12-54. Datum A must be flat within 0.004 inches. Datum B must be perpendicular to datum A within 0.004 inches. Datum C must be perpendicular to datums A and B within 0.004 inches. All dimensions are in inches.

The size of the 5/8-inch diameter holes has a tolerance of +.003 and -.000 inches. The two 5/8-inch holes must be located within a diameter of 0.002 inches relative to their true position at their maximum material condition based upon datums A, B, and C in that order.

The tolerance for the ¾-inch diameter hole is +.008 and -0.000 inches. The diametric positional tolerance for the ¾-inch hole is .002 inches at MMC relative to datums C, B, then A.

Be sure to indicate which dimensions are basic dimensions. Also, convert all basic dimensions to 3-place decimals to match the accuracy of the geometric position tolerance.

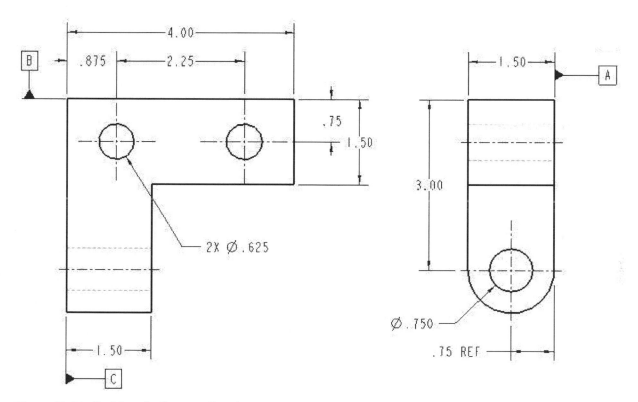

Figure 12-54 Problem 2—Support Bracket

12.3 Draw the "**spacer**" made from 1018 HR steel shown (Figure 12-55). Create a 2-view engineering drawing at ½-scale, drawing number 123003. The back of the spacer must be flat within 0.001 inches and will become datum A. The front surface must be parallel to the back surface within 0.001 inches. The left ¾-inch hole is datum B and it needs to be perpendicular to the back surface within 0.00 inches at its maximum material condition (MMC). The right side slot is datum C and it must be positioned exactly at MMC relative to datums A, and B at MMC, in that order. All dimensions are in inches.

The center hole must be positioned relative to datums A, B at MMC, and C at MMC in that order within 0.002 inches at MMC. The outside surface must have a profile tolerance of 0.003 inches <u>all around</u>.

The ¾ -inch diameter hole and slot have an upper limit of 0.754 and a lower limit of 0.746 inches. The 2¼-inch diameter hole has an upper limit of 2.260 and a lower limit of 2.240 inches. The thickness of the spacer must be between 0.376 and 0.374 inches.

All basic dimensions should have 3-decimal places to match the accuracy of the geometric tolerances.

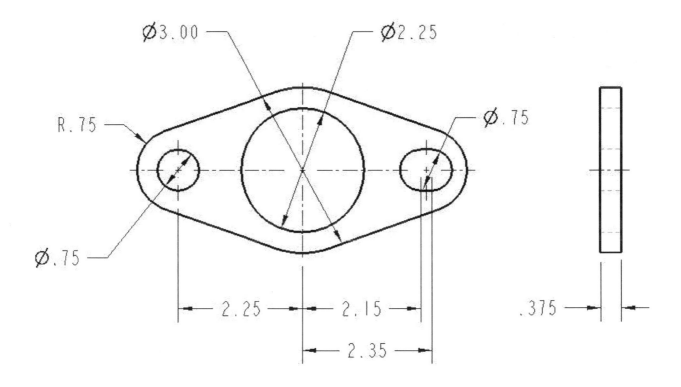

Figure 12-55 Problem 3—Spacer

12.4 Create a 2-view engineering drawing at ¾-scale, drawing number 124004, of the "**flanged_bushing**" made from bronze shown in Figure 12-56. The axis of the 1 3/8-inch diameter hole is datum A. The 1 7/8-inch outside diameter must have a total runout tolerance of 0.0004 inches relative to datum A. The right side of the flange is datum B and must be perpendicular to datum axis A within 0.002 inches. The left side of the flange must be parallel to datum B within 0.002 inches. All dimensions are in inches.

The round between the 2 ¾-inch diameter and the 1 7/8-inch diameter has a radius of 0.15 inches. The 1 3/8-inch diameter hole has an upper limit of 1.3756 and a lower limit of 1.3750 inches. The 1 7/8-inch outside diameter of the tube has an upper limit of 1.8758 and a lower limit of 1.8753 inches. The outer diameter of the ring has an upper limit of 2.755 and a lower limit of 2.745 inches. The thickness of the ring must be between .245 and .255 inches. The overall length of the 1 7/8-inch diameter tube including the round must be between 3.000 and 2.996 inches.

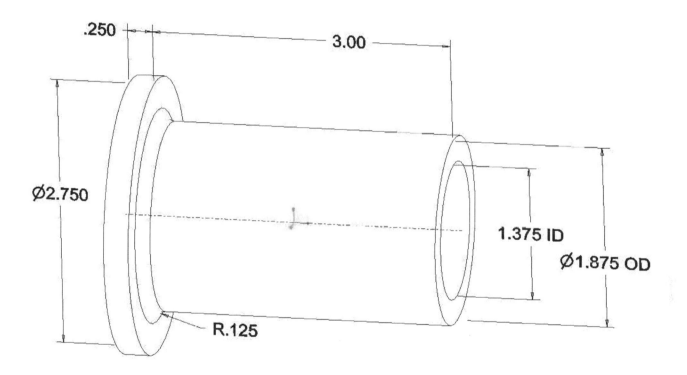

Figure 12-56 Problem 4—Flanged_Bushing

12.5   Create a 2-view engineering drawing at ½-scale for the "**slotted_hub** " shown in Figure 12-57, drawing number 125005. All dimensions are in inches. This part is to be made from 1040 CD Steel. The 1.50-inch diameter hole is the datum A with an upper limit of 1.5010 and a lower limit of 1.5000 inches. An .06 x 45° chamfer must be added to both ends on this 1.50-inch diameter hole. All dimensions are in inches.

The front face (marked B in the figure below) must have a total runout geometric tolerance of 0.0005 inches relative to datum A and is labeled datum B. The back surface must be parallel to datum B within 0.002 inches. The 3-inch diameter hub has an upper limit of 3.0009 and a lower limit of 3.0003 inches. This 3-inch diameter also has a total runout geometric tolerance of 0.0005 inches relative to datum A. The ¾-inch high slot has an upper limit of 0.752 and a lower limit of 0.750 inches. It has a geometric positional tolerance of 0 at maximum material condition (MMC) and is relative to datum A at MMC and datum B. This ¾-inch high slot then becomes datum C.

A geometric profile tolerance of 0.004 inches applies to the back side of the slot and is relative to datums A, B, and C in that order. With that said, the 2.50-inch dimension becomes a basic dimension.

The three ½-inch holes have an upper limit of 0.520 and a lower limit of 0.510 inches. The three holes must be positioned within an 0.003-inch diameter of their true position relative to datums A, B, and C in that order. With that said, the 60° angle and the 4¼-inch bolt circle become basic dimensions with 3-decimal places to match the accuracy of the geometric tolerance.

The remaining dimensions will use the default tolerances found in the title block.

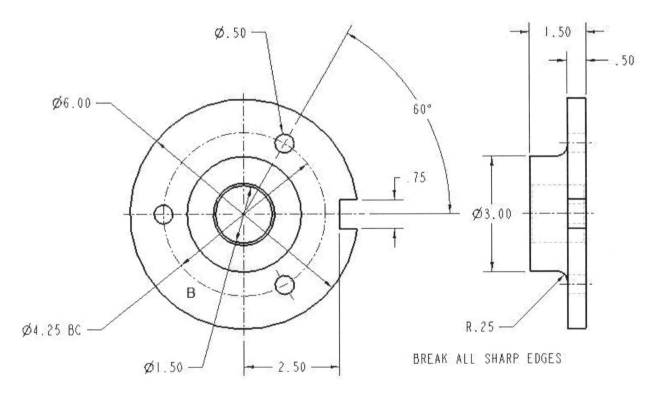

Figure 12-57 Problem 5—Slotted Hub

12.6 Determine the number of orthographic views necessary for the 1/8-inch thick "**roller_bracket**" shown in Figure 12-58, and then create the appropriate engineering drawing at full scale, drawing number 126006. All dimensions are in inches. Add a general note to break all sharp edges. Part made from 1020 HR steel. The ¼ diameter hole has limits of .258 inches and .250 inches. The ½-inch diameter hole has limits of 0.504 and 0.496 inches. The diameter of the slot uses the default tolerance from the title block. All dimensions are in inches.

The lower front surface (marked A) must be flat within 0.010 inches and is primary datum A. The ¼-inch diameter hole must perpendicular to datum A within 0.003 inches and will be designated at datum B. The ¼-inch height of the slot is datum C. The slot has a geometric positional tolerance at maximum material condition of zero relative to datums A and B. The back side of the upper portion with the ½-inch diameter hole must be parallel to datum A within 0.015 inches. The ½-inch diameter hole must be positioned within a diameter of 0.015 inches relative to datums A, B, and C in that order.

With that said, the horizontal 1.625-inch, the horizontal 1.375-inch, and the vertical 1.50-inch dimensions become basic dimensions with 3-decimal places to match the accuracy of the geometric tolerances.

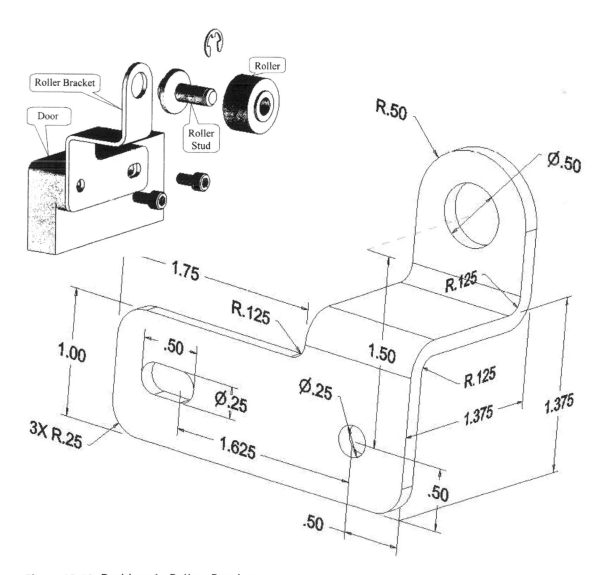

Figure 12-58 Problem 6—Roller_Bracket

12.7   Draw the "**spacer_mm**" made from 1018 HR steel shown (Figure 12-59). Create a 2-view engineering drawing at ½-scale, drawing number 217700. The back of the spacer must be flat within 0.05 millimeters and will become datum A. The front surface must be parallel to the back surface within 0.05 millimeters. The left 20-mm hole is datum B and it needs to be perpendicular to the back surface within 0.0 millimeters at its maximum material condition (MMC). The right side slot is datum C and it must be positioned within a diameter of zero millimeters at MMC relative to datums A, and B at MMC, in this order. All dimensions are in millimeters.

The center hole must be positioned relative to datums A, B at MMC, and C at MMC in that order within 0.05 millimeters at MMC. The outside surface must have a profile tolerance of 0.1 millimeters <u>all around</u>.

The 20-mm diameter hole and slot have an upper limit of 20.1 and a lower limit of 0.19.9 millimeters. The 56-mm diameter hole has an upper limit of 56.8 and a lower limit of 55.4 inches. The thickness of the spacer must be between 10.05 and 9.95 millimeters.

Be sure to indicate which dimensions are basic dimensions.

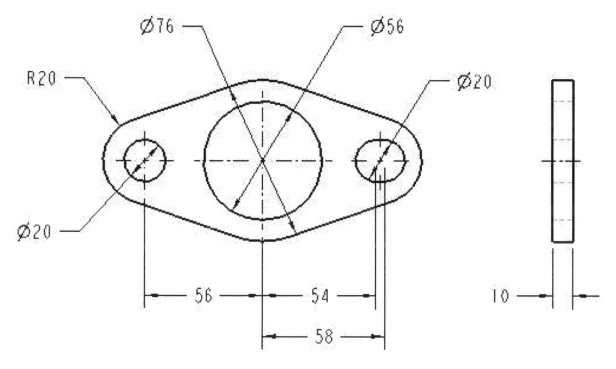

Figure 12-59 Problem 7—Spacer_mm (metric)

CREO SIMULATE AND FEA

Objectives

▶ Introduction to Finite Element Analysis (FEA)

▶ Solve simple beam deflection problems

▶ Create simple beam shear and moment diagrams

▶ Solve 3D stress analysis problems

▶ Determine if the FEA solution makes sense

▶ Creo Simulate and FEA Explored

One aspect of finite element analysis (FEA) is the numerical estimation of stresses in a 3D part by applying real-world loads and boundary conditions. Creo is a suite of programs; one of which is Creo Simulate (formerly Mechanica), which can solve FEA problems. This section focuses on an introduction to structural analysis using Creo Simulate. This chapter will not discuss the theory behind FEA.

It is important to design a mathematical model for FEA that will reflect the actual real-world conditions. The CAD model for producing the part is not the model needed to solve for the real-world stresses in the part. Simple models for FEA work best. Complex models work best for the production of the part. Keep this difference in mind when creating the FEA model.

FEA methods break the part into small regions, then apply the governing equations to each region. FEA can be divided into three categories: 1-dimensional using line elements, 2-dimensional using plane elements, and 3-dimensional using volume elements. Although 3D models can be used for all FEA problems, simplifying the model to 1D or 2D, when appropriate, reduces computer time without loss of accuracy. In order to verify the FEA results, some preliminary hand calculations should be done to estimate the stresses in a few locations on the part, then compare the FEA results with these estimates.

Creo Parametric lets you create the model which can be moved into Simulate for analysis. Simulate lets you define the FEA elements for the part, add the material properties, add the applied loads and the constraints, generate and solve the system of equations, then review the results. The following is a simplified procedure for using Simulate to solve for the reaction loads, deflections, and stresses in a part.

Overview of Required Steps

Step 1:   Create a folder in your working directory for the FEA part and analysis.

Step 2:   Start Creo Parametric.

Step 3:   Set the working directory to the newly created FEA folder.

Step 4:   **File>New…** (name the part).

Step 5:   **File>Prepare>Model Properties.**

Step 6:   Set the Material type and the units of measure.

Step 7:   Close on the Model Properties window.

Step 8:   Construct the FEA part.

Step 9:   *Save* the part.

Step 10:   Pick the Applications tab.

Step 11:   Select the *Simulate* icon in the Applications ribbon.

Step 12:   Material Assignment Assign a material to the part. (Since Simulate can solve for stresses in an assembly of parts, each part must be assigned a material property inside Creo Simulate.)

Step 13:   Create or sketch a region on any surface that needs to reflect a load or constraint area, such as a force load is applied to the last 2 inches of a 15-inch handle.

Step 14:   Assign the real-world loads to the part in the appropriate regions. The loads can be force and moment loads applied to a surface, region, edge, or point; pressure loads or bearing loads applied to a surface; gravity loads, temperature loads, centrifugal loads, or preload applied to the part.

Step 15:   Assign the displacement constraints, planar constraints, pin constraints, ball constraints, or symmetry constraints to reflect the real-world conditions.

Step 16:   Create a new analysis by prescribing the analysis type, the type of convergence wanted, the desired output, the load set, and the constraint set.

Step 17: 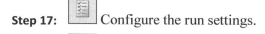 Configure the run settings.

Step 18: Run the Solver using single-pass convergence.

Step 19: Check for errors in the model setup.

Step 20: Fix any errors.

Step 21: Run the solver again using multi-pass convergence.

Step 22: Display the study analysis status. Watch the solution as it progresses.

Step 23: Wait for the solver to obtain a solution.

Step 24: Select the type of results to display.

Step 25: Display the results.

Step 26: Look at the different types of results.

Step 27: Verify if the results make sense.

Step 28: Refine the model if necessary and repeat any necessary steps.

Step 29: Draw conclusions.

Step 30: Redesign the part if necessary and repeat the procedure.

Step 31: Move on to production with the information obtained.

▶ Simulate and FEA Practice

Design Intent
Design the 1040 CD steel, constant diameter shaft in Figure 13-1 given the approximate constant loading shown. The slope of the roller bearings at O and B cannot exceed .06° for good bearing life. (There is about a 20% loss of life in roller bearings for every .06 degrees of neutral axis slope beyond .06 degrees.[1]) Initial estimates led the design engineer to a 1.25-inch diameter shaft. Use Simulate to determine the slope of the steel shaft at points O and B. If the slope requirement is not met, then increase the shaft's diameter in 0.25-inch increments until the slope requirement is met. We are not concerned with the twist of the shaft at this time.

---

[1] Shigley & Mischke, "Mechanical Engineering Design," McGraw-Hill Publishers, 5th edition, 2002, p. 473.

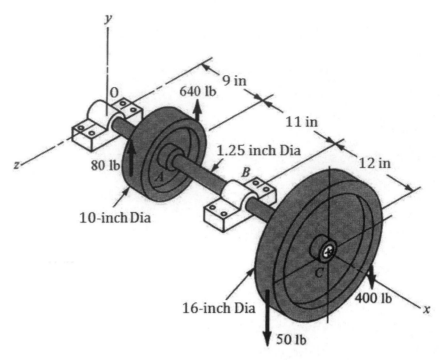

Figure 13-1　Shaft Design Based upon Shaft Deflection

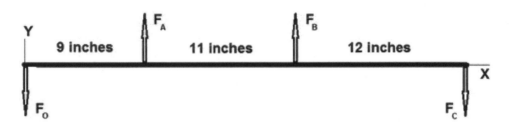

Figure 13-2　Free Body Diagram of Shaft with Assumes Force Directions

Before we begin using Simulate (formally Mechanica), let's calculate the reaction forces at bearings O and B, then calculate the deflections at pulleys A and C. Finally, calculate the estimated slopes of the shaft at O and B. This will give us a comparison when we do the finite element analysis (FEA) with Simulate.

From statics, $\Sigma F_y = 0$ and $\Sigma M_z = 0$, but first we need to draw the free body diagram (FBD). See Figure 13-2.

From Figure 13-1:

$F_A = 640 + 80 = 720$ lbf.
$F_C = 400 + 50 = 450$ lbf.

From Figure 13-2:

$\Sigma M_z = 0 = 9 * F_A + 20 * F_B - 32 * F_C$

$\boxed{F_B = 396 \text{ lbf.}}$

$\Sigma F_y = 0 = -F_O + F_A + F_B - F_C$
$F_O = 720 + 396 - 450$

$\boxed{F_O = 666 \text{ lbf.}}$

For deflections, $Y_O = 0$ inches $= Y_B$.

Using the superposition method of applying F_A, then F_C, we obtain the following.

$$y_A = \frac{-F_A \cdot L_{AB} \cdot x_A \cdot (x_A^2 + L_{AB}^2 - L_{OB}^2)}{6 \cdot E \cdot I \cdot L_{OB}} + \frac{F_C \cdot L_{BC} \cdot x_A \cdot (L_{OB}^2 - x_A^2)}{6 \cdot E \cdot I \cdot L_{OB}}$$

$$y_A = \frac{-720 \cdot 11 \cdot 9 \cdot (9^2 + 11^2 - 20^2)}{6 \cdot 30 \times 10^6 \cdot \dfrac{\pi(1.25^4)}{64} \cdot 20} + \frac{450 \cdot 12 \cdot 9 \cdot (20^2 - 9^2)}{6 \cdot 30 \times 10^6 \cdot \dfrac{\pi(1.25^4)}{64} \cdot 20}$$

$$y_A = \frac{14.11 \times 10^6 + 15.50 \times 10^6}{4.314 \times 10^6} = .069 \text{ inches}$$

Can you solve for the deflection at point C?

$$y_C = ? = _____ \text{ inches}$$

Most textbooks do not have the slope curves for beams as a function of the distance along the beam, so we will use the Deflection Method using Singularity Functions.[2] Assume the origin is at the left end and the variable "x" is positive toward the right. The slope and deflection analysis follows. The moment equation becomes:

$$M = -666 x^1 + 720 \langle x-9 \rangle^1 + 396 \langle x-20 \rangle^1 - 450 \langle x-32 \rangle^1$$

$$EI\theta = \int M dx = -333 x^2 + 360 \langle x-9 \rangle^2 + 198 \langle x-20 \rangle^2 - 225 \langle x-32 \rangle^2 + C_1$$

$$EIy = \int \theta dx = -111 x^3 + 120 \langle x-9 \rangle^3 + 66 \langle x-20 \rangle^3 - 75 \langle x-32 \rangle^3 + C_1 x^1 + C_2$$

$y = 0 @ x = 0$

$C_2 \rightarrow 0$

$y = 0 @ x = 20$

$C_1 \rightarrow +36414$

$$y(@ x = 9) = \frac{-111 \cdot 9^3 + 36414 \cdot 9 + 0}{30 \times 10^6 \cdot \dfrac{\pi(1.25^4)}{64}} = .069 \text{ inches}$$

$$y(@ x = 32) = \frac{-111 \cdot 32^3 + 120 \cdot (32-9)^3 + 66 \cdot (32-20)^3 + 36414 \cdot 32 + 0}{30 \times 10^6 \cdot \dfrac{\pi(1.25^4)}{64}} = .250 \text{ inches}$$

$$\theta(@ x = 20) = \frac{-333 \cdot 20^2 + 360 \langle 20-9 \rangle^2 + 36414}{30 \times 10^6 \cdot \dfrac{\pi(1.25^4)}{64}} = -.015 \text{ radians} = -.85°, (\text{magnitude} > .06°)$$

$$\theta(@ x = 0) = \frac{36414}{30 \times 10^6 \cdot \dfrac{\pi(1.25^4)}{64}} = .010 \text{ radians} = .58°, (\text{magnitude} > .06°)$$

---

[2] Shigley & Mischke, "Mechanical Engineering Design," McGraw-Hill Publishers, 5th edition, 2002, pp. 98-101.

First, we will verify these results using Simulate, then we will redesign the constant diameter shaft to meet the bearing slope specifications using Simulate.

Step 1:   Create the folder "Beam_Analysis" inside your working directory.

Step 2:   Start Creo Parametric.

Step 3:   Set your working directory to the "Beam_Analysis" folder. **Note**: your working directory name must contain only alphanumerics and underscores.

Step 4: File>New.

Step 5:   With **Part** and **Solid** selected, name it "design_shaft". Make sure **Use default template** is checked. Pick **OK.**

Step 6: File>Prepare>Model Properties.

Step 7:   Change the units to **IPS** (inch-pound-seconds). *Pick* **Yes** to convert units.

Step 8:   Change the material to **"Legacy-Materials → steel.mtl** by *double-clicking* on it. *Pick* **Yes** if asked to convert units to your model's units.

Step 9:   Double-click on **➡STEEL** located in the "Materials in Model" window at the bottom of the Materials window to bring up the Material Definition window. At the top of the window, change the name to **"1040_CD_STEEL"**. Under the Structural tab, change Poisson's ratio to 0.292, and Young's Modulus to "3.0e+07" psi. In the Material Limits section, change the Tensile Yield Stress to 71 ksi and the Tensile Ultimate Stress to 85 ksi. Under the Miscellaneous tab, clear the Hardness and Hardness Type if present. *Pick* **OK** to close the Material Definition window.

Step 10: *Pick* **OK** to close the Materials window.

Step 11: Pick **Close** to close the Model Properties window.

Step 12:   Select the **Datum Point** tool from the top of the screen. With **New Point** highlighted under the Placement tab in the Datum Point window, pick the FRONT Datum plane in the model tree or the graphics window. Select the Offset references area in the Datum Point window. Pick the RIGHT datum plane, then hold down <Ctrl> and pick the TOP datum plane. Set the offset values to zero. Select **PNT0** under the Placement tab. Rename it as "O." See Figure 13-3.

Step 13: With New Point highlighted under the Placement tab in the Datum Point window, pick the FRONT Datum plane in the model tree or the graphics window again. Select the Offset references area in the datum point window. Pick the RIGHT datum plane, then hold down <Ctrl> and pick the TOP datum plane. Set the RIGHT offset value to "9.00" and the TOP offset value to zero. Select **PNT0** under the Placement tab. Rename it as "A." Be sure it goes in the +x direction. If it isn't, enter −9.00 for the RIGHT offset.

Step 14: With New Point highlighted under the Placement tab in the Datum Point window, pick the FRONT Datum plane in the model tree or the graphics window again. Select the Offset references area in the datum point window. Pick the RIGHT datum plane, then hold down <Ctrl>

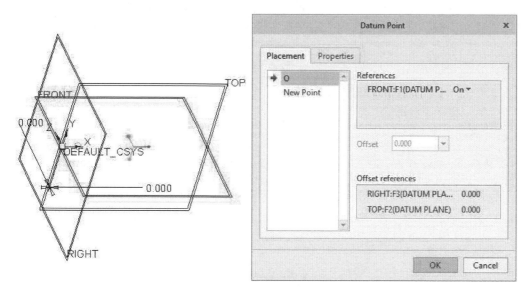

Figure 13-3 O Datum Point

and pick the TOP datum plane. Set the RIGHT offset value to "20.00" and the TOP offset value to zero. Select **PNT0** under the Placement tab. Rename it as "B."

Step 15: With New Point highlighted under the Placement tab in the Datum Point window, pick the FRONT Datum plane in the model tree or the graphics window again. Select the Offset references area in the datum point window. Pick the RIGHT datum plane, then hold down <Ctrl> and pick the TOP datum plane. Set the RIGHT offset value to "32.00" and the TOP offset value to zero. Select PNT0 under the Placement tab. Rename it as "C." See Figure 13-4.

Step 16: Pick **OK** to close the Datum Point window.

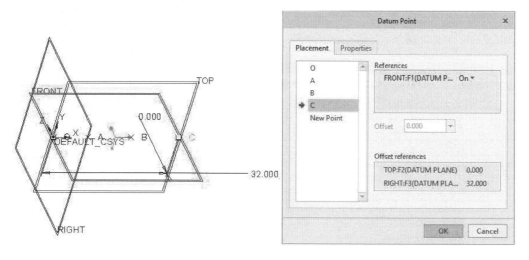

Figure 13-4 Four Datum Points

Step 17: Select the *Sketch* tool from the top of the screen. Pick the FRONT datum plane in the graphics window. Orient the RIGHT datum plane toward the Right. Pick **Sketch.**

Step 18: Pick the *Sketch View* icon if needed to orient the sketch plane so that it is parallel with the display screen.

Step 19: References Select *References* from the top left corner of the Sketch ribbon.

Step 20: Place the cursor on top of datum point O, then press the LMB to select it. (You may have to *right-click* several times in the area of the point before the point becomes highlighted, then press the LMB.)

Step 21: Place the cursor on top of datum point A, and select it.

Step 22: Repeat this procedure for datum points B and C.

Step 23: Pick **Close** to close the References window.

Step 24: Starting at the origin draw three horizontal, straight lines along the FRONT datum plane between points O and A, A and B, and B and C.

Step 25: Add *reference* dimensions to verify their lengths as 9 inches, 11 inches, and 12 inches. See Figure 13-5. Forces can only be applied at line segment endpoints. Simulate will subdivide each line segment into five equal parts while doing calculations. This should be adequate for our purposes.

Step 26: Exit sketcher and keep the sketch.

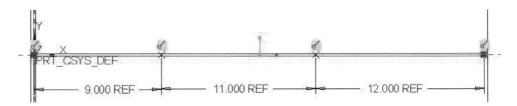

Figure 13-5 Three Line Segments Representing the Shaft and Force Placement

These three lines created along the x-axis in Creo Parametric represent a CAD model, not an FEA model. However, Simulate will use this CAD model when it creates its FEA model.

Step 27: Select the Applications tab. Select the *Simulate* icon. **Note:** your working directory name must contain only alphanumerics and underscores.

Step 28: Select the Refine Model tab. Select **Beam Sections.** See Figure 13-6.

Step 29: Pick the **New** button in the Beam Sections window.

Step 30: Under the Section tab, select **Solid Circle** as the type.

Step 31: A circle representing the cross-section shows up with a box for the radius, R. Type ".625" since the original shaft diameter was 1.25 inches. Pick **OK** to close the Beam Section Definition window.

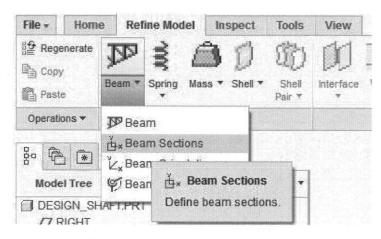

Figure 13-6   Beam Sections

Step 32: Pick **Close** to close the Beam Sections window.

Step 33: 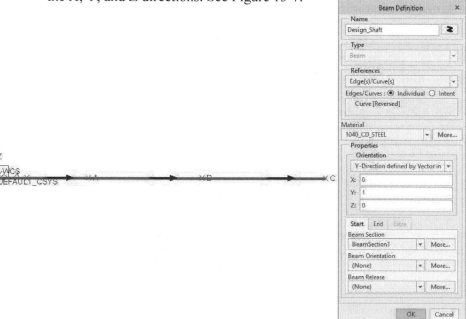 Select the *Beam* icon from the top of the screen.

Step 34: Set the name to "Design_Shaft." Set the References to **Edge(s)/Curve(s).**

Step 35: Pick the 3-line segment CAD model in the graphics window.

Step 36: If the colored arrows (hard to see) are pointed toward the left, they need to be changed toward the right. Click on one of the colored arrows to change their direction to the right if necessary.

Step 37: Choose **1040_CD_STEEL** from the pop-up menu as the material. **(Part Material)** is also a valid option.

Step 38: Confirm that the Y-Direction defined by Vector in WCS is "0, 1, 0," for the X, Y, and Z directions. See Figure 13-7.

Figure 13-7   Beam Definition

Step 39: Pick **OK** to close the Beam Definition window.

Step 40:  Select the ***Displacement Constraint*** under the Home tab.

Step 41: Change the References selection to **Points,** then pick bearing point O, which is located at the origin. Translation is fixed about the X, Y, and Z axes. Rotation is fixed about the X-axis and Y-axis. See Figure 13-8. Pick **OK** to close the Constraint window.

Step 42: *Select* the ***Displacement Constraint*** again.

Step 43: Verify that Constraint2, the 2nd constraint, is a member of MyConstraintSet.

Step 44: Change the References selection in the lower left corner of the screen to **Points** again, then pick bearing point B, which is located at x equals 20 inches.

Step 45: Set the X Translation, Z Translation, and all Rotations to Free. Leave Y Translation fixed. See Figure 13-9. Pick **OK** to close the second Constraint.

Step 46: *Select* the ***Force/Moment Load*** icon.

Step 47: Change the References selection to **Points** again, then pick pulley point A where x equals 9 inches.

Step 48: Set the Y component of the Force to "720" lbf.

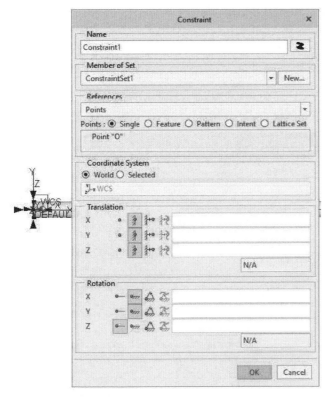

Figure 13-8   Constraint at Bearing O

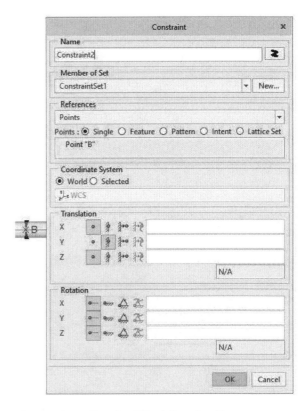

Figure 13-9   Constraint at Bearing B

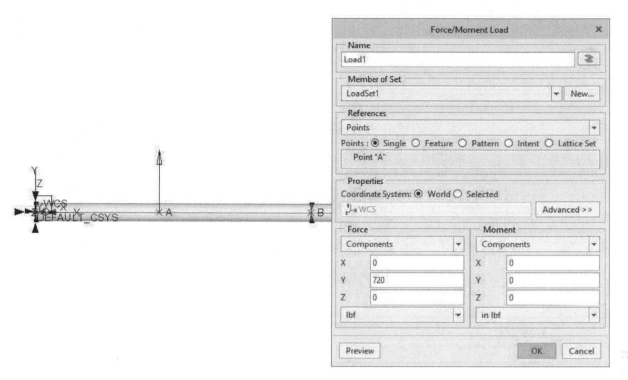

Figure 13-10   Load at Pulley A

Step 49: *Pick* the **Preview** button to verify the applied load at pulley A. See Figure 13-10. *Pick* **OK** to accept the first load.

Step 50: *Select* the **Force/Moment Load** icon again.

Step 51: Verify that LOAD2, the 2nd load, is a member of MyLoadSet.

Step 52: Change the Reference selection to **Points** again, then *pick* the far right end of the rightmost line segment, which is located at pulley C where x equals 32 inches.

Step 53: Set the Y component of the Force to "-450" lbf.

Step 54: *Pick* the **Preview** button to verify the applied load at pulley C. See Figure 13-11. *Pick* **OK** to accept the second load.

Step 55: *Select* the **Named View List** icon at the top of the screen. *Select* FRONT from the pop-up menu.

Step 56: Material Assignment *Select* the **Material Assignment** icon from the top of the

Step 57: In the Name area, type "1040_CD_Steel." *Pick* **OK.** (We already set the material properties for this steel before we created the CAD sketch.) A material tag should show up below the shaft. See Figure 13-12.

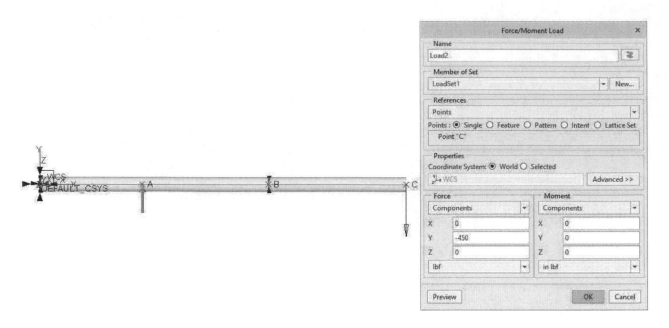

Figure 13-11   Load at Pulley C

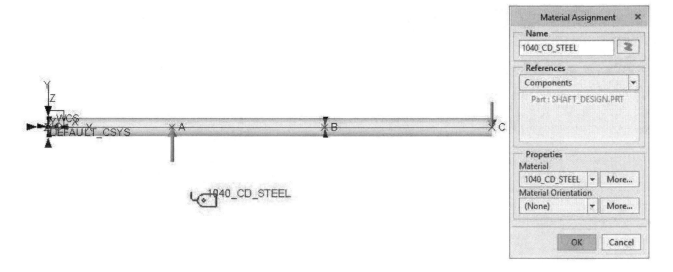

Figure 13-12   Material Assignment Tab

Step 58: **File>Save.** Pick **OK.** (We have done a lot of work that we don't want to lose.)

Step 59: [icon] Select the *Analyses and Studies* icon in the [Home] tab ribbon.

Step 60: In the Analyses and Design Studies window select **File>New Static…**

Step 61: Verify that MyConstraintSet is checked in the constraints area and MyLoadSet is checked in the Loads area. Change the name to "Beam_Analysis." Type, "Shaft Design Analysis" for the description. Under the Convergence tab, set the method to **Multi-Pass Adaptive.** The Polynomial order should be set to a minimum of 1 and a maximum of 6. The Percent Convergence should be set to 10. Pick **OK.** See Figure 13-13.

Step 62:  Select the *green flag* to start the analysis.

Step 63: When the question, "Do you want to run interactive diagnostics?" appears, pick **Yes** (Figure 13-14).

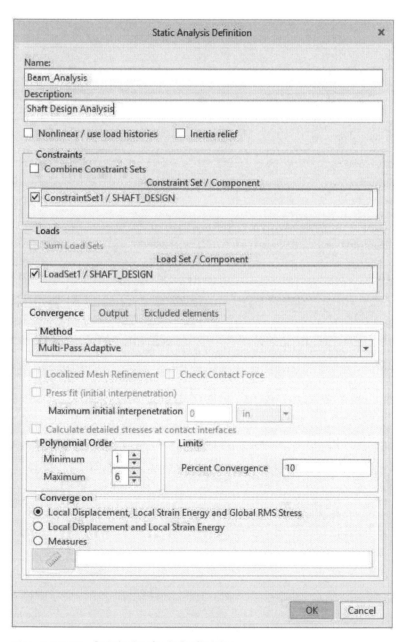

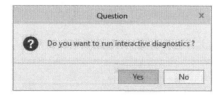

Figure 13-13 Static Analysis Definition   **Figure 13-14**   Run Diagnostics?

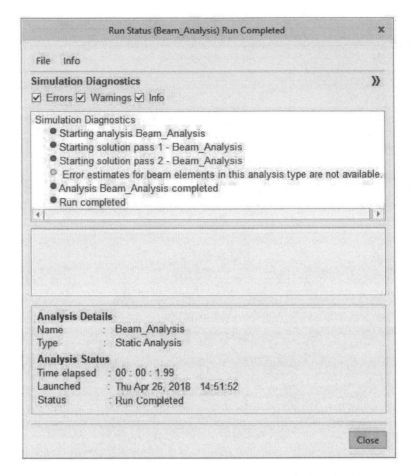

Figure 13-15   Diagnostics: Analysis Beam_Analysis window

Step 64: After verifying the run has completed, pick **Close** to close the Diagnostics: Analysis Beam_Analysis window (Figure 13-15). Don't worry about the warning about the "Error estimates for beam elements in this analysis type are not available" in the Diagnostics window.

Step 65: 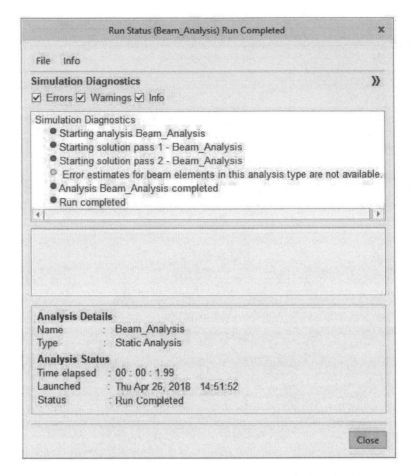 Select the ***Review results of a Design Study or FEA Analysis*** icon in the Analyses and Design Studies window.

Step 66: Change the name to "Forces." Change the Display type to **Model.**

Step 67: Under the Quantity tab, select **Reactions at Point Constraints** from the pop-up menu. Select **Force** and **lbf** as the Secondary Quantity. Select **Y** as the Component. See Figure 13-16.

Step 68: Under the Display Options tab, check the **Deformed** box (Figure 13-17).

Step 69: Pick **OK and Show.**

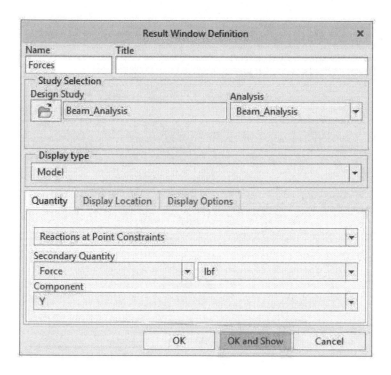

Figure 13-16   Result Window Definition—Quantity Tab

Figure 13-17   Result Window Definition—Display Options Tab

Figure 13-18   Constraint Forces at Bearings O and B

Step 70: Verify the constraint forces are -666 lbf at O and 396 lbf at B (Figure 13-18).

Step 71: Select the ***Edit the selected definition*** icon.

Step 72: Select **Graph** for the Display type.

Step 73: Under the Quantity tab, select **Shear & Moment.** Check **Vy** only. Make sure P, Vz, Mx, My, and Mz are unchecked.

Step 74: Set Graph abscissa relative to **Curve Arc Length.**

Step 75: Set Graph Location to Beams. (Be sure Beams is selected.)

Step 76: Select the arrow icon. In the graphics window, pick the CAD beam. It will turn a different color, possibly red.

Step 77: *Pick* **OK** in the small Select window or press MMB to accept the selection.

Step 78: If the little plus sign is at the right end of the beam, as shown in Figure 13-19, pick the **Toggle** button. If you are not sure, pick **Toggle** and watch the beam.

Step 79: When the little plus sign is at the left end of the beam, pick **OK.**

Step 80: Pick **OK and Show** to see the Shear diagram (Figure 13-20).

Step 81: Select the ***Edit the selected definition*** icon again.

Step 82: Select **Graph** for the Display type.

Figure 13-19   Information Window

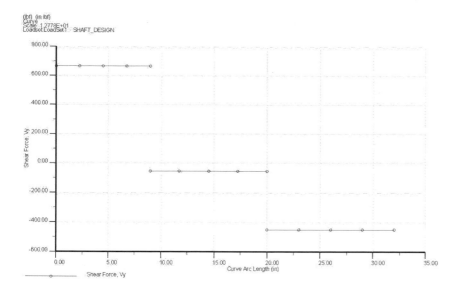

Figure 13-20   Shear Diagram

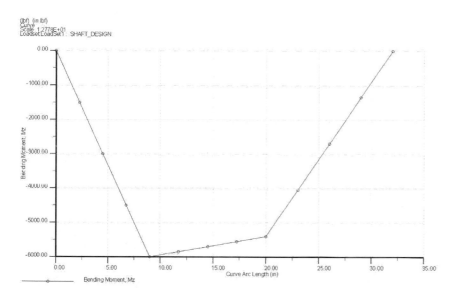

Figure 13-21   Moment Diagram

Step 83: Under the Quantity tab, select **Shear & Moment.** Check **Mz** only. Make sure P, Vy, Vz, Mx, and My are unchecked.

Step 84: Pick **OK and Show** to see the Shear diagram (Figure 13-21).

Step 85: Select the *Edit the selected definition* icon again.

Step 86: Select **Graph** for the Display type.

Step 87: Under the Quantity tab, select **Rotation** and **rad.** In the Components pop-up menu, select **Z.**

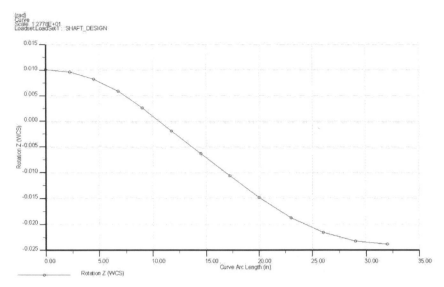

Figure 13-22   Beam Slope using 1.25-inch Diameter Shaft

Step 88: Pick **OK and Show** to see the Rotation about the Z-axis diagram (Figure 13-22).

At x equals 0.0 inches, the rotation is .010 radians or .58 degrees. At x equals 20 inches, the rotation is −.015 radians or −.86 degrees. Both of these results agree with the hand calculations done previously. We have now verified the FEA model using Creo Simulate. Since the bearing deflections are greater than .06 degrees, the shaft diameter must be increased.

Step 89: Pick the *X* in the upper right corner of the window to close the window.

Step 90: When the message window appears asking, "Do you want to exit from Creo Simulate Results and save the current results window?" pick the **Save** button.

Step 91: In the Save Result window, type the filename "original_diameter", then *pick* Save.

Step 92: Pick **Close** to close the Analyses and Design Studies window.

Step 93: Select the ｜Refine Model｜tab. Select **Beam Sections.**

Step 94: Make sure BeamSection1 is highlighted, then pick the **Edit...** button.

Step 95: Change the radius of the solid cross-section to ".75" inches. Pick **OK.**

Step 96: Pick **Close** to close the Beam Sections window.

Step 97: File>Save.

Step 98: *Select* the ｜Home｜ tab, then *select* the ***Analyses and Studies*** icon.

Step 99: With **Beam_Analysis** highlighted in the Analyses and Design Studies window, select the ***green flag*** to start the analysis again.

Step 100: When the question, "Output files for "Beam_Analysis" already exists. They must be removed before starting the design study. Do you want to delete the files?" appears, pick **Yes.**

Step 101: When the question, "Do you want to run interactive diagnostics?" appears, pick **Yes.**

Step 102: After verifying the run has completed, pick **Close** to close the Diagnostics: Analysis Beam_Analysis window.

Step 103: Select the ***Review results of a Design Study or FEA Analysis*** icon in the Analyses and Design Studies window. Type "Forces" for the name.

Step 104: Change the Display type to **Graph.**

Step 105: Under the Quantity tab, select **Rotation** and **rad.** In the Components pop-up menu, select **Z.**

Step 106: Set Graph abscissa relative to **Curve Arc Length.** Set Graph Location to **Beams. (Be sure Beams is selected.)**

Step 107: Select the arrow icon. In the graphics window, pick the CAD beam. It will turn a different color, possibly red. Pick **OK** in the small Select window or press MMB to accept the selection.

Step 108: If the little plus sign is at the right end of the beam, pick the **Toggle** button.

Step 109: When the little plus sign is at the left end of the beam, pick **OK.**

Step 110: Pick **OK and Show** to see the Rotation about the Z-axis diagram (Figure 13-23).

The slope at bearing O is .0049 radians. The slope at bearing B is −.0071 radians. The maximum acceptable shaft slope at the bearings is .001 radians or .06 degrees. We need a larger diameter shaft.

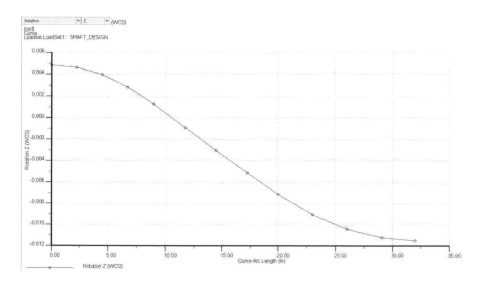

Figure 13-23 Beam Slope using 1.50-inch Diameter

Step 111:   *Close* the window. Do not save the results window.

Step 112:   Repeat the previous steps 92 through 110 using a radius of .875 inches. This is a diameter of 1.75 inches. Slope at bearing O is _____. radians. Slope at bearing B is _____ radians. Is this acceptable? If not, increase the shaft diameter and repeat the process.

Step 113:   *Close* the window. Do not save the results window.

Step 114:   Repeat the previous steps 93 through 110 using a radius of 1.00 inches. This is a diameter of 2.00 inches. Slope at bearing O is _____. radians. Slope at bearing B is _____ radians. Is this acceptable? If not, increase the shaft diameter and repeat the process.

Step 115:   *Close* the window. Do not save the results window.

Step 116:   Repeat the previous steps 93 through 110 using a radius of 1.125 inches. This is a diameter of 2.25 inches. Slope at bearing O is _____ radians. Slope at bearing B is _____ radians. Is this acceptable? If not, increase the shaft diameter and repeat the process.

Step 117:   *Close* the window. Do not save the results window.

Step 118:   Repeat the previous steps 93 through 110 using a radius of 1.25 inches. This is a diameter of 2.50 inches. Slope at bearing O is _____. radians. Slope at bearing B is_____ radians. Is this acceptable? Yes… Yea!

Step 119:   *Save* the results window as "New_Diameter", then *pick* **OK.**

Step 120:   *Pick* ***Close*** to close the Analyses and Design Studies window.

Step 121:   *Pick* ***Close*** in the Home tab.

Step 122:   **File>Save.** *Pick* **OK** if asked.

Step 123:   **File>Manage File>Delete Old Versions.** *Pick* the ***green checkmark.***

Step 124:   **File>Close.** *Exit from Creo Parametric.*

If we had thought about the design problem more carefully, we might have arrived at a solution much sooner. For example, the slope equation, $\theta = \text{constant}/\text{diameter}^4$, tells us the diameter is related to the slope through the fourth power. If we rearrange the equation, $\text{constant} = \theta * \text{diameter}^4 = \theta_{new} * \text{diameter}_{new}^4$. Solving for the new diameter leads to: $\text{diameter}_{new} = (\theta/\theta_{new})^{.25} * \text{diameter}$.

$\text{diameter}_{new} = (.86 \text{ degrees}/.06 \text{ degrees})^{.25}*(1.25) = 2.43 \text{ inches} \rightarrow \boxed{2.50 \text{ inches}}$
or
$\text{diameter}_{new} = (.015 \text{ radians}/.001 \text{ radians})^{.25}*(1.25) = 2.46 \text{ inches} \rightarrow \boxed{2.50 \text{ inches}}$

End of Simulate and FEA Practice

► Simulate and FEA Exercise

Design Intent

The lever shown in Figure 13-24 needs to be redesigned to use less 1018 HR steel, thus lowering its cost. The hole at the left end is 1.50 inches in diameter with a slot cut for a square key. The outer diameter is 3.00 inches. The lever is a constant 1.00-inch thick. A maximum force of 300 pounds is applied to the last, flat 2.00 inches of the 15-inch lever (center of the hole to far right end). The height of the lever near the hub is 2.00 inches. Using a safety factor of 5 for 1018 HR steel, the allowable von Mises stress is 6400 psi. What is the weight savings for the redesigned tool?

Figure 13-24 Original Lever and Suggested Redesigned Lever

For a 1.50-inch diameter shaft, the square key should be $\frac{3}{8} \times \frac{3}{8} \times 1.00$. The distance to the top of the keyway from the opposite side of the hole is 1.67 inches.

Step 1: Create a new folder in your working directory called "Lever_Design."

Step 2: Start Creo Parametric.

Step 3: Set the working directory to "Lever_Design."

Step 4: **File>New.**

Step 5: With **Part** and **Solid** selected, name the part "Lever." With **Use default template** checked, pick **OK.**

Step 6: **File>Prepare>Model Properties.**

Step 7: Change the units to **IPS** (inch-pound-seconds), and then change the material to "**Legacy-Materials → steel.mtl**" by *double-clicking* on it. If asked to convert the units to your model's units, *pick* **Yes.** *Double-click* on the name ►STEEL in the "Materials in Model" window and change the name to "**1018_HR_STEEL**". Change its density to 0.0007324 lbf.sec^2/in^4, Poisson's ratio to 0.292, Young's modulus to "3.0e+07" psi, yield strength to 32 ksi, its ultimate strength to 58 ksi, and Under the $\boxed{\text{Miscellaneous}}$ tab, clear the Hardness and Hardness Type if present.

Step 8: *Pick* **OK** twice, then *pick* **Close** to close the Model Properties window.

Step 9: *Select* the Extrude tool, then *pick* the FRONT datum plane as the sketching plane.

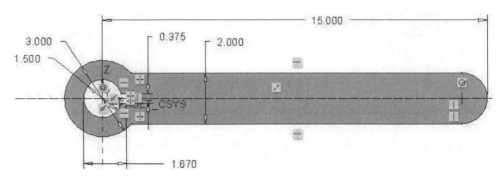

Figure 13-25   Sketch of Lever

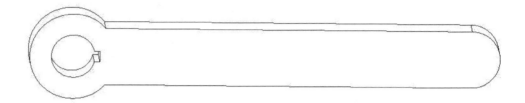

Figure 13-26   Lever is 1.00-inch Thick

Step 10:   *Pick* the ***Sketch View*** icon <u>if necessary</u> to orient the sketch plane so that it is parallel with the display screen. Sketch the lever as shown in Figure 13-25. The origin is at the center of the hole. Dimension according to design intent.

Step 11:   Exit sketcher, then extrude the lever to 1.00-inch thick. See Figure 13-26.

This completes the 3D CAD model so save the model. Now let's move to Simulate to set up and perform a static stress analysis.

Step 12: File>Save. *Pick* **OK.**

Step 13:   *Select* the   Applications   tab. *Select* the ***Simulate*** icon.

Step 14:   *Select* ***Material Assignment*** icon from the top of the screen. For the material name, type "1018_HR_Steel". This material should be selected in the Material box since we already made this material available. *Pick* **OK.** The material tag should appear on or below the lever.

The load is applied in a 2-inch area near the end of the lever. Because of this, we need to create a surface region on the top surface of the lever.

Step 15:   *Select* the   Refine Model   tab.

Step 16:   *Select* the ***Surface Region*** icon from the top of the screen to create a simulated surface region on the last, flat 2.00 inches of the lever.

Step 17:   *Select* the red word **References** in the upper left corner.

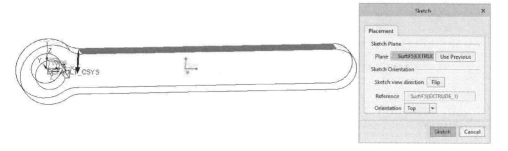

Figure 13-27 Sketch Surface

Step 18: *Pick* the **Define…** button to the right of the sketch box.

Step 19: *Select* the top surface of the lever. Orient the back surface toward the Top. See Figure 13-27. *Pick* the **Sketch** button.

Step 20: 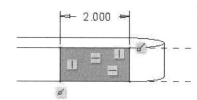 *Pick* the **Sketch View** icon <u>if necessary </u>to orient the sketch plane so that it is parallel with the display screen.

Step 21: *Select* **References.** *Pick* the lower surface of the lever and the line between the flat surface of the lever and the rounded end. See the dashed lines in Figure 13-28. Pick **Close.**

Step 22: Draw a 2.000-inch long rectangle at the right end of the lever using the sketch references as boundaries. See Figure 13-29.

Step 23: Exit sketcher.

Step 24: *Pick* the top surface of the lever as the surface placement. See Figure 13-30.

Step 25: 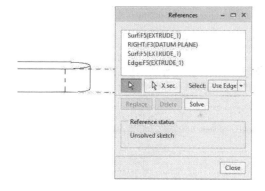 *Pick* the **green checkmark** to accept the surface region.

Step 26: *Select* the Home Tab.

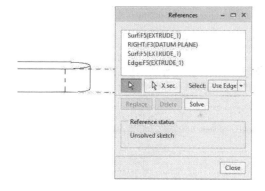

Figure 13-28 Sketch References

Figure 13-29 Rectangular Region

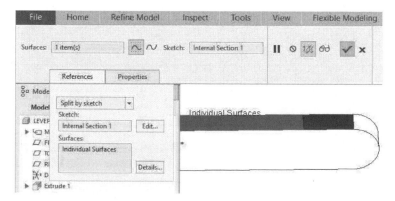

Figure 13-30 Surface Region

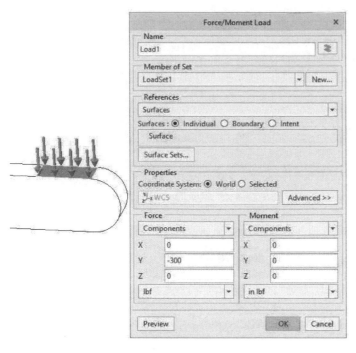

Figure 13-31   Apply Vertical Force to Lever

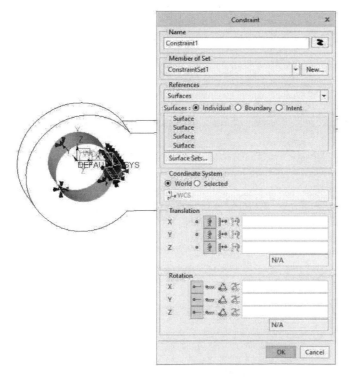

Figure 13-32   First Constraint with Free Z Rotation

Step 27: _Select_ the **Force/Moment Load** icon. _Pick_ the surface region that was just created. Type "-300" for the Y component of the force. See Figure 13-31. Pick **OK.**

Step 28: _Select_ the **Displacement Constraint** icon. _Pick_ the inside surface of the hole. Hold down <Ctrl> and _pick_ the three surfaces that make up the keyway slot. See Figure 13-32. Pick **OK.**

Step 29: _Select_ the **Analyses and Studies** icon.

Step 30: In the Analyses and Design Studies window, _select_ **File>New Static...**

Step 31: In the Static Analysis Definition window, change the name to "Initial_Design." Make sure ConstraintSet1 and LoadSet1 are checked.

Step 32: _Select_ the Convergence tab. _Select_ **Single-Pass Adaptive** for the method. _Pick_ **OK** to close the window.

Step 33: _Select_ the **Start Run** Flag. _Pick_ the **Yes** button when the question, "Do you want to run interactive diagnostics?" appears.

Step 34: After a few moments/minutes the Diagnostics: Analysis Initial_design window will appear. See Figure 13-33. **Close** this window.

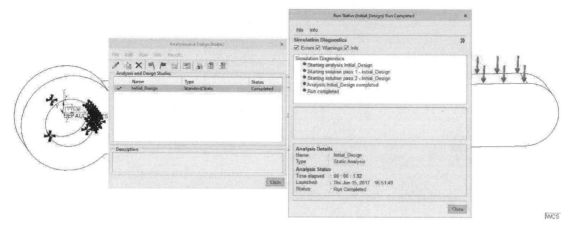

Figure 13-33 Run Diagnostics Window

Step 35: 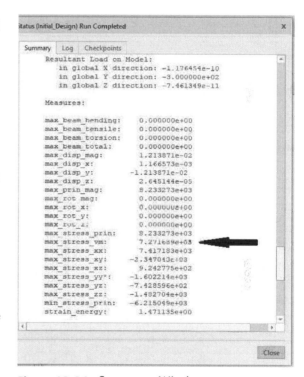 *Select* the **Display Study Status** icon. Scroll through the listing and examine the measures section. The maximum von Mises stress is at 7270 psi. See Figure 13-34. Pick **Close.**

Step 36: *Select* the **Review Results of Design Study or FEA** icon.

Step 37: When the Result Window Definition window appears, name it "vonMises" and the title as "von Mises Stresses." The display type should be **Fringe.** Under the Quantity tab, make sure **Stress, psi,** and **von Mises** are selected.

Step 38: Under the Display Options tab, *check* the **Continuous Tone** and the **Deformed** boxes.

Step 39: *Pick* **OK and Show.** The following stress distribution figure appears. See Figure 13-35. The highest stress is where the round head meets the lever. Because of the sharp corner, the stresses are higher than normal. Overall, everything looks reasonable.

Figure 13-34 Summary Window

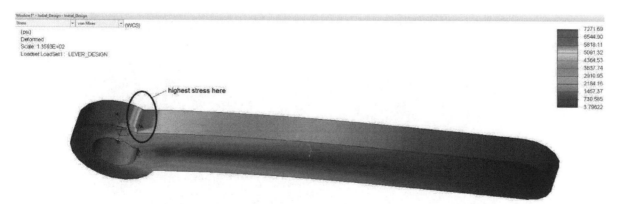

Figure 13-35 von Mises Stress Distribution

Step 40: ▭ ▭ ⊠ *Pick* the X in the upper right corner of the window to close the results window. Do not save the current results window.

Because we are interested in redesigning the lever/handle shape, we need to reduce the stress concentration factor at the sharp edge. Placing a radius in this corner should do the trick.

Step 41: **Close** the Analyses and Design Studies window.

Step 42: **File>Save.**

Step 43: ⊠ *Select* the **Close** icon in the Home ribbon to exit from Creo Simulate and reenter Creo Parametric.

Step 44: *Select* the **Round** tool from the top of the screen. *Pick* the sharp edge, then hold down the <Ctrl> key and *pick* the other sharp edge on the opposite side. Set the radius value to "2.00" inches. *Pick* the **green checkmark.**

Step 45: **File>Save.**

Step 46: **File>Manage File>Delete Old Versions.** *Pick* **Yes** or press **MMB.**

Step 47: *Select* the Applications tab. Select the **Simulate** icon.

Step 48: *Select* the **Analyses and Studies** icon.

Step 49: *Select* the **Configure run Settings** icon. With Memory Allocation (MB) checked, set the value to ½ of the computer's total memory. For example, if the computer has 4 Gb of RAM, set the value to 2048 Mb. *Pick* **OK.**

Step 50: *Select* the **Start Run** Flag.

Step 51: *Pick* **Yes** for the question, "Output files for "Initial_Design" already exist. They must be removed before starting the design study. Do you want to delete the files?"

Step 52: *Pick* the **Yes** button for the question, "Do you want to run interactive diagnostics?"

Step 53: After a few moments/minutes the Diagnostics: Analysis Initial_design window will appear again. **Close** this window.

Step 54: *Select* the **Display Study Status** icon. Scroll through the listing and examine the measures section. Maximum von Mises stress = _____ psi.

Step 55: Scroll to locate the mass of the lever. Record it here: $Mass_o$ = _____ $lbf.sec^2/inch$. The tool steel's density is .0007324 $lbf.sec^2/inch^4$ = .283 $lbm./in^3$. (The Lever mass should be .02353 $lbf.sec^2/inch$.)

Step 56: *Pick* **Close** to close this window.

"Window1" - Initial_Design - Initial_Design
Stress   von Mises   (WCS)
(psi)
Deformed
Scale 1 3929E+02
Loadset:LoadSet1 : LEVER_DESIGN

5838.39
5255.24
4672.09
4088.93
3505.78
2922.62
2339.47
1756.32
1173.16
590.007
6.85303

Figure 13-36 von Mises Stresses with 2-inch Rounds Present

Step 57:  *Select* the ***Review Results of Design Study*** icon.

Step 58: When the Result Window Definition window appears, name it "vonMises" and the title as "von Mises Stresses." The display type should be **Fringe.** Under the Quantity tab, make sure **Stress, psi,** and **von Mises** are selected.

Step 59: Under the Display Options tab, *check* the **Continuous Tone** and the **Deformed** boxes.

Step 60: *Pick* **OK and Show.** The following stress distribution figure appears. See Figure 13-36. The highest stress is still where the round head curves into the lever, but the stress is reduced. Overall, everything looks reasonable.

Let's assume the allowable stress is 6.40 ksi (32 ksi/5 = S_y/SF), and the FEA model shows a maximum stress less than 5.84 ksi. Before we modify the lever it is important to calculate a more accurate von Mises stress using the Multi-Pass Adaptive. (Single-pass adaptive was a quick check.)

Step 61: *Pick* the **View Max** icon to show a maximum stress of 5650 psi at the rounded area.

Step 62: *Pick* the **X** in the upper right corner of the screen to close the results window. Do not save the current results window.

Step 63: With the ***Analyses and Studies*** window still open, *double-click* on the **Initial_Design** so we can edit it.

Step 64: *Select* the Convergence tab.

Step 65: *Select* **Multi-Pass Adaptive** for the method. Set the polynomial order to a minimum of "1" and a maximum of "9." Set the percent convergence to "5." *Pick* **OK** to close the window.

Step 66: *Select* the ***Start Run*** Flag. *Pick* the **Yes** button for both questions.

Step 67: *Select* the ***Display Study Status*** icon. Watch the (FEA) Finite Element analysis proceed.

Step 68: After a few moments/minutes the Diagnostics: Analysis Initial_design window will appear. Close this window.

Step 69: Scroll through the listing and examine the measures section. This time the maximum von Mises stress seems to be approximately 5.68 ksi with a convergence of 0.9%. Everything looks OK. Let's proceed. Because the stresses are very low near the far end of the lever, we know that we can reduce the height of the lever in this area.

Step 70: *Close* all small, open windows.

Step 71: ☒ *Select* the red *Close* icon in the Home ribbon to exit from Creo Simulate and reenter Creo Parametric.

Step 72: *Pick* **Extrude1** in the model tree. *Select* the *Edit Definition* icon from the pop-up menu.

Step 73: *Pick* the Placement tab in the upper left corner of the window. *Pick* the **Edit...** button.

Step 74: Delete the ▬ constraints (if present) for the horizontal lines of the lever.

Step 75: Create a dimension for the radius at the lever's end as shown in Figure 13-37. Set this radius to .75 inches. Make sure the 0.75-inch radius and straight portions of the lever are tangent at their intersections. Make sure the intersection between the lever and the round hub is symmetric about the horizontal centerline. The thickness of the lever where it meets the rounded end should remain at 2.00 inches. Be sure to maintain design intent.

Step 76: Exit from sketch and accept the changes.

Step 77: *Pick* the **green checkmark** to accept the changes.

Step 78: **File>Save.**

Step 79: **File>Manage File>Delete Old Versions.** *Pick* **Yes** or *press* **MMB**.

Step 80: *Select* the Applications tab. *Select* the *Simulate* icon.

Step 81: ▨ *Select* the *Analyses and Studies* icon.

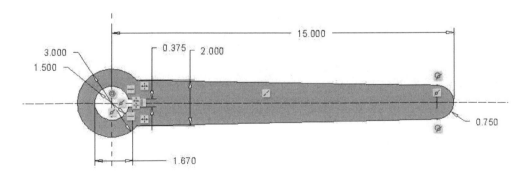

Figure 13-37   Radius Modification

Step 82: [flag icon] *Select* the **Start Run** Flag with Initial_Design highlighted.

Step 83: *Pick* **Yes** for the question, "Output files for "Initial_Design" already exist. They must be removed before starting the design study. Do you want to delete the files?"

Step 84: *Pick* the **Yes** button for the question, "Do you want to run interactive diagnostics?"

Step 85: After a few minutes the Diagnostics: Analysis Initial_design window will appear again. Close this window.

Step 86: [icon] *Select* the **Display Study Status** icon. Scroll through the listing and examine the measures section. The maximum von Mises stress, max_stress_vm, is listed as 6.03 ksi, which is less than the allowable stress specified. The model shows 5.86 ksi maximum.

Step 87: Scroll through the listing to locate the mass of the lever, then record it here. Mass$_{new}$ = _____ lbf.sec^2/inch. Carbon steel density is .0007324 lbf.sec^2/inch^4 = .283 lbm./in^3.

Step 88: Calculate the mass/weight reduction according to the equation.

$$\%\text{Reduction} = 100\%\left(\frac{Mass_{original} - Mass_{new}}{Mass_{original}}\right) = \underline{\qquad}$$

Step 89: *Pick* **Close.**

Step 90: [icon] *Select* the **Review Results of Design Study** icon.

Step 91: When the Result Window Definition window appears, name it "vonMises" and the title as "von Mises Stresses." The display type should be **Fringe.** Under the Quantity tab, make sure **Stress, psi,** and **von Mises** are selected.

Step 92: Under the Display Options tab, *check* the **Continuous Tone** and the **Deformed** boxes.

Step 93: *Pick* **OK and Show.** The following stress distribution figure appears. See Figure 13-38. The highest stress is still where the round head curves into the lever and the stress is more evenly distributed. Overall, everything looks reasonable.

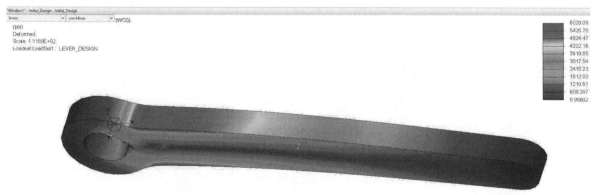

Figure 13-38 von Mises Stresses with Tapered Lever

Step 94: *Pick* the **X** in the upper right corner of the screen to close the results window. Do not save the current results window.

Step 95: *Close* the Analysis and Design Studies window.

Step 96: **File>Save.**

Step 97: **File>Manage File>Delete Old Versions.** Pick **Yes.**

Simulate is capable of optimizing a design based on a single goal of minimizing or maximizing a function while making the design follow a set of constraint equations. This part of the exercise is optional. You can stop at this point, or continue as described below. Rather than reset the design back to its original shape, let's optimize the design starting at this point. See Figure 13-39. We will minimize the mass while keeping the maximum von Mises stress less than 6.4 ksi or less than 3% of the yield strength for tool steel, and the maximum lever deflection less than .025 inches. (Note that yield strength for W1 Tool steel 7320 is 218 ksi, and we have a safety factor of approximately 34.)

Simulate has the ability to run a sensitivity analysis to determine which model parameters have the greatest effects, but we will skip this step. We will vary just the radius on the far end of the lever from .75 inches to .25 inches.

Step 98: *Select* the **Analyses and Studies** icon if the Analyses and Design Studies window is not still open.

Step 99: **File>New Optimization Design Study...** in the Analyses and Design Studies window.

Step 100: Change the name to "Reduced_Mass." Add a description, "Reduce the mass by varying the radius at the end of the lever."

Step 101: *Select* **Optimization** for the type. Set the Goal to **Minimize and total_mass.**

Step 102: *Select* the **Design Limits Measure** icon. In the Measures window, select **max_disp_mag.** *Pick* **OK.** Set the maximum displacement magnitude to less than (<) "**0.025**" inches.

Step 103: *Select* the **Design Limits Measure** icon again. In the Measures window, *select* **max_stress_vm.** *Pick* **OK.** Set the von Mises stress to less than (<) **6400** lbf/in^2.

Figure 13-39 Tapered Lever Design

Step 104: 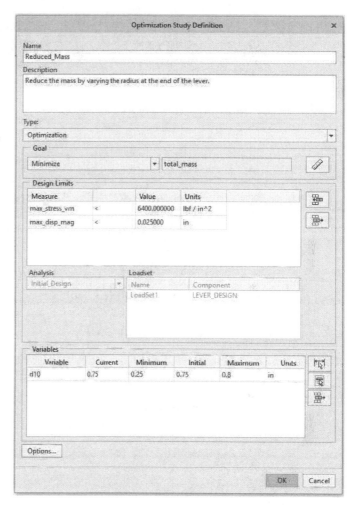 *Pick* the **Select dimension from model** icon. *Pick* the lever in the graphics window with the LMB. When the dimensions for the lever appear, *pick* the R.75-inch dimension at the far end of the lever. Set the minimum value to .25 inches. Set its maximum value to .80 inches. Set the Initial value to 0.75 inches. See Figure 13-40.

Step 105: *Pick* the **Options...** button. *Check* **Repeat P-Loop Convergence** and **Remesh after each shape update.** *Pick* **Close.** See Figure 13-41.

Step 106: *Pick* **OK** to close the Optimization Study Definition window.

Step 107: With the **Reduced_Mass** study highlighted, *pick* the **Start run** icon.

Step 108: Answer **Yes** to delete existing files and run diagnostics.

Step 109: *Select* the **Display Study status** icon to watch the optimization proceed.

Step 110: After a few minutes the Diagnostics: Analysis Reduced_mass window will appear again. **Close** this window.

Step 111: Scroll through the Run Status window looking for Best Design Found: Parameters: (radius) d10 = _____ inches. Goal: mass = _____ lbf.sec^2/in. Record the values. **Close** the window.

Step 112: *Select* the **Review Results of Design Study** icon.

Step 113: When the Result Window Definition window appears, name it "vonMises" and the title as "von Mises Stresses." The display type should be **Fringe.** Under the Quantity tab, make sure **Stress, psi,** and **von Mises** are selected.

Step 114: Under the Display Options tab, *check* the **Continuous Tone** and the **Deformed** boxes.

Figure 13-40 Optimization Study Definition

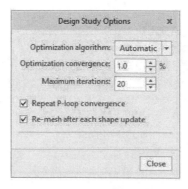

Figure 13-41 Design Study Options

Figure 13-42   von Mises Stresses for Optimized Lever

Step 115: *Pick* **OK and Show.** The following stress distribution figure appears. See Figure 13-42.

Step 116: **Info>View Max.** The highest stress is still where the round head curves into the lever at approximately _____ ksi; however, the stress is more evenly distributed. Overall, everything looks good.

Step 117: *Select* the ***Edit the Selected definition*** icon.

Step 118: In the Result Window Definition window, *select* the Quantity tab. *Select* **Displacement, in,** and **Magnitude.** *Pick* **OK and Show.** See Figure 13-43.

Step 119: *Pick* the **View Max** icon. Note where the maximum deflection is _____ inches. Is the max deflection where you expected it to be?

Step 120: *Select* the ***Edit the Selected definition*** icon. Set the Display type to Graph. *Select* the Quantity tab. *Select* Measure.

Figure 13-43   Optimized Lever Displacement

Step 121: 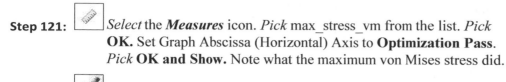 *Select* the **Measures** icon. *Pick* max_stress_vm from the list. *Pick* **OK.** Set Graph Abscissa (Horizontal) Axis to **Optimization Pass.** *Pick* **OK and Show.** Note what the maximum von Mises stress did.

Step 122: *Select* the **Edit the Selected definition** icon again. Set the Display type to **Graph.** Select the Quantity tab. Select **Measure.**

Step 123: *Select* the **Measures** icon. *Pick* **max_disp_mag** from the list. *Pick* **OK.** Set Graph Abscissa (Horizontal) Axis to **Optimization Pass.** Change the Title at the top of the window to "Maximum Displacement." Change the name to "Displacement." *Pick* **OK and Show.** Note what the maximum displacement did on the graph. Which constraint governed the final optimized lever design: Stress or Displacement? _____

Step 124: Calculate the optimized mass/weight reduction using the equation.

$$\% \text{ Reduction} = 100\% \left(\frac{Mass_{original} - Mass_{new}}{Mass_{original}} \right) = \underline{\hspace{2cm}}$$

Step 125: *Pick* the **X** in the upper right corner of the screen to close the results window. Do not save the current results window.

Now we need to update our part model to reflect the optimized design.

Step 126: **Info>Optimize History** in the Analyses and Design Studies window.

Step 127: Answer Yes (y) to the question, "Do you want to review the next shape?" by *picking* the **green checkmark** or *pressing* the MMB.

Step 128: Repeat the above step as long as it continues to get to the last design, the final design.

Step 129: *Close* the Analyses and Design Studies window.

Step 130: Select the **Close** icon to exit from Creo Simulate and reenter Creo Parametric.

Step 131: File>Save.

Step 132: File>Manage File>Delete Old Versions. *Pick* **Yes.**

Step 133: *Select* **No Hidden** from the graphics toolbar. Orient the part and Refit it to the window.

Step 134: *Select* **Extrude1** from the model tree. *Pick* the **Edit Dimensions** icon.

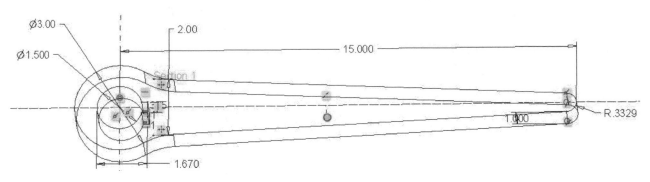

Figure 13-44   Final Lever Design

Step 135: *Double-click* with LMB on the far right radius of the handle to see more digits. It initially should have appeared as R.3329. See Figure 13-44. If you wanted to display more digits by default, select the R.333 dimension with the LMB. In the Dimension tab, change the number of decimal places to four. *Pick* **OK.**

Step 136: File>Print... Pick **OK** twice.

Step 137: File>Close.

Step 138: *Exit from Creo Parametric.*

End of Simulate and FEA Exercise

Review Questions

1. How does an FEA model differ from a CAD model?

2. Is it possible for an FEA solution to give you the exact stresses in a part? Why or why not?

3. What is the maximum edge order available in Creo SIMULATE?

4. What measures are typically used in Creo SIMULATE to monitor convergence?

5. What is a design study?

6. What restrictions are there on 2D FEA models?

7. Why would you use a 2D FEA model, when applicable, over a 3D FEA model?

8. Where and how do you set the units for FEA models?

9. How do you assign material properties to an FEA model?

10. What is the difference between single-pass and multi-pass adaptive analysis?

11. What is the command that shows where and what the maximum stress is on an FEA model?

12. How do you examine the stresses at the current cursor location on the FEA model?

13. What are the minimum material properties necessary to conduct a static stress analysis?

14. What information is necessary to optimize a part using Creo SIMULATE?

15. How do you create a shear or moment diagram for a simple beam in Creo SIMULATE?

16. Besides a load set, an FEA model needs _____ .

Simulate and FEA Problems

13.1   Perform a beam analysis similar to the one done in the practice section by creating datum points, then connecting line segments. The beam can be made from steel or aluminum. (Be sure to include both materials for this part.) For steel, Poisson's ratio equals 0.3 and Young's modulus equals 30×10^6 psi (Figure 13-45). Use 2% convergence and IPS units for the FEA analysis.

   a.   Determine the left and right vertical reaction forces. Point O restricts the movement of the beam in the horizontal direction. Both points O and B are free to rotate about the Z-axis.

   b.   Plot the Shear and Moment diagrams.

   c.   Determine the maximum von Mises stress in the beam and its location.

   d.   Determine the maximum deflection in the beam and its location.

   e.   Repeat parts a) through d) using aluminum with Poisson's ratio equal to 0.334 and Young's modulus equal to 10.6×10^6 psi.

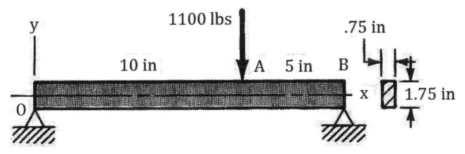

Figure 13-45   Problem 13.1—FEA Beam Analysis 1

13.2   Perform a beam analysis similar to the one done in the practice section by creating datum points, then connecting line segments. The beam can be made from steel or cast iron. (Be sure to include both materials for this part.) For steel, Poisson's ratio equals 0.3 and Young's modulus equals 30×10^6 psi (Figure 13-46). Use 2% convergence and IPS units for the FEA analysis.

   a.   Determine the left and right vertical reaction forces. Point O restricts the movement of the beam in the horizontal direction. Both points O and A are free to rotate about the Z-axis.

   b.   Plot the Shear and Moment diagrams.

   c.   Determine the maximum von Mises stress in the beam and its location.

   d.   Determine the maximum deflection in the beam and its location.

   e.   Repeat parts a) through d) using gray cast iron with Poisson's ratio equal to 0.211 and Young's modulus equal to 14.5×10^6 psi.

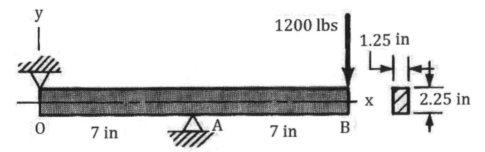

Figure 13-46   Problem 13.2—FEA Beam Analysis 2

13.3 Perform a stress and defection analysis in the 3D part shown in Figure 13-47. Do this analysis similar to the one done in the exercise section. The latching spring is made from steel with Poisson's ratio equal to 0.3 and Young's modulus equal to 30×10^6 psi. A 2.5-pound force is applied in the .125-inch radius channel center 4.1094 inches from the wall. Create a surface region in the channel for the force. The latching spring is attached to the wall using two ¼-20 UNC cap screws. Assume the surface area of the latching spring that touches the wall has its directions and rotations fixed. Use 2% convergence and IPS Units for the FEA analysis.

a. Determine the maximum von Mises stress in the latching spring and its location.
b. Determine the maximum deflection in the latching spring and its location.
c. Using a safety factor of at least 3, what carbon content steel would you recommend? (For example 1010 CD steel, 1040 HR steel, etc.)

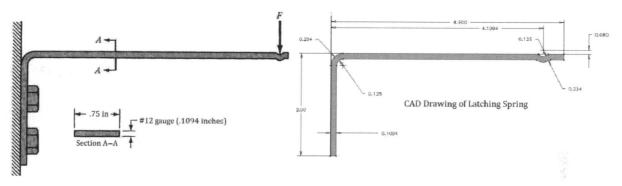

Figure 13-47   Problem 13.3—3D Solid FEA Analysis 1

13.4 Do this analysis similar to the one done in the exercise section. The shaft is made from 1040 CD steel with Poisson's ratio equal to 0.3, Young's modulus equal to 30×10^6 psi, and its yield strength equal to 71 ksi. The applied forces are applied over a ½-inch wide region centered as shown in Figure 13-48. Create a surface region on the shaft for each force and the two bearing reactions. Assume the left bearing restricts the X-direction motion. A .03-inch radius is located at each section where the shaft changes diameter. Use 2% convergence and IPS units for the FEA analysis.

a. Determine the maximum von Mises stress in the shaft and its location.
b. Determine the maximum deflection in the shaft and its location.
c. Using a safety factor of 2, is the shaft a safe design?

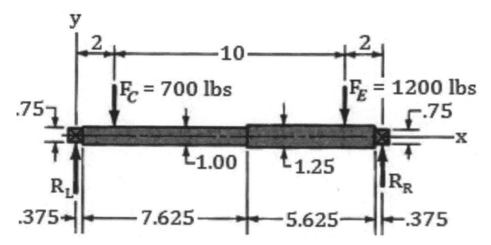

Figure 13-48   Problem 13.4—3D Solid FEA Analysis 2

13.5　Do this analysis similar to the one done in the exercise section. The shaft is made from 1030 CD steel with Poisson's ratio equal to 0.3, Young's modulus equal to 30×10^6 psi, and its yield strength equal to 64 ksi. The applied forces are applied over a ½-inch wide pulley face centered as shown in Figure 13-49. The pulley hub is also ½-inch wide. Create a surface region on the pulley for the two forces and on the shaft for the two 1-inch wide bearing reactions. Assume the bearing O restricts the X-direction motion. For calculation purposes, $T_1 = 8 * T_2$. Use 2% convergence and IPS units for the FEA analysis.

   a.　Determine the maximum von Mises stress in the shaft and its location.
   b.　Determine the maximum deflection in the shaft and its location.
   c.　What is the maximum slope of the centerline of the shaft at the two bearings?

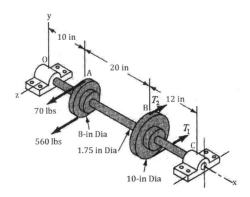

Figure 13-49　Problem 13.5—3D Solid FEA Analysis 3

13.6　Do this analysis similar to the one done in the exercise section. The shaft is made from 1050 CD steel with Poisson's ratio equal to 0.3, Young's modulus equal to 30×10^6 psi, and its yield strength equal to 84 ksi. The applied forces are applied over a ½-inch region centered as shown in Figure 13-50. Create a surface region on the shaft for the two forces and for the two 1-inch wide bearing reactions at A and B. Assume the bearing A restricts the X-direction motion. The shaft has diameters of 2.0 inches and 1.0 inches. Force 1 is 150 pounds vertical and force 2 is 250 pounds horizontal. Distances are: $a_1 = 3$ inches, $b_1 = 9$ inches, and $a_2 = b_2 = 6$ inches. There is a .12-inch radius at the step changes in the shaft. Use 2% convergence and IPS units for the FEA analysis.

   a.　Determine the maximum von Mises stress in the shaft and its location.
   b.　Determine the maximum deflection in the shaft and its location.
   c.　What is the maximum slope of the centerline of the shaft at the two bearings?
   d.　If the slope exceeds .001 radians at either bearing, what is the smallest standard size diameter that will meet this requirement? (Leave the bearing diameters at 1.0 inch.)

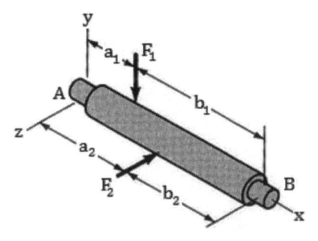

Figure 13-50　Problem 13.6—3D Solid FEA Analysis 4

13.7 A 25-mm diameter, solid steel rod is firmly attached to a wall as in Figure 13-51. The crank lever is located 150 mm from the wall. The handle is located 125 mm from the 25-mm diameter rod. Place 3-mm rounds where the handle meets the solid rods to simulate welds. A force of (0, -800, 300) N is applied to the end of the crank. Assume Poisson's ratio is 0.285 and Young's Modulus is 207 GPa. Use 5% convergence and IPS units for the FEA analysis.

a. What is the magnitude of the deflection at the far end of the handle?
b. Use Dynamic Query to determine the maximum von Mises stress at the wall?
c. Does the FEA analysis agree with the closed-form equations that you learned in the mechanical design course? (Calculate the maximum bending stress at the wall? Calculate the maximum shear stress at the wall. Calculate the maximum von Mises stress at the wall?)

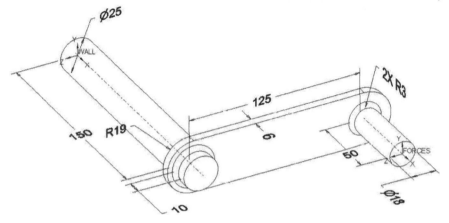

Figure 13-51 Problem 13.7—3D Solid FEA Analysis 5

13.8 A 25-mm diameter, 100 mm long solid steel cantilever rod is firmly attached to a wall as shown in Figure 13-52, then a 5000 N force is applied to the end of the rod. Assume Poisson's ratio is 0.285 and Young's Modulus is 207 GPa. Using a 3% convergence, determine the following. Use 3% convergence and IPS units for the FEA analysis.

a. What is the deflection at the end of the solid rod in the y-direction?
b. What is the maximum von Mises stress at the wall?
c. Does the FEA analysis agree with the closed-form equations that you learned in the strength of materials course? (Calculate the deflection of the rod at its end? Calculate the maximum bending stress at the wall?)

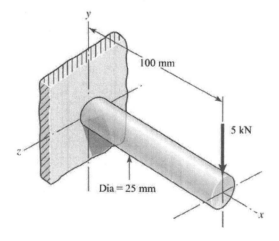

Figure 13-52 Problem 13.8—3D Solid FEA Analysis 6

13.9 A force of 3000 pounds is applied to each end of a ⅜-inch × 1-inch rectangular plate that is 6.00 inches long. See Figure 13-53. There is a .25-inch diameter hole in the middle of the plate. The plate is made from 1018 CD steel with Poisson's ratio equal to 0.3, Young's modulus equal to 30×10^6 psi, and a yield strength equal to 54 ksi. Use 2% convergence and IPS units for the FEA analysis. All dimensions are in inches.

    a. Determine the maximum von Mises stress in the rectangular plate and its location.

    b. Calculate the stress at the hole using the corresponding stress concentration factor, then compare it   with the results from part a).

    c. Determine the change in length in the rectangular plate because of the load.

    d. Vary the diameter of the hole from .10 inches to .50 inches to minimize the maximum von Mises Stress. What diameter provides the minimum stress?

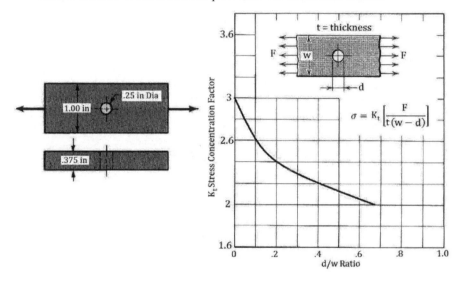

Figure 13-53 Problem 13.9—3D Solid FEA Analysis and Optimization

PARAMETERS FOR DRAWINGS

Name	Definition
&d#	Displays a dimension in a drawing note, where # is the dimension ID.
&ad#	Displays an associative dimension in a drawing note, where # is the dimension ID.
&rd#	Displays a reference dimension in a drawing note, where # is the dimension ID.
&p#	Displays an instance number of a pattern in a drawing note, where # is the pattern ID.
&g#	Displays a gtol in a drawing note, where # is the gtol ID.
&<param_name>	Displays a user-defined parameter value in a drawing note.
&<param_name>:att_cmp	An object parameter that indicates the parameters of the component to which a note is attached.
&<param_name>:att_edge	An object parameter that indicates the parameters of the edge to which a note is attached.
&<param_name>:att_feat	An object parameter that indicates the parameters of the feature to which a note is attached.
&<param_name>:att_mdl	An object parameter that indicates the parameters of the model to which a note is attached.
&<param_name>:att_pipe_bend	An object parameter that indicates the parameters of the pipe bend to which a note is attached.
&<param_name>:att_spool	An object parameter that indicates the parameters of the spool to which a note is attached.
&<param_name>:EID_<edge_name>	An object parameter that references edges.
&<param_name>:FID_<feat_ID>	An object parameter that includes a feature parameter in a note by ID.
&<param_name>:FID_<FEAT_NAME>	An object parameter that includes a feature parameter in a note by name.
&<param_name>:SID_<surface_name>	An object parameter that references surfaces.
&angular_tol_0_0	Specifies the format of angular tolerance values in a note from one to six decimal places.
&sheet_number	Displays a drawing label indicating the current sheet number.
&sheet_name	Displays a drawing label indicating the current sheet name.
&det_scale	Displays a drawing label indicating the scale of a detailed view. You cannot use this parameter in a drawing note. Pro/ENGINEER creates this parameter with a view and places it in notes automatically. You can modify its value, but you cannot call it out in another note.

Name	Definition
&dtm_name	Displays datum names in a drawing note, where name is the name of a datum plane. The datum name in the note is read-only, so you cannot modify it; unlike dimensions, a datum name does not disappear from the model view if included in a note. The system encloses its name in a rectangle, as if it were a set datum.
&dwg_name	Displays a drawing label indicating the name of the drawing.
&format	Displays a drawing label indicating the format size (for example, A1, A0, A, B, and so forth).
&linear_tol_0_0	Specifies the format of dimensional tolerance values in a note from one to six decimal places.
&model_name	Displays a drawing label indicating the name of the model used for the drawing.
&parameter:d	Adds drawing parameters to a drawing note, where parameter is the parameter name and :d refers to the drawing.
&pdmdb	Displays the database of origin of the model.
&pdmrev	Displays the model revision.
&pdmrev:d	Displays the revision number of the model (where :d refers to the drawing).
&pdmrl	Displays the release level of the model.
&scale	Displays a drawing label indicating the scale of the drawing
&sym(<symbolname)	Includes a drawing symbol in a note, where symbolname is the name of the symbol.
&todays_date	Displays a drawing label indicating the date on which the note was created in the form dd-mm-yy (for example, 2-Jan-92). You can edit it as any other nonparametric note, using Text Line or Full Note. If you include this symbol in a format table, the system evaluates it when it copies the format into the drawing. To specify the initial display of the date in a drawing, use the configuration file option "todays_date_note_format."
&total_sheets	Displays a drawing label indicating the total number of sheets in the drawing.
&type	Displays a drawing label indicating the drawing model type (for example, part, assembly, etc.).
&view_name	Displays a drawing label indicating the name of the view. You cannot use this parameter in a drawing note. Pro/ENGINEER creates it with a view and places it in notes automatically. You can modify its value, but you cannot call it out in another note.
&view_scale	Displays a drawing label indicating the name of a general scaled view. You cannot use this parameter in a drawing note. Pro/ENGINEER creates it with a view and places it in notes automatically. You can modify its value, but you cannot call it out in another note.
Pro/REPORT System Parameters	
&asm.mbr.comp....	Retrieves information about the component from the model data and displays it in the report table.
&asm.mbr.cparam....	Retrieves a given component parameter.
&asm.mbr.cparams....	Lists information pertaining to all component parameters for the current model.
&asm.mbr.name	Displays the name of an assembly member. To show tie wraps and markers, the region attributes must be set to Cable Info.
&asm.mbr.param....	Displays information about parameters in an assembly member.

Name	Definition
&asm.mbr.type	Displays the type (part or assembly) of an assembly member.
&asm.mbr.User Defined	Lists the specified user-defined parameter for the respective assembly components. Note that "&asm.mbr." can be used as a prefix before any user-defined parameter in an assembly member.
&fam....	Retrieves Family Table information about the model.
&harn....	Shows cable harness parameters for 3-D harness parts and flat harness assemblies.
&lay....	Retrieves layout information about the model.
&mbr....	Retrieves parameters about a single component.
&mdl....	Retrieves information about a single model.
&prs....	Retrieves process-specific report parameters used to create reports on the entire process sequence.
&rpt....	Displays information about each record in a repeat region.
&weldasm....	Retrieves welding information about the model.
&asm.mbr.cblprm....	Lists values for a given cabling parameter.
&asm.mbr.cblprms....	Lists values for cabling and wire parameters.
&asm.mbr.connprm....	Lists parameters for connector pins in flat harness assemblies.
&asm.mbr.location...	Lists the location callouts in a specified view or all views of the drawing in session.
&asm.mbr.pipe....	Shows pipeline, pipe segment, and Pro/REPORT bend information parameters.
&asm.mbr.generic.name....	Lists the generic name information for a Family Table instance in a table.
&asm.mbr.topgeneric.name....	Lists the top generic name information for a Family Table instance in a table when working with a nested Family Table.

DRILL AND TAP CHART

Major Dia		Threads (Thd/in)	Tap Drill Size		Clearance Hole	
Size	(inch)		Drill	Decimal	Drill	Decimal
0	0.06	80	3/64	0.0469	50	0.07
1	0.073	64	53	0.0595	46	0.081
		72	53	0.0595		
2	0.086	56	50	0.07	41	0.096
		64	50	0.07		
3	0.099	48	47	0.0785	35	0.11
		56	45	0.082		
4	0.112	40	43	0.089	30	0.1285
		48	42	0.0935		
5	0.125	40	38	0.1015	29	0.136
		44	37	0.104		
6	0.138	32	36	0.1065	25	0.1495
		40	33	0.113		
8	0.164	32	29	0.136	16	0.177
		36	29	0.136		
10	0.19	24	25	0.1495	7	0.201
		32	21	0.159		
12	0.216	24	16	0.177	1	0.228
		28	14	0.182		
		32	13	0.185		
1/4	0.25	20	7	0.201	H	0.266
		28	3	0.213		
		32	7/32	0.2188		
5/16	0.3125	18	F	0.257	Q	0.332
		24	I	0.272		
		32	9/32	0.2812		

Major Dia		Threads (Thd/in)	Tap Drill Size		Clearance Hole	
Size	(inch)		Drill	Decimal	Drill	Decimal
3/8	0.375	16	5/16	0.3125	X	0.397
		24	Q	0.332		
		32	11/32	0.3438		
7/16	0.4375	14	U	0.368	15/32	0.4687
		20	25/64	0.3906		
		28	Y	0.404		
1/2	0.5	13	27/64	0.4219	17/32	0.5312
		20	29/64	0.4531		
		28	15/32	0.4688		
9/16	0.5625	12	31/64	0.4844	19/32	0.5938
		18	33/64	0.5156		
		24	33/64	0.5156		
5/8	0.625	11	17/32	0.5312	21/32	0.6562
		18	37/64	0.5781		
		24	37/64	0.5781		
11/16	0.6875	24	41/64	0.6406	23/32	0.6562
3/4	0.75	10	21/32	0.6562	25/32	0.7812
		16	11/16	0.6875		
		20	45/64	0.7031		
13/16	0.8125	20	49/64	0.7656	27/32	0.8438
7/8	0.875	9	49/64	0.7656	29/32	0.9062
		14	13/16	0.8125		
		20	53/64	0.8281		
15/16	0.9375	20	57/64	0.8906	31/32	0.9688
1	1	8	7/8	0.875	1 1/32	1.0313
		12	15/16	0.9375		
		20	61/64	0.9531		

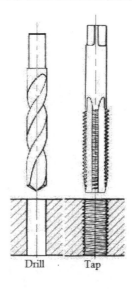

Drill   Tap

SURFACE ROUGHNESS CHART

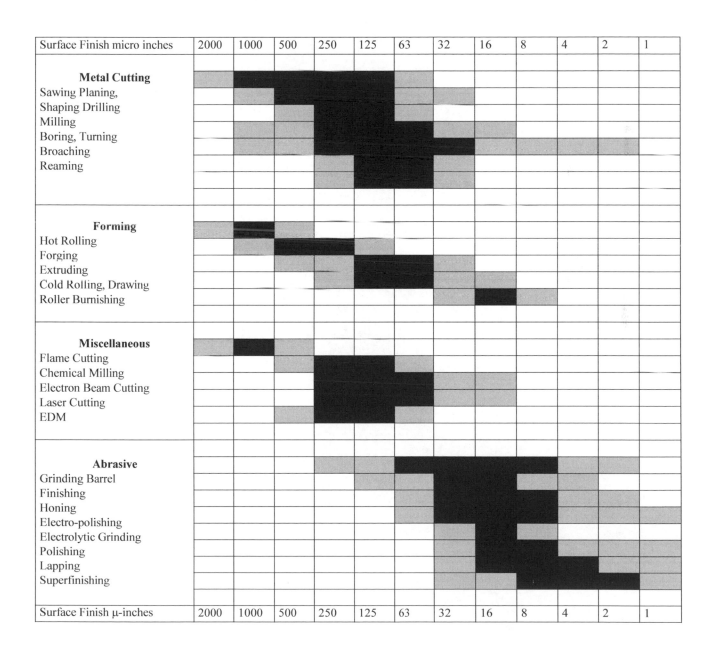

Surface Finish micro inches	2000	1000	500	250	125	63	32	16	8	4	2	1
Metal Cutting												
Sawing Planing,												
Shaping Drilling												
Milling												
Boring, Turning												
Broaching												
Reaming												
Forming												
Hot Rolling												
Forging												
Extruding												
Cold Rolling, Drawing												
Roller Burnishing												
Miscellaneous												
Flame Cutting												
Chemical Milling												
Electron Beam Cutting												
Laser Cutting												
EDM												
Abrasive												
Grinding Barrel												
Finishing												
Honing												
Electro-polishing												
Electrolytic Grinding												
Polishing												
Lapping												
Superfinishing												
Surface Finish μ-inches	2000	1000	500	250	125	63	32	16	8	4	2	1

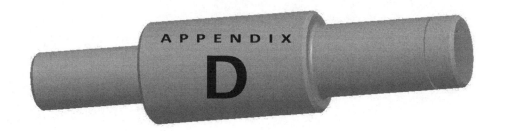

CLEVIS PIN SIZES

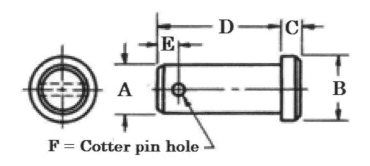

F = Cotter pin hole

Pin Dia (inch) A	Head Dia (inches) B	Hd Height (inches) C	Min. Length (inches) D	Pin Locate (inches) E	Hole Size (inches) F
0.188	0.31	0.06	0.59	0.11	0.078
0.250	0.38	0.09	0.80	0.12	0.078
0.312	0.44	0.09	0.97	0.16	0.109
0.375	0.50	0.12	1.09	0.16	0.109
0.500	0.62	0.16	1.42	0.22	0.141
0.625	0.81	0.20	1.72	0.25	0.141
0.750	0.94	0.25	2.05	0.30	0.172
1.000	1.19	0.34	2.62	0.36	0.172

Pin Dia (mm) A	Head Dia (mm) B	Hd Height (mm) C	Min. Length (mm) D	Pin Locate (mm) E	Hole Size (mm) F
4	6	1	16	2.2	1
6	10	2	20	3.2	1.6
8	14	3	24	3.5	2
10	18	4	28	4.5	3.2
12	20	4	36	5.5	3.2
16	25	4.5	44	6	4
20	30	5	52	8	5
24	36	6	66	9	6.3

NUMBER AND LETTER DRILL SIZES

Drill Size Number	Diameter (inches)	Diameter (mm)		Drill Size Number	Diameter (inches)	Diameter (mm)		Drill Size Number	Diameter (inches)	Diameter (mm)
1	0.228	5.79		28	0.141	3.58		55	0.052	1.32
2	0.221	5.61		29	0.136	3.45		56	0.047	1.19
3	0.213	5.41		30	0.129	3.28		57	0.043	1.09
4	0.209	5.31		31	0.120	3.05		58	0.042	1.07
5	0.206	5.23		32	0.116	2.95		59	0.041	1.04
6	0.204	5.18		33	0.113	2.87		60	0.040	1.02
7	0.201	5.11		34	0.111	2.82		61	0.039	0.99
8	0.199	5.05		35	0.110	2.79		62	0.038	0.97
9	0.196	4.98		36	0.107	2.72		63	0.037	0.94
10	0.194	4.93		37	0.104	2.64		64	0.036	0.91
11	0.191	4.85		38	0.102	2.59		65	0.035	0.89
12	0.189	4.80		39	0.100	2.54		66	0.033	0.84
13	0.185	4.70		40	0.098	2.49		67	0.032	0.81
14	0.182	4.62		41	0.096	2.44		68	0.031	0.79
15	0.180	4.57		42	0.094	2.39		69	0.029	0.74
16	0.177	4.50		43	0.089	2.26		70	0.028	0.71
17	0.173	4.39		44	0.086	2.18		71	0.026	0.66
18	0.170	4.32		45	0.082	2.08		72	0.025	0.64
19	0.166	4.22		46	0.081	2.06		73	0.024	0.61
20	0.161	4.09		47	0.079	2.01		74	0.023	0.58
21	0.159	4.04		48	0.076	1.93		75	0.021	0.53
22	0.157	3.99		49	0.073	1.85		76	0.020	0.51
23	0.154	3.91		50	0.070	1.78		77	0.018	0.46
24	0.152	3.86		51	0.067	1.70		78	0.016	0.41
25	0.150	3.81		52	0.064	1.63		79	0.015	0.38
26	0.147	3.73		53	0.060	1.52		80	0.014	0.36
27	0.144	3.66		54	0.055	1.40				

Drill Size Letter	Diameter (inches)	Diameter (mm)		Drill Size Letter	Diameter (inches)	Diameter (mm)		Drill Size Letter	Diameter (inches)	Diameter (mm)
A	0.234	5.94		J	0.277	7.04		S	0.348	8.84
B	0.238	6.05		K	0.281	7.14		T	0.358	9.09
C	0.242	6.15		L	0.290	7.37		U	0.368	9.35
D	0.246	6.25		M	0.295	7.49		V	0.377	9.58
E	0.250	6.35		N	0.302	7.67		W	0.386	9.80
F	0.257	6.53		O	0.316	8.03		X	0.397	10.08
G	0.261	6.63		P	0.323	8.20		Y	0.404	10.26
H	0.266	6.76		Q	0.332	8.43		Z	0.413	10.49
I	0.272	6.91		R	0.339	8.61				

F

SQUARE AND FLAT KEY SIZES

Shaft Diameter		Key Size		Key Size Depth (inches)
Over (inch)	Including (inch)	w (inch)	h (inch)	
5/16	7/16	3/32	3/32	3/64
7/16	9/16	1/8	3/32	3/64
		1/8	1/8	1/16
9/16	7/8	3/16	1/8	1/16
		3/16	3/16	3/32
7/8	1 1/4	1/4	3/16	3/32
		1/4	1/4	1/8
1 1/4	1 3/8	5/16	1/4	1/8
		5/16	5/16	5/32
1 3/8	1 3/4	3/8	1/4	1/8
		3/8	3/8	3/16
1 3/4	2 1/4	1/2	3/8	3/16
		1/2	1/2	1/4
2 1/4	2 3/4	5/8	7/16	7/32
		5/8	5/8	5/16
2 3/4	3 1/4	3/4	1/2	1/4
		3/4	3/4	3/8
3 1/4	3 3/4	7/8	7/8	7/16
3 3/4	4 1/2	1	1	1/2

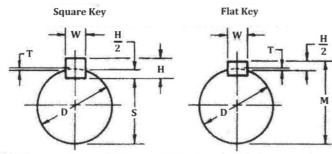

Square Key   Flat Key

C = Allowance for parallel keys = .005 inches
W = Nominal key width (Inches)

$$S = D - \frac{H}{2} - T = \frac{D - H + \sqrt{D^2 - W^2}}{2} \qquad T = \frac{D - \sqrt{D^2 - W^2}}{2}$$

$$M = D - T + \frac{H}{2} + C = \frac{D + H + \sqrt{D^2 - W^2}}{2} + C$$

| Shaft Diameter | | Key Size | | Key Size |
Over (mm)	Including (mm)	w (mm)	h (mm)	Depth (mm)
6	8	2	2	1.0
8	10	3	3	1.5
10	12	4	4	2.0
12	17	5	5	2.5
17	22	6	6	3.0
22	30	7	7	3.5
		8	7	3.5
30	38	8	8	4.0
		10	8	4.0
38	44	9	9	4.5
		12	8	4.0
44	50	10	10	5.0
		14	9	4.5
50	58	12	12	6.0
		16	10	5.0

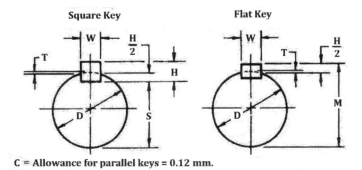

Square Key   **Flat Key**

C = Allowance for parallel keys = 0.12 mm.
W = Nominal key width (millimeters)

$$S = D - \frac{H}{2} - T = \frac{D - H + \sqrt{D^2 - W^2}}{2} \qquad T = \frac{D - \sqrt{D^2 - W^2}}{2}$$

$$M = D - T + \frac{H}{2} + C = \frac{D + H + \sqrt{D^2 - W^2}}{2} + C$$

SCREW SIZES

Nominal Size (inch)	Slot Width (inches)	Flat & Oval Head		Round Head		Hexagon Head	
		A (inches)	B (inches)	C (inches)	D (inches)	E (inches)	F (inches)
0 (.060)	0.023	0.119	0.035	0.113	0.053	-	-
1 (.073)	0.026	0.146	0.043	0.138	0.061	-	-
2 (.086)	0.031	0.172	0.051	0.162	0.069	0.125	0.050
3 (.099)	0.035	0.199	0.059	0.187	0.078	0.187	0.055
4 (.112)	0.039	0.225	0.067	0.211	0.086	0.187	0.060
5 (.125)	0.043	0.252	0.075	0.236	0.095	0.187	0.070
6 (.138)	0.048	0.279	0.083	0.260	0.103	0.250	0.080
8 (.164)	0.054	0.332	0.100	0.309	0.120	0.250	0.110
10 (.190)	0.06	0.385	0.116	0.359	0.137	0.312	0.120
12 (.216)	0.067	0.438	0.132	0.408	0.153	0.312	0.155
1/4	0.075	0.507	0.153	0.472	0.175	7/16	0.172
5/16	0.084	0.635	0.191	0.590	0.216	1/2	0.219
3/8	0.094	0.762	0.230	0.708	0.256	9/16	0.250
7/16	0.094	0.812	0.223	0.750	0.328	5/8	0.297
1/2	0.106	0.875	0.223	0.813	0.355	3/4	0.344
9/16	0.118	1	0.260	0.938	0.410	13/16	0.359
5/8	0.133	1.125	0.298	1.000	0.438	15/16	0.422
3/4	0.149	1.375	0.372	1.250	0.547	1 1/8	0.500

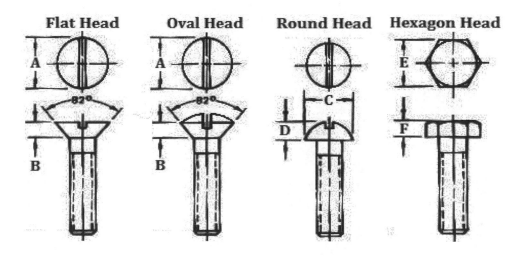

Flat Head   Oval Head   Round Head   Hexagon Head

Nominal Size (mm)	Slot Width (mm)	Flat & Oval Head		Round Head		Hexagon Head	
		A (mm)	B (mm)	C (mm)	D (mm)	E (mm)	F (mm)
3	1	5.6	1.6	6	2.4	5.5	2
4	1.3	7.5	2.2	8	3.2	7	2.8
5	1.5	9.2	2.5	9.8	4	8.5	3.5
6	1.7	11	3	11.8	4.7	10	4
8	2.1	14	4	14.8	6	13	5.5
10	2.6	18	5	19.2	7.6	15	7
12	3	23	6.4	24.5	9.7	18	8
14	3.2	26	7.2	27.8	11	21	9.3
16	3.2	29	8	30.8	12.2	24	10.5
20	4.2	35	9	37.2	14.8	30	13.1
24	4.8	41	11.6	43.5	17	36	15.6
30	6	51	14.4	54.2	22	46	19.5
36	7	60	17	63.8	25	55	23.4

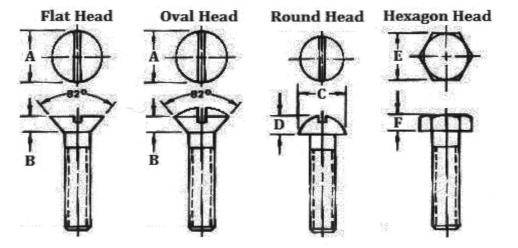

Flat Head **Oval Head** **Round Head** **Hexagon Head**

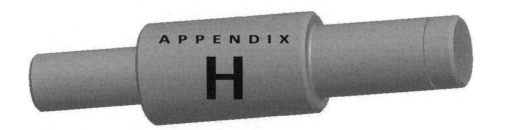

APPENDIX

H

NUT SIZES

Major Dia Nominal Size (inch)	Distance across Flats (inches)	Height	
		Regular (inches)	Thick (inches)
1/4	7/16	7/32	9/32
5/16	1/2	17/64	21/64
3/8	9/16	21/64	13/32
7/16	11/16	3/8	29/64
1/2	3/4	7/16	9/16
9/16	7/8	31/64	39/64
5/8	15/16	35/64	23/32
3/4	1 1/8	41/64	13/16
7/8	1 5/16	3/4	29/32
1	1 1/2	55/64	1
1 1/8	1 11/16	31/32	1 5/32
1 1/4	1 7/8	1 1/16	1 1/4
1 3/8	2 1/16	1 11/16	1 3/8
1 1/2	2 1/4	1 9/32	1 1/2

Major Dia Nominal Size (mm)	Distance across Flats (mm)	Height	
		Regular (mm)	Thick (mm)
4	7	-	3.2
5	0.4	4.7	5.1
6	0.5	5.2	5.7
8	0.6	6.8	7.5
10	0.7	8.4	9.3
12	0.8	10.8	12.0
14	0.9	12.8	14.1
16	0.9	14.8	16.4
20	1.1	18.0	20.3
24	1.3	21.5	23.9
30	1.5	25.6	28.6
36	1.7	31.0	34.7

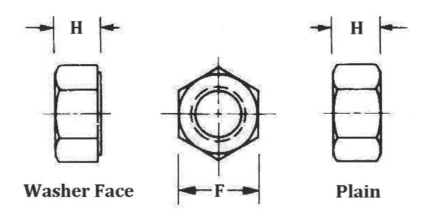

Washer Face   **Plain**

SETSCREW SIZES

Shaft Diameter		Setscrew Diameter (inches)	Seating Torque (in.lb.)
Over (inch)	Including (inch)		
–	–	0.125	10
–		0.138	10
–		0.164	20
5/16	7/16	0.190	36
7/16	9/16	0.250	87
9/16	7/8	0.375	290
7/8	1 1/4	0.500	450
1 1/4	1 3/8	0.625	620
1 3/8	1 3/4	0.750	2400
1 3/4	2 1/4	1.000	7200
2 1/4	2 3/4	1.250	15000

Shaft Diameter		Setscrew Diameter (mm)	Seating Torque (N.m)
Over (mm)	Including (mm)		
–	–	1.4	1
–	–	2	1.1
–	–	3	2.3
6	8	4	4
8	10	6	10
10	12	8	25
12	17	10	40
17	22	12	50
22	30	14	75
30	38	16	100
38	44	18	280
44	50	20	400

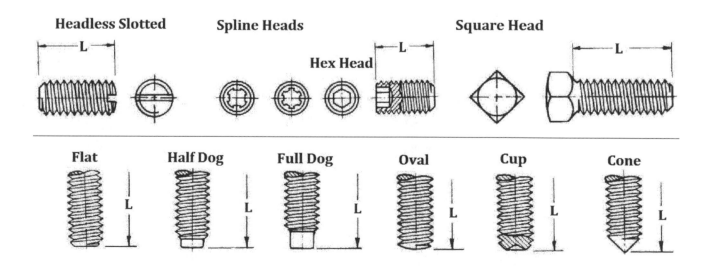

Headless Slotted   **Spline Heads**   **Square Head**

Hex Head

Flat   **Half Dog**   **Full Dog**   **Oval**   **Cup**   **Cone**

WASHER SIZES

Flat Washers			
Screw Size No. or Inch	Inside Diameter (inches)	Thickness (inches)	Outside Diameter (inches)
6 (.138)	5/32	0.049	3/8
8 (.164)	3/16	0.049	7/16
10 (.190)	7/32	0.049	1/2
12 (.216)	1/4	0.065	9/16
1/4	9/32	0.065	5/8
5/16	11/32	0.065	11/16
3/8	13/32	0.065	13/16
7/16	15/32	0.065	59/64
1/2	17/32	0.095	1 1/16
9/16	19/32	0.095	1 5/32
5/8	21/32	0.095	1 5/16
3/4	13/16	0.134	1 15/32
7/8	15/16	0.134	1 3/4
1	1 1/16	0.134	2
1 1/8	1 1/4	0.134	2 1/4
1 1/4	1 3/8	0.165	2 1/2
1 3/8	1 1/2	0.165	2 3/4
1 1/2	1 5/8	0.165	3
1 3/4	1 7/8	0.180	4
2	2 1/8	0.180	4 1/2

Lock Washers			
Screw Size No. or Inch	Inside Diameter (inches)	Thickness (inches)	Outside Diameter (inches)
4 (.112)	0.120	0.025	0.209
5 (.125)	0.133	0.031	0.236
6 (.138)	0.148	0.031	0.250
8 (.164)	0.174	0.040	0.293
10 (.190)	0.200	0.047	0.334
12 (.216)	0.227	0.056	0.377
1/4	0.262	0.062	0.489
5/16	0.320	0.078	0.586
3/8	0.390	0.094	0.683
7/16	0.455	0.109	0.779
1/2	0.518	0.125	0.873
9/16	0.582	0.141	0.971
5/8	0.650	0.156	1.079
3/4	0.770	0.188	1.271
7/8	0.905	0.219	1.464
1	1.042	0.250	1.661
1 1/8	1.172	0.281	1.853
1 1/4	1.302	0.312	2.045
1 3/8	1.432	0.344	2.239
1 1/2	1.561	0.375	2.430

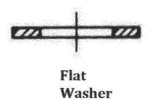

Flat Washer

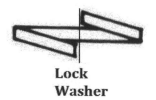

Lock Washer

Flat Washers			
Screw Size (mm)	Inside Diameter (mm)	Thickness (mm)	Outside Diameter (mm)
2	2.5	0.35	5
3	3.5	0.55	7
4	4.7	0.9	9
5	5.5	1.1	10
6	6.6	1.8	12
8	8.9	1.8	16
10	10.8	2.2	20
12	13.3	2.7	24
14	15.2	2.7	28
16	17.2	3.3	30
18	19.2	3.3	34
20	21.8	3.3	37
22	23.4	4.3	42
24	25.6	4.3	44
27	28.8	4.3	50
30	32.4	4.3	56
36	38.3	5.6	66

Lock Washers			
Screw Size (mm)	Inside Diameter (mm)	Thickness (mm)	Outside Diameter (mm)
2	2.1	0.5	4.4
3	3.1	0.8	6.2
4	4.1	0.9	7.6
5	5.1	1.2	9.2
6	6.1	1.6	11.8
8	8.2	2	14.8
10	10.2	2.2	18.1
12	12.2	2.5	21.1
14	14.2	3	24.1
16	16.2	3.5	27.4
18	18.2	3.5	29.4
20	20.2	4	33.6
22	22.5	4	35.9
24	24.5	5	40
27	27.5	5	43
30	30.5	6	48.2
36	33.5	6	55.2

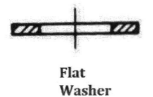

**Flat
Washer**

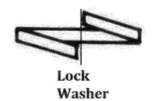

**Lock
Washer**

APPENDIX K

RETAINING RING SIZES

Nominal Shaft Diameter (inch)	Minimum Groove Diameter (inches)	Maximum Groove Diameter (inches)	Retaining Ring Diameter (inches)	Retaining Ring Thickness (inches)
1/4	0.218	0.032	0.311	0.025
5/16	0.274	0.032	0.376	0.025
3/8	0.333	0.032	0.448	0.025
1/5	0.447	0.042	0.581	0.025
5/8	0.560	0.042	0.715	0.035
3/4	0.673	0.049	0.845	0.042
7/8	0.786	0.049	0.987	0.042
1	0.897	0 049	1.127	0.042
1 1/8	1.009	0.060	1.267	0.050
1 1/4	1.122	0.060	1.410	0.050
1 3/8	1.233	0.060	1.550	0.050
1 1/2	1.346	0.060	1.691	0.050
1 3/4	1.571	0.072	1.975	0.062

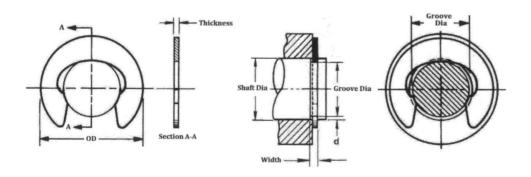

Nominal Shaft Diameter (mm)	Minimum Groove Diameter (mm)	Maximum Groove Diameter (mm)	Retaining Ring Diameter (mm)	Retaining Ring Thickness (mm)
8	6.9	0.8	10.0	0.6
10	8.9	0.8	12.2	0.6
12	10.8	0.8	14.4	0.6
14	12.6	1.2	16.3	1.0
16	14.4	1.2	18.5	1.0
18	16.2	1.2	20.4	1.2
20	17.9	1.4	22.6	1.2
22	19.7	1.4	25.0	1.2
24	21.5	1.4	27.1	1.2
25	22.4	1.4	28.3	1.2
30	26.8	1.5	33.7	1.5
35	31.3	1.8	39.4	1.5
40	25.8	1.8	45.0	1.5
45	40.3	1.8	50.6	1.5
50	44.8	2.4	56.4	2.0

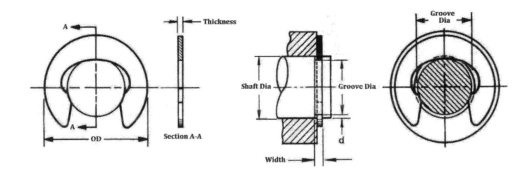

BASIC HOLE TOLERANCE

Basic Hole Symbol	Description of Basic Hole Fit
H11/c11	Loose running fit—for wide commercial tolerances or allowances on external members.
H9/d9	Free running fit—not for use where accuracy is essential, but good for large temperature variations, high running speeds, or heavy journal pressures.
H8/f7	Close running fit—for running on accurate machines and for accurate location at moderate speeds and journal pressures.
H7/g6	Sliding fit—not intended to run freely, but to move and turn freely and locate accurately.
H7/h6	Locational clearance fit—provides snug fit for locating stationary parts; but can be freely assembled and disassembled.
H7/k6	Location transition fit—for accurate location, a compromise between clearance and interference.
H7/n6	Locational transition fit—for more accurate location where greater interference is permissible.
H7/p6	Locational interference fit—for parts requiring rigidity and alignment with prime accuracy of location, but without special bore pressure requirements.
H7/s6	Medium drive fit—for ordinary steel parts or shrink fits on light sections, the tightest fit usable with cast iron.
H7/u6	Force fit—suitable for parts which can be highly stressed or for shrink fits where heavy pressing forces are impractical.

BASIC SHAFT TOLERANCE

Basic Shaft Symbol	Description of Basic Shaft Fit
C11/h11	Loose running fit—for wide commercial tolerances or allowances on external members.
D9/h9	Free running fit—not for use where accuracy is essential, but good for large temperature variations, high running speeds, or heavy journal pressures.
F8/h7	Close running fit—for running on accurate machines and for accurate location at moderate speeds and journal pressures.
G7/h6	Sliding fit—not intended to run freely, but to move and turn freely and locate accurately.
H7/h6	Locational clearance fit—provides snug fit for locating stationary parts; but can be freely assembled and disassembled.
K7/h6	Location transition fit—for accurate location, a compromise between clearance and interference.
N7/h6	Locational transition fit—for more accurate location where greater interference is permissible.
P7/h6	Locational interference fit—for parts requiring rigidity and alignment with prime accuracy of location, but without special bore pressure requirements.
S7/h6	Medium drive fit—for ordinary steel parts or shrink fits on light sections, the tightest fit usable with cast iron.
U7/h6	Force fit—suitable for parts which can be highly stressed or for shrink fits where heavy pressing forces are impractical.

TOLERANCE ZONES

Basic Size Up to (inches)	c (inches)	d (inches)	f (inches)	g (inches)	h (inches)	k (inches)	n (inches)	p (inches)	s (inches)	u (inches)
0.12	-0.0024	-0.0008	-0.0002	-0.0001	0.0000	0.0000	0.0002	0.0002	0.0006	0.0007
0.24	-0.0028	-0.0012	-0.0004	-0.0002	0.0000	0.0000	0.0003	0.0005	0.0007	0.0009
0.40	-0.0031	-0.0016	-0.0005	-0.0002	0.0000	0.0000	0.0004	0.0006	0.0009	0.0011
0.72	-0.0037	-0.0020	-0.0006	-0.0002	0.0000	0.0000	0.0005	0.0007	0.0011	0.0013
0.96	-0.0043	-0.0026	-0.0008	-0.0003	0.0000	0.0001	0.0006	0.0009	0.0014	0.0016
1.20	-0.0043	-0.0026	-0.0008	-0.0003	0.0000	0.0001	0.0006	0.0009	0.0014	0.0019
1.60	-0.0047	-0.0031	-0.0010	-0.0004	0.0000	0.0001	0.0007	0.0010	0.0017	0.0024
2.00	-0.0051	-0.0031	-0.0010	-0.0004	0.0000	0.0001	0.0007	0.0010	0.0017	0.0028
2.60	-0.0055	-0.0039	-0.0012	0.0004	0.0000	0.0001	0.0008	0.0013	0.0021	0.0034
3.20	-0.0059	-0.0039	-0.0012	-0.0004	0.0000	0.0001	0.0008	0.0013	0.0023	0.0040
4.00	-0.0067	-0.0047	-0.0014	-0.0005	0.0000	0.0001	0.0009	0.0015	0.0028	0.0049
4.80	-0.0071	-0.0047	-0.0014	-0.0005	0.0000	0.0001	0.0009	0.0015	0.0031	0.0057
5.60	-0.0079	-0.0057	-0.0017	-0.0006	0.0000	0.0001	0.0011	0.0017	0.0036	0.0067
6.40	-0.0083	-0.0057	-0.0017	-0.0006	0.0000	0.0001	0.0011	0.0017	0.0039	0.0075
7.20	-0.0091	-0.0057	-0.0017	-0.0006	0.0000	0.0001	0.0011	0.0017	0.0043	0.0083
8.00	-0.0094	-0.0067	-0.0020	-0.0006	0.0000	0.0002	0.0012	0.0020	0.0048	0.0093
9.00	-0.0102	-0.0067	-0.0020	-0.0006	0.0000	0.0002	0.0012	0.0020	0.0051	0.0102
10.00	-0.0110	-0.0067	-0.0020	-0.0006	0.0000	0.0002	0.0012	0.0020	0.0055	0.0112
11.20	-0.0118	-0.0075	-0.0022	-0.0007	0.0000	0.0002	0.0013	0.0022	0.0062	0.0124
12.60	-0.0130	-0.0075	-0.0022	-0.0007	0.0000	0.0002	0.0013	0.0022	0.0067	0.0130
14.20	-0.0142	-0.0083	-0.0024	-0.0007	0.0000	0.0002	0.0015	0.0024	0.0075	0.0154
16.00	-0.0157	-0.0083	-0.0024	-0.0007	0.0000	0.0002	0.0015	0.0024	0.0082	0.0171

Basic Size Up to (mm)	c (mm)	d (mm)	f (mm)	g (mm)	h (mm)	k (mm)	n (mm)	p (mm)	s (mm)	u (mm)
3	-0.060	-0.020	-0.006	-0.002	0.000	0.000	0.004	0.006	0.014	0.018
6	-0.070	-0.030	-0.010	-0.004	0.000	0.001	0.008	0.012	0.019	0.023
10	-0.080	-0.040	-0.013	-0.005	0.000	0.001	0.010	0.015	0.023	0.028
14	-0.095	-0.050	-0.160	-0.006	0.000	0.001	0.012	0.018	0.028	0.033
18	-0.095	-0.050	-0.160	-0.006	0.000	0.001	0.012	0.018	0.028	0.033
24	-0.110	-0.065	-0.020	-0.007	0.000	0.002	0.015	0.022	0.035	0.041
30	-0.110	-0.065	-0.020	-0.007	0.000	0.002	0.015	0.022	0.035	0.048
40	-0.120	-0.080	-0.025	-0.009	0.000	0.002	0.017	0.026	0.043	0.060
50	-0.130	-0.080	-0.025	-0.010	0.000	0.002	0.017	0.026	0.043	0.070
65	-0.140	-0.100	-0.030	-0.010	0.000	0.002	0.020	0.032	0.053	0.087
80	-0.150	-0.100	-0.030	-0.012	0.000	0.002	0.020	0.032	0.059	0.102
100	-0.170	-0.120	-0.036	-0.012	0.000	0.003	0.023	0.037	0.071	0.124
120	-0.180	-0.120	-0.036	-0.014	0.000	0.003	0.023	0.037	0.079	0.144
140	-0.200	-0.145	-0.043	-0.014	0.000	0.003	0.027	0.043	0.092	0.170
160	-0.210	-0.145	-0.043	-0.014	0.000	0.003	0.027	0.043	0.100	0.190
180	-0.230	-0.145	-0.043	-0.015	0.000	0.003	0.027	0.043	0.108	0.210
200	-0.240	-0.170	-0.050	-0.015	0.000	0.004	0.031	0.050	0.122	0.236
225	-0.260	-0.170	-0.050	-0.015	0.000	0.004	0.031	0.050	0.130	0.258
250	-0.280	-0.170	-0.050	-0.015	0.000	0.004	0.031	0.050	0.140	0.284
280	-0.300	-0.190	-0.056	-0.017	0.000	0.004	0.034	0.056	0.158	0.315
315	-0.330	-0.190	-0.056	-0.017	0.000	0.004	0.034	0.056	0.170	0.350
355	-0.360	-0.210	-0.062	-0.018	0.000	0.004	0.037	0.062	0.190	0.390
400	-0.400	-0.210	-0.062	-0.018	0.000	0.004	0.037	0.062	0.208	0.435

INTERNATIONAL TOLERANCE GRADES

Basic Size Up to (inches)	IT6 (inches)	IT7 (inches)	IT8 (inches)	IT9 (inches)	IT11 (inches)
0.12	0.0002	0.0004	0.0006	0.0010	0.0024
0.24	0.0003	0.0005	0.0007	0.0012	0.0030
0.40	0.0004	0.0006	0.0009	0.0014	0.0035
0.72	0.0004	0.0007	0.0011	0.0017	0.0043
1.20	0.0005	0.0008	0.0013	0.0020	0.0051
2.00	0.0006	0.0010	0.0015	0.0024	0.0063
3.20	0.0007	0.0012	0.0018	0.0029	0.0075
4.80	0.0009	0.0014	0.0021	0.0034	0.0087
7.20	0.0010	0.0016	0.0025	0.0039	0.0098
10.00	0.0011	0.0018	0.0028	0.0045	0.0114
12.60	0.0013	0.0020	0.0032	0.0051	0.0126
16.00	0.0014	0.0022	0.0035	0.0055	0.0142

Basic Size Up to (mm)	IT6 (mm)	IT7 (mm)	IT8 (mm)	IT9 (mm)	IT11 (mm)
3	0.006	0.010	0.014	0.025	0.060
6	0.008	0.012	0.018	0.030	0.075
10	0.009	0.015	0.022	0.036	0.090
18	0.011	0.018	0.027	0.043	0.110
30	0.013	0.021	0.033	0.052	0.130
50	0.016	0.025	0.039	0.062	0.160
80	0.019	0.030	0.046	0.074	0.190
120	0.022	0.035	0.054	0.087	0.220
180	0.025	0.040	0.063	0.100	0.250
250	0.029	0.046	0.072	0.115	0.290
315	0.032	0.052	0.081	0.130	0.320
400	0.036	0.057	0.089	0.140	0.360

REFERENCES

Giesecke, Mitchell, Spencer, Hill, & Loving (1970). *Engineering Graphics* (2nd printing). New York: Macmillan Company.

Giesecke, Mitchell, Spencer, Hill, & Loving (1975). *Engineering Graphics* (2nd ed.). New York: Macmillan Company.

Jensen, Helsel, and Short (2002). *Engineering Drawing and Design* (6th ed.). New York: McGraw-Hill Publishers.

Madsen, David A. (1999). *Geometric Dimensioning and Tolerancing* (6th ed.). Tinley Park, IL: The Goodheart-Willcox Company, Inc.

Marelli, Richard S. & McCuistion, Patrick J. (2001). *Geometric Tolerancing, A Text-Workbook* (3rd ed.). New York: McGraw-Hill Publishers.

Shih, Randy H. (2011). *Parametric Modeling with Creo Parametric, An Introduction to Creo Parametric 1.0.* Mission, KS: SDC Publications.

Thiagu, Palaniappan (2011). *Creo 1.0 Quick Reference Guide.* Phoenix, AZ: TriStar Inc.

INDEX

A

Alpha characters, 34
American Standards Association
  (ASA), 203
Angle, 42
Annotation display, 47
Arc, 31–32, 54
Assembly, 287, 287–338. *See also*
  Assembly drawings
  exercise, 307–332
  practice, 293–306
  predefined constraints, 291–292
  problems, 334–338
  user-defined constraints, 289–291
  using extra constraints, 292
Assembly drawings, 287, 339–368
  defined, 339
  exercise, 352–364
  parts identification on, 288
  practice, 343–352
  problems, 366–368
  subassembly, 339, 340
A-size format sheet, 222–231
A-size template, 232–239
Axial pattern tool, 170
Axis ends ellipse, 32
Axis pattern, 161

B

Baseline, 43
Basic hole tolerance, 513

Basic shaft tolerance, 515
Bell crank mechanism, 22
Block creation, 75–78
Border, 225
Blueprinting, 200
Brainstorming, 18–19, 21–23

C

CAD file folder, 7
CASASP-D (Computer-Aided
  Simulation and Analysis of
  Sculptured Parts in 3D), 1
Cathode-ray tube, 1
Center and ends arc, 31
Center and point circle, 31
Centerline, 199
Center rectangle, 31
Chamfer fillet, 33
Chamfer trim fillet, 33
Child feature, 75
Circle, 31, 51, 100
Circular fillet, 33
Circular trim fillet, 33
Clevis pin sizes, 497
Coincident axes, 289
Coincident constraint, 40
Coincident surfaces, 289
Common system parameters, 229
Computer-Aided Design (CAD),
  history of, 1–2
Concentric arc, 32
Concentric circle, 31, 54